Darwinism Applied

DARWINISM APPLIED

Evolutionary Paths to Social Goals

JOHN H. BECKSTROM

Human Evolution, Behavior, and Intelligence
Seymour W. Itzkoff, Consulting Editor

Westport, Connecticut
London

Library of Congress Cataloging-in-Publication Data

Beckstrom, John H.
    Darwinism applied : evolutionary paths to social goals / John
H. Beckstrom.
        p.   cm.—(Human evolution, behavior, and intelligence, ISSN
1063-2158)
    Includes bibliographical references and index.
    ISBN 0–275–94568–5 (alk. paper)
    1. Sociobiology. 2. Social Darwinism. 3. Social values.
4. Social policy. I. Title. II. Series.
GN365.9.B43 1993
304.5—dc20   93–2863

British Library Cataloguing in Publication Data is available.

Library of Congress Catalog Card Number: 93–2863
ISBN: 0–275–94568–5
ISSN: 1063–2158

First published in 1993

Praeger Publishers, 88 Post Road West, Westport, CT 06881
An imprint of Greenwood Publishing Group, Inc.

Printed in the United States of America

∞™

The paper used in this book complies with the
Permanent Paper Standard issued by the National
Information Standards Organization (Z39.48–1984).

10 9 8 7 6 5 4 3 2 1

Contents

Acknowledgments

This book contains considerable distillation and refinement of technical scientific information so as to make it accessible to the educated person who has little natural science background. Thus, I asked for reactions and comments on early drafts from a wide variety of people ranging from professors of anthropology, biology, sociology and law to relatives and running companions. I received much appreciated responses from the following: Ronald Allen, Garland Autrey, Linda Beckstrom, Page Beckstrom, Robert W. Bennett, Laura Betzig, Cynthia Bowman, Anthony D'Amato, Clinton Francis, Mary Ann Glendon, John T. Harney, William Irons, Seymour Itzkoff, Gary Johnson, Jane E. Larson, Gary Lawson, Edward Lev, Randolph Neese, Joanne Silver, Peter Strahlendorf and Randy Thornhill.

Innumerable people, past and present, have contributed to the body of science reported on in the book. A large portion of them are cited in Notes and Bibliography. In the interests of a concise presentation, I have mentioned such people only occasionally in the text—when they figured most prominently in the particulars on which I concentrate. Probably the individual whose work is most reflected in a general manner throughout the book, but is not named in the text, is Richard D. Alexander.

I want to thank Northwestern University School of Law for its support of this project.

Introduction: Proper Use of Evolutionary Science in Social Affairs

This book describes how new insights into human behavior obtained from evolutionary science may be used to help achieve social goals such as the reduction of child abuse and incest. We will see sometimes surprising methods of doing so that have recently been suggested by science. Advice of this sort available today will be discussed in the early part of the book. Later, we will discuss help that may be on the horizon for goals as ambitious as reducing the risk of international war.

Here, at the outset, we need to make it clear that evolutionary learning cannot be used to *justify* social goals. It cannot, for example, justify the goal of reducing the incidence of child abuse, or—bear with me on this—justify proposals such as removing society's restraints on child abusers. Early attention to this point is necessary so that we do not confuse the proper use of evolutionary learning with its improper use as exemplified by the turn of the century ideology called Social Darwinism.[1]

Although Social Darwinists did not advocate the removal of restraints on child abusers, they took an analogous position.[2] They believed that essentially unrestrained, winner-take-all competition should be permitted in economic and social arenas. They arrived at this conclusion by first noting that Charles Darwin and others had exposed the historical evolutionary process in which those humans best equipped for life's struggles, against nature and other humans, have tended to survive and thrive. Therefore, it was argued, unrestrained competition was "natural" and collective social action like welfare programs for the poor and property taxation should not interfere with it.[3] This position was nonsense, of course. One may as well argue that floods and diseases are "natural," so they should not be interfered with; or that child abuse has always been with us[4] so it should not be restrained. The fact that something exists and has existed for a very long time does not call for society's endorsement. To say that it does is to commit the "naturalistic fallacy."[5] "Is does not equal ought," as the saying goes among philosophers.[6]

So references to the factual processes of evolution—how living things got where they are and what they are like at present—do not support arguments that human society ought to stay the same. Neither do they support arguments that it ought to change.[7] The propriety of any societal goal is largely a matter of personal values and tastes. Evolutionary science cannot supply values and tastes. It may explain the facts of how humans came to have certain values and tastes, but it cannot justify them or any societal goals that hinge on them.[8]

For these reasons the ideology called Social Darwinism deservedly died out. But because it was the first publicized attempt to apply evolutionary learning to social matters, its wrongheadedness left an unfortunate legacy. It led many to assume that evolutionary insights have no proper place in social affairs. The time has come when that wrong impression could result in serious losses for society if not corrected. Large leaps forward in evolutionary learning in the last few decades of the twentieth century have brought us to the beginning of an era where science may be able to offer social planners advice on how to reduce or even eliminate a large array of social problems.

As I said, evolutionary science cannot be used "normatively"—it cannot be used to determine what social goals ought to be pursued. But once social planners, public or private, have used their values and tastes to select social goals, modern evolutionary science may then step in to provide factual guides toward achieving those goals—facilitative guides.[9] The repositories of evolutionary science learning can act like travel agents. They cannot tell you where to go,[10] but they can give you information about the costs and benefits of various destinations and help you to get there once you finalize your decision.[11] Let me illustrate this concept with an analogy from health science, which is more familiar to most of us than evolutionary science.

Look at the first entry in Figure 1.

If you approach health scientists and tell them you are thinking about living as long as possible and would like to know the costs of doing so, they can give you such factual cost information. For example, they can tell you about the inevitable physical and mental deterioration, with attending discomforts, that occur with increasing age. Although it is not

Figure 1.
Human Goals: Their Costs and Facilitative Guides

Ultimate (Target) Goals	Costs	Facilitative Guides (Sub-Goals)			
1. Long life	(Physical and mental deterioration)	Avoid heart attacks	by	Losing weight	by	Dieting and exercising
2. Continuance and proliferation of human species	(Oppression and elimination of other life forms)	Promote variety in human characteristics	by	Preserving or expanding present human characteristics	by	Promoting equal reproductive opportunities in present population

in their field particularly, they might also point out benefits—such as wisdom and extended opportunities for life experiences—that can accompany advanced age.

Now, assume that you have weighed the costs and benefits of long life and are indecisive as to whether to go for it. If you were to ask health scientists whether there is anything in their theory or factual discoveries that can help you make your decision, their answer should be no. Here scientific information is of no use. Your decision hinges largely on personal values and tastes. (Some may prefer to live indulgently and die young before deterioration and senility set in.)

Next, let us assume you have gotten over the subjective, personal decision hurdle. You have looked to your religion, your parents, your bank account, or whatever source, and you have definitely decided to pursue long life. Health scientists can now help you again. For example, they can recommend that you avoid heart attacks, which kill many early in life. Avoiding heart attacks may be regarded as a subsidiary or secondary or proximate "goal" on the way to the ultimate goal of a long life. To avoid confusion, however, let us use "goal" only to refer to ultimate target objectives. We will call the steps that may help to achieve those objectives "facilitative guides."

It is important to distinguish clearly in this way, because when say I say science cannot tell us whether or not to seek a goal, some may wonder, then, what doctors are doing when they flatly tell us to avoid heart attacks and to do that by losing weight and to lose weight by dieting and exercising. The answer is that they are giving us facilitative guides on the *assumption* that we have the goal of long life—because almost all people do. If you say to the doctors, "but I haven't decided whether to try to live long or not," they should admit that, as scientists, they cannot help you with that decision. But once you have decided to aim for long life, make another appointment, because they can then give you advice—facilitative guidelines.

These distinctions regarding what science can and cannot do for us have not been clear in the writing of some authors who have applied evolutionary learning to human social affairs. Writers have suggested "primary values," "principles," or "standards" for human societies when it might seem as if they were announcing goals that evolutionary learning tells us we *ought* to work toward. However, on close inspection, it often appears that the authors are *assuming* that the reader and they agree on an ultimate or target goal for which the announced "principles," and such are merely facilitators. They make the assumption because disagreements with the goal are almost unheard of.

For example, both Roger Masters[12] and Edward O. Wilson[13] have suggested that evolutionary learning points to preserving the vast differences that one finds among humans in physical characteristics, behavior, and culture. At first glance, in context, their suggestions may seem like "is equals ought" statements. A closer reading reveals that the writers are saying that preserving human diversity is a way to accomplish ultimate goals they have in mind. Masters suggests that preserving diversity can help ensure the survival of the human species. (When an unforeseen global disaster occurs, those with special characteristics resistant to the onslaught may survive and carry on.) He is *assuming* survival of the species as a target goal—without being fully aware that he is doing so, perhaps[14]—and telling us that science suggests preservation of human diversity as a way of facilitating the achievement of that goal (see Figure 1, entry 2). Similarly, Wilson is *assuming* a target goal of assuring continuance in the population of people with extraordinary physical and mental capacities—superb athletes and geniuses. He says that, because of the intricacies of genetic reproduction, "the human gene pool creates hereditary genius in many ways in many places only to have it come apart the next generation." Thus, he suggests that the best way today to help ensure the continued appearance of exceptional individuals in the human population is not just to promote reproduction by people displaying the extraordinary characteristics, but for all members of the diverse population to continue reproducing themselves.

If we were to ask professors Masters and Wilson whether science tells us that the human species *ought* to be preserved or that extraordinary athletes and thinkers *ought* to continue to be produced by the species, I fully expect they would say no.[15] The answers to such questions do not come out of fact, which is what natural scientists strive to uncover. Rather, they come out of opinion derived from personal values and tastes. Opinion cannot be definitely demonstrated; it cannot be "nailed down" like fact. You and I may believe that the human species ought to continue indefinitely and that appropriate steps should be taken toward that end. Another individual may believe that the human species ought to be eliminated when it seriously threatens the existence of other life forms on earth. Natural science may be able to explain why we and that other individual happen to hold the opinions we do, and it may also be able to give us facilitative (factual) guidelines toward accomplishing the goals those opinions represent. But it cannot tell us whether either opinion is right, true, proper, or just, and ought to be pursued as a goal.

THE PLAN OF THIS BOOK

Having established ground rules for the proper use of evolutionary science in human social affairs, let us move to my principal purpose for writing this book. I want to introduce people who have social goals to a modern, rapidly developing aspect of evolutionary science that could help to achieve many of those goals. This evolutionary perspective has been variously named—neo Darwinism, evolutionary human behavior, bicultural science—but it has most often been called sociobiology. It deals with the evolved biological basis for social behavior in all organisms including humans.

There is no longer any doubt that as a result of the processes of evolution human behavior is, to some extent, influenced by the genes we carry in our bodies. A project of human sociobiology is to map the extent to which the influence of our genes—let us call it genetic programming[16]—operates in conjunction with learning, culture and other aspects of human environments to produce our social behavior.

Based on facts known about the behavior of all the earth's animals, human physiology and the history of evolution, evolutionists have hypothesized genetically influenced processes or mechanisms in the human mind and body that lie behind behavior. Many of these behavioral mechanism hypotheses have received empirical support; others are still quite speculative.[17]

The plan of this book is to sort through these behavior mechanism hypotheses, concentrating on those that are most solidly supported, and show how social goals might be accomplished by utilizing these mechanisms. The goals I have selected as illustrations are mostly noncontroversial. Virtually everyone would subscribe to the reduction of child abuse, rape, incest and war, for example. But controversial social goals may also be served in the way I will outline. Pure natural science is factual and nonpartisan. Social planners of any persuasion can get useful insights from natural science. For example, eventually the behavioral mechanisms bearing on patriotic activity should be more clearly identified than they are at present, so that scientifically appropriate steps to increase patriotism in a society may be known. This prospect will be offensive, or at least distasteful, to those who believe excessive patriotism is a cause of wars between nations. Very well then, the same relevant behavioral mechanisms should be able to be influenced in a reverse manner to *decrease* patriotism. Decreasing and increasing patriotism may be seen as opposing political goals. But facts about the workings of behavioral mechanisms are apolitical. Pure natural science is apolitical. The manner in which natural science information is used may be political, but the information itself is not.[18]

The natural science information I will be discussing will have uses for individuals in their daily lives, as well as for the projects of professional social planners. For example, Chapter 3 suggests a rather simple method of helping to ensure that one's male and female children do not have incestuous relations after puberty. Readers might apply it in their own families. Social planners may also be interested in such information if reducing incest in their society is on their agenda. They might, for example, mount a publicity campaign to inform individual families. Inevitably, information given to individuals in book form will be available for such social planning. Do we want to give organized social forces more or better information with which to influence society? My answer is yes.

I may be mildly offended by the idea of being personally influenced (led) away from conduct such as child abuse and rape. (I can avoid them myself, right?) As a participant in society, however, I want to see it as well equipped as possible to handle these phenomena. If you agree with me on that, I expect you want to include me in the group that is led away from such conduct. And I will agree with you there. I have to be included.

People who believe that society should be as well equipped as possible to encourage or discourage conduct that is collectively considered desirable or undesirable should find this book of particular interest. But what about libertarians who disagree and do not want collective interference with individual actions even when they agree with the targeted social goals? The message of this book is for them, too. Through introduction to sociobiology, they may better detect and resist behavioral molding by future social programs—or present programs: Some behavioral mechanisms that scientists are attempting to illuminate are likely being influenced today, in subtle ways, by current social programs.[19] The existence of some of these mechanisms and their susceptibility to influence may be intuitively sensed by many people,[20] including those who are presently making and enforcing society's rules.

In any case, I am convinced that all available information about the workings of human behavior should be generally known. I would like to see all the players have full knowledge of the terrain on which the game is being played. That has to be good for us all in the end.

In this book we will, as I said, be assuming certain illustrative social goals and suggesting various facilitative programs. Readers may find new ideas—for example, reducing the risk of incest between male and female siblings after puberty by maximizing their daily contact before puberty. Other suggestions may not be new ideas for many readers—for example, reducing the incidence of rape by increasing and publicizing the assured penalty for rapists. In such cases the principal value of the scientific insight could be to provide theoretical support from evolutionary learning for arguments that the program should be instituted. For instance, opponents of increasing publicized assured penalties for rapists could—and probably would—claim that evidence of the measure's deterrence value is lacking or insufficient. If so, theoretical support from evolutionary learning could add weight to whatever inconclusive empirical evidence exists that an increase in publicized assured penalties would have deterrence value.

Let me now give the reader a very broad introductory idea of the mechanics of translating evolutionary science into suggestions for promoting social goals.

Sociobiology stipulates that human behavior results from the interaction of (1) genetic programming we all carry, (2) our life histories of learning, and (3) the environment we find ourselves in at the time of the behavior. Our environment is composed of natural elements (heat, wind, etc.), cultural elements (clothes, institutions, etc.) and social elements (other people) that are ever changing and unpredictable. Our life histories and the learning we acquire in them are unique and diverse. On the other hand, genetic programming which *inclines* us toward certain types of behavior, is fixed at birth and in many respects uniform in the human species. Therefore, to the extent that genetic programming exists in the genetic/life history/environmental mix that produces our behavior, the behavior is somewhat predictable. It becomes especially predictable when one looks for *typical* behavior in large population groupings. In such groupings environmental quirks and extremes will be counterbalanced, and the personal, marginal idiosyncrasies of individuals will be negligible in the big population picture. Let me illustrate this thesis by citing behavior that one person can exhibit in isolation from others—nonsocial behavior.

If the temperature were to fall by 20 degrees in Sweden in January, the Swedes would typically put on more clothes or turn up the heat. A few would not respond for various reasons, such as they want to appear tough or are too ill to notice the change or they already had more heat on than they needed. But more clothes or heat will be the typical reaction to this environmental change. You can bet on it, because (1) we know that clothes and heat produced from fuel are learned, cultural devices that twentieth-century Swedes use as means of adjusting body temperature, (2) the maintenance of a certain body temperature is important for survival, and (3) evolutionary learning tells us that all humans are genetically programmed to behave in ways that promote their survival.

Some such behavioral programming is almost automatic, requiring little or no thought or planning; it is essentially involuntary—for example, eye-blinking. Other human genetic programming is looser and more general, calling on the brain and other bodily systems to respond as if to messages like, "do whatever is most appropriate, in your circumstances, to be thermally comfortable (keep the body at about 98.6° fahrenheit)." In response to such a genetically generated message, given today's Swedish culture, Swedes will typically put on more clothes or turn up the heat when the temperature falls 20 degrees.

So, the typical behavior of the isolated individual in response to climate change is somewhat predictable in large population groupings. Sociobiology tells us that social behavior (relations *between* people) in response to various stimuli can often be similarly predicted through genetic programming in humans aimed at survival and reproduction. For example, sociobiologists will predict that if the assured punishment for rape were generally increased and the increase was publicized around the United States, a decrease in the incidence of rape would occur in the country.[21] Some potential rapists would remain undeterred, for one reason or another, by knowledge of an assured penalty increase, but enough would be discouraged that a decrease in the incidence of rape nationwide would result. Such is the prediction.

The evolutionary learning that prompts such predictions from scientists is complicated. In the space of this book I cannot hope to give the reader a complete and firm grasp of the subject; several large volumes would be required for that. I will, however, summarize enough of the science in lay terms to constitute an insight into why the attainment of the social goals we will address may be facilitated by the methods suggested. The science is particularly concentrated in Chapters 1 and 5, which constitute evolutionary biology primers.

WHEN CAN THE SCIENTIFIC INFORMATION BE APPLIED TO SOCIAL PROBLEMS?

Sociobiology is a relatively young and developing scientific discipline. Many sociobiological hypotheses are still in the early stages of formation and/or lack empirical substantiation. In Chapter 8 we will look at some of this "soft information" that may eventually be firmed up to provide useful guides for achieving social goals.

In Chapters 2, 3, 4, 6 and 7, however, we will deal with sociobiological propositions about the predictability of human behavior that are already quite solid.[22] They are solid enough so that if I were personally, or as a social planner, concerned with the problems we will discuss, knowing what I know about the relevant science, I would consider using the information today in order to help solve those problems. Indeed, in the chapter dealing with reducing sibling incest, we will see empirical data on *actual behavior* that I believe is substantial enough to justify action based on what the data alone tell us, even if there were no sociobiological theory in accord with it.

This question of when sociobiological theory might be used in practical affairs is important and momentous, so let me take the time to draw an analogy to the use of newly developed medicines. When someone "hypothesizes" a new medicine as an aid in combatting a disease, the decision to use it on patients can hinge on three factors: (1) the results of tests as to effectiveness and side effects, (2) the availability of alternatives and (3) the condition of the patient. If the patient's condition is serious enough and no reliable alternatives are available, a doctor may reasonably choose to administer the medicine, even though testing is incomplete or inconclusive.[23] If the patient is on the verge of death, one may reasonably choose to administer the medicine in the absence of tests if no reliable alternatives exist.

Because many of society's problems have no clear, easy solutions, concerned people will be looking for new ways to relieve the patient. I will leave it to them to decide whether the social problems we will discuss in the first seven chapters are serious and pressing. I will also leave it to concerned people to weigh alternative treatments. I will simply offer methods of treating the problems that have been suggested by recent evolutionary learning. In the next to last chapter, I will outline problem treatments that are on the horizon. I would not yet utilize them, unless my patient was on the verge of death and there was no alternative.

Sociobiology is not a cure-all, however, and social problem solving can be complex. So as not to waste time and resources, advice taken from evolutionary learning should be as clear and pragmatic as possible with real goals in mind. In a final chapter, as a caution to those who would too quickly locate solutions to social problems in evolutionary learning, I will illustrate "blind alleys" that commentators have gone down.

This book is written for the educated nonscientist. I have aimed to reach as broad a readership as possible without distorting the science involved by oversimplification. Virtually any aspect of evolutionary biology I mention has much more technical and detailed treatment in other books. And there is a vast number of human behavioral implications of sociobiology not mentioned in the book because they bear no direct relevance to the particular social goals we will discuss. Any reader interested in more information is invited to consult the chapter notes at the end of the book.[24] There I cite authorities (which key into the Bibliography) for my textual statements and have often added extensive substantive comments and quotations.

Chapter One

Evolutionary Biology Primer I: Aid-Giving Behavior

This chapter presents an overview of some basic evolutionary biology. The biology I will be describing is most often called human sociobiology when it concerns social relations—interactions between people—and we will be focusing on such interactions in this book.

I guarantee that the reader to whom this subject is new will be interested, and perhaps even surprised or perplexed.[1] We will be dealing with new perspectives provided by modern science on why others treat us as they do and why we treat them as we do. We know we are complicated creatures, so, as we might expect, the science turns out to be complicated. Much study in many recent source books is required for a thorough insight into it.[2] But, at the cost of leaving some readers unsatisfied, the essence can be distilled. I will do that here[3] and at other places in the book.

I will present what I take to be the consensus view of the many scientists from various academic disciplines[4] who are concentrating on sociobiology. It is a developing mix of theory (actually hundreds of interrelated hypotheses radiating from a core) and supporting fact. Much of my narrative will be stated in a matter-of-fact way. I will only occasionally interrupt with observations on the extent to which various parts of the theory have been substantiated. However, I can assure the reader that the contents of this chapter are at the best developed and substantiated core of sociobiology.

The ideal place to start with evolutionary learning is as far back as knowledge reaches on the history of the earth. For that comprehensive treatment I highly recommend Richard Dawkins's *The Blind Watchmaker*.

Our focus in this book is on evolved human behavior, however, so in this brief overview we will skip ahead to the point where human beings came on the scene. In that setting let us begin by examining what is meant by "natural selection."[5] It is the key concept in the evolutionary process, and it can be briefly summarized as follows: Since life first appeared on earth, individual organisms, from amoebas to humans, have been in a kind of competition against the forces of nature, including others of their species. Those with certain genetically influenced characteristics have tended to survive the competition and reproduce more successfully than those without such characteristics.[6] That is generally what laypeople have in mind when they tell you that natural selection means "survival of the fittest." But most people cannot go beyond that definition and describe the mechanics of natural selection. It is essential to grasp those mechanics before the rest of evolutionary learning can be fully appreciated. So let us look at an illustration.

Although we cannot know the exact details of early human history, we can construct plausible scenarios that illustrate the natural selection process. That process is rooted in the fact that mutations—random structural changes—occasionally occur in the genes humans carry inside the cells that make up their bodies.[7] Now, let us assume that an early ancestor of ours developed one of these mutant genes which, either alone or more probably in combination with other genetic materials,[8] gave her the dexterity and inspiration to use a simple tool like a sharp stone found in her environment. This capacity permitted her to survive during hard times while others around her, who were not thus genetically endowed, were dying off. Because she prospered longer than the others, she had more offspring than they did. Moreover, a large percentage of her descendants inherited exact copies of that special gene from her, so they, in turn, were more successful in staying alive and reproducing than were others around them. Eventually, after a great many generations had passed, essentially every human alive carried this particular gene. This happened because those who did not carry the gene were at a competitive disadvantage in dealing with adverse environments, which lessened their lifetime reproductive activity compared to those who did carry it. So they died leaving relatively few offspring. Therefore, as the gene was spreading throughout the population, people without it were slowly washed out of the "gene pool," which represents the total combined genetic complement of a population at any given time.

In the above scenario, the behavioral characteristics that the special gene promoted were being *selected* by the process of natural selection. Indeed, the gene itself was being selected; selection was acting at the level of the gene. The gene was surviving from generation to generation because the behavioral characteristics it prompted worked well in the environments its carriers inhabited. It was only incidental that the individual humans who carried the gene had, during their brief lifetimes, high survival value and were "fit" in terms of reproductive capacity.[9] This generation-to-generation gene survival is an evolutionary insight that science has only recently provided.[10] Darwin knew something survived as represented in the phrase "survival of the fittest,"[11] but he was unclear as to exactly what it was. Genes, the chromosomes on which they are found and the DNA of which they are composed, were there but undiscovered in his day.[12]

Evolution is, of course, a continuing process. As the environments of our ancestors have changed and increased in range through the ages, vast numbers of genes bearing on behavioral traits that adapted well to varying environments have arisen by mutation, spread by reproduction and settled into the gene pool that represents all humans living today. Thus, we are all quite well "programmed" by the genes we carry to survive and reproduce in the environments in which we find ourselves.

Some behavior in response to the environment, such as eye-blinking, is genetically "hard-wired" (relatively inflexible). Blinking moistens and helps clear the surface of the eye as air and dust come into contact with it. No great advantage would have accrued to a genetic program that promoted complex calculation or thought before blinking. Hence, any such neurologically expensive program that popped up by mutation in the gene pool would have been selected against by the economics of nature.[13] Other behavior does have complex behavioral mechanisms behind it and is quite flexibly programmed.[14] This is because calculating flexibility has often had survival and reproductive value, since it permits adjustments to rapid and diverse changes in unpredictable environments. As wind, rain, snow, disease-causing organisms, predators, and competitors have appeared and disappeared, those who could respond most flexibly were often better adapted for the struggle. It has been advantageous for individuals to be able to learn by trial and error how best to respond to a given environmental circumstance. Programming for this learning behavior has therefore been selected for in the evolutionary process leading to today's human beings. Thus, the genetic programming that enabled that remote ancestor of ours to remember and reapply the skill of using a sharp stone as a tool was retained and proliferated in the gene pool.

Now, theoretically, sharp stone use might have spread throughout the population by the tedious route of each individual stumbling onto it by trial and error. Most certainly it did not happen that way. Instead, individuals in whom genetic programming developed that equipped them to learn by observing others and imitating the use of sharp stones, fire, warm clothing, the wheel and books had an advantage. They had an advantage on an advantage, so to speak. Thus vicarious, as well as firsthand, learning capacity proliferated in the gene pool.

Notice that with vicarious learning, culture entered into the human picture. Useful devices for survival and reproduction could now be passed rapidly, in a moment's time, from person to person without having to wait for the slower, generation-to-generation reproductive process by which genes and genetic programming are inherited. Keep in mind, however, that the *capacity to learn* from accumulated culture is genetically influenced. And that capacity, *at any given moment* in evolutionary time, has outward limits.[15] At present, for example, starting with mother's milk, we can go on to learn to ingest hamburgers, artichokes and red hot peppers. However, if seed salespeople were to promote lawn grass as food—which would be a cultural innovation—very few, if any, of us could "learn" that. Unlike sheep, we are not programmed at this point in our evolutionary history to digest adequately and gain nutrition from grass.

By putting a minimum of two people in our picture (one to learn from the other), we have also moved into the social dimension of human life—let us stay there. Picture a crowd and then focus on one person. That individual (a gene carrier) has a package of evolved genetic material inherited from countless remote ancestors and directly from its parents. During the sexual reproduction process, the thousands of genes that each parent contains are shuffled together and reissued in such a way that each emerging person's gene package is unique—not exactly like that of any other individual (unless the person has an identical twin).[16] This fact, plus the principle of natural selection, suggests that people will be programmed to act in their self-interest. If natural selection means anything, it should mean that organisms *not* designed to look out for their interests will soon expire and leave no descendants. Everybody alive is a descendant, of course. Thus, one might assume that essentially everybody alive would be programmed to compete in the world—behave—in ways that promote the survival and reproduction of their own particular genetic package. How, then, can one account for aid-giving between individuals? Here sociobiology has provided an insight that even some of the discipline's most noted detractors have admired.[17] To obtain this insight we need first to narrow our focus to a single gene.

Any single gene one person carries (aside from rare mutations)[18] is and has been carried by other individuals as well. Each carrier has an identical copy of the gene inherited from some ancestor and reproduced from generation to generation. Wherever a gene appears it is *identical* in each incarnation. This fact leads to a key abstract concept that must be appreciated if the rest of evolutionary biology is to be understood, and it has no exact parallel in human experience. So let us go slow here. It is extremely useful to employ anthropomorphic devices (pretend that things nonhuman are human) and metaphors to convey this concept, so bear with me while I do. This is the abstract concept that we need to illustrate: *All the identical copies of a gene in whatever bodies they appear are of equal value to the genetic material they represent.* If we could ask a sandy beach which of its grains of sand were most important to its existence, the answer should be that they are all equally important. Now, let us give the grains of sand life, put the self-consciousness of the whole beach into the mind of each grain, place each grain in a container and spread the containers out over a large part of the earth. If we could do all that, we would fairly closely parallel the relationship that exists between any gene we contain and the copies of it that are contained in other people.

Our genes may not have minds as such, but they do constitute living matter. And any gene you contain is equally as "interested" in its identical copy in another person as it is in itself. For that reason, a genetic mechanism, whether due to one gene or a combination, that promotes aid-giving to others who contain the same genetic material could be selected in the evolutionary process. The economics of natural selection will have this happening when any loss in reproductive capacity (fitness) to the aid-giver is sufficiently offset by a gain in fitness to others who contain the same genetic material. I will illustrate this proposition with a simplified example involving you rescuing people who contain high percentages of the same genetic material you do—your brothers and sisters. First, however, let me lead into the example with a few words about genetic overlap between such close relatives.

All humans alive today share, in common, the vast majority of their genes. But over and above those genes that prevail in the entire species, additional gene commonality exists between close relatives because they received their genes from close ancestors whom they have in common.[19] Let us call these genes that are common in a family, but less common in the species population as a whole, "familial" genes.

Geneticists express the genetic relatedness between relatives in terms of fractions, like one-half for siblings. In the computation of these fractional relationships the focus is on the familial genes.[20] The fraction for parent and offspring is always one-half: you have one-half of your mother's genes and one-half of your father's genes. For all other relationships the figure represents an average. If you had 100 brothers and sisters, approximately 50 of them would contain less than half of your familial genes and 50 of them more than half, for an average of one-half. Similarly, half-siblings are related by one-fourth, as are grandparents and grandchildren and the aunt/uncle–niece/nephew combinations. Cousins are related by one-eighth. A family tree showing genetic relatedness is set out in Figure 2.

Now, let us get to our rescue illustration. Assume you find yourself in a situation where you could either save yourself or three of your siblings. If you save yourself and let your siblings die, you have saved 100 percent of your familial gene copies. Remember, however, that all copies of a gene are of equal value to the genetic material they represent and each of your siblings has copies representing about 50 percent of your familial genes. Thus, if you sacrifice yourself and save your three siblings, you have saved about 150 percent (3 x 50) of the genetic material represented in your copies of your familial genes. In other words, the siblings, combined, would be better equipped to reproduce *your* genetic material than you alone would be. For this reason, sociobiologists suggest that any genetic programming prompting self-sacrifice in similar aid-giving situations would proliferate in the gene pool. Sociobiologists assume that aid-giving tendencies following this principle were selected in humans ages ago and exist in us today.[21] The phenomenon, first elaborated by William D. Hamilton, is usually called kin selection by natural scientists, and its existence in the nature has been well substantiated, in particular by elegant experiments with social insects.[22] The evolutionary calculus of aid-giving is much more complicated than indicated in the above sibling-saving illustration. For instance, if your siblings were beyond reproductive age and you were still within it, the calculations you would be assumed to make, unconsciously at least, would ease the balance toward your favor. But, *everything else being equal*, sociobiologists would predict that you would sacrifice yourself for your three siblings because of the high percentage of familial genes you share in common with them.

The effects of kin selection on aid-giving behavior are manifested in countless everyday settings short of the extreme "save your siblings" example we have used. Assume I have an indivisible piece of a life-sustaining resource that I do not presently need and several people surrounding me are in need of it. These people are related to me in varying degrees. Sociobiologists would predict that, if I give it away, I would give it to my child before I would give it to my niece, everything else being equal. This is because my children (one-half related) would, when mating with people unrelated to me, pass on one-half of their genes—and an average of one-fourth of my familial genes—to each grandchild ($\frac{1}{2}$ of $\frac{1}{2} = \frac{1}{4}$), whereas

Figure 2.
The Family Tree

(The fractions after each group of relatives indicate the degree of genetic relatedness of anyone in that group to Ego.)

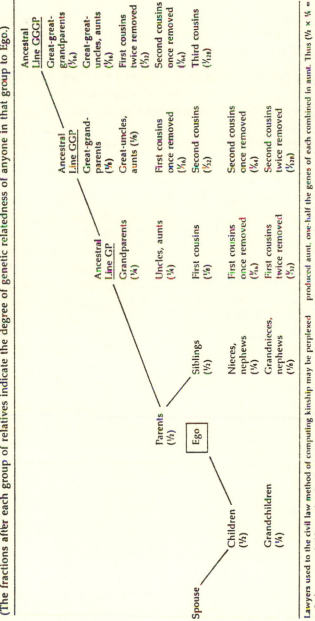

Lawyers used to the civil law method of computing kinship may be perplexed to find parents and siblings related to Ego in the same degree, as are grandparents and uncles/aunts. In the civil law method one counts up from Ego to the common ancestor, then down to the kin in question. This results, e.g., in aunts being considered, legally, as more remote "blood" kin than grandparents, though genetically they are the same degree distant. The genetic explanation is this: both grandparents are ancestors related to Ego by ¼. When they together produced aunt, one-half the genes of each combined in aunt. Thus (½ × ¼ = ⅛) + (½ × ¼ = ⅛) = ¼. Similar combinations occur when any two ancestors mate together, as when Ego's parents produced his siblings. When a relative mates with a (comparative) genetic stranger, the result is different. E.g., when aunt (related by ¼) mates with a man unrelated to Ego, the genes of Ego found in the resulting cousin include one-half the familial genes aunt has in common with Ego. Thus cousin and Ego are related by ⅛.

my nieces (one-fourth related) would reproduce an average of only one-eighth of those genes ($\frac{1}{2}$ of $\frac{1}{4} = \frac{1}{8}$) in each offspring. Similarly, sociobiologists predict, all else being equal, I would aid my niece before my cousin (one-eighth related) whose children would only have roughly one-sixteenth of my familial genes ($\frac{1}{2}$ of $\frac{1}{8} = \frac{1}{16}$). My cousin, however, would normally get my aid before a comparative stranger would.

Why do we aid strangers at all if natural selection has programmed each of us to act for the benefit of the gene package we carry? Here is where a second recent theoretical insight bearing on sociality comes into the picture. It is usually called reciprocal altruism and was first elaborated by Robert Trivers.[23]

The principle of reciprocal altruism assumes that evolutionary processes have selected for genetic programming that promotes aid-giving when an aid-giver has reasonable assurance that there will eventually be at least a substantially equivalent return for the benefit of his or her genetic material. This is the sophisticated scientific version of "I scratch your back and you scratch mine." Thus, it is assumed that genetic programming would be selected in the evolutionary process if it prompted the giving of life-promoting aid by A to B when the cost to A was small compared to B's need and the likelihood was high that B—or someone representing B—would return the favor. The forms of reciprocal altruism can be varied and complex. The key to it all is that natural selection has favored participation in reciprocal social behavior that tends to promote the reproduction of the participant's genetic package. Reciprocated returns for aid given may be experienced in the short or long run and may be experienced by persons other than the aid-giver, such as children, grandchildren or nieces and nephews who contain high degrees of genetic overlap with the aid-giver.

Commercial life insurance is a somewhat complex manifestation of the reciprocal altruism principle. In the life insurance scheme, one pays an amount of money into a fund that will benefit strangers. However, there is a good chance that a larger amount will someday, on the death of the insured, be paid out of the fund to (usually) people who are in a good position to promote the continuance and proliferation in the gene pool of representatives of the insured's gene package. The capacity to enter into all such arrangements may well have genetic underpinnings in that our genes equip us to learn and utilize the arrangements when they can promote the survival and reproduction of our genetic material.[24]

With all this discussion of genetic programming, we should not lose sight of the cultural input to human behavior. I mentioned rudimentary culture earlier when discussing our remote ancestors, their use of sharp stones and the evolution of their ability to learn from others. Today culture surrounds us, of course, in all its complex, multidimensional manifestations, ready to be learned and applied (as appropriate) in the pursuit of survival and reproduction. The commercial life insurance I just mentioned is a fairly complex cultural creation; so are mousetraps, the refrigerator, representative government, Zen Buddhism, child-care centers and the Australian crawl stroke.They are there waiting to be learned and applied to the extent that our genetic programming permits.

Many readers will be familiar with what was called the "nature-nurture debate." The central issue of this debate was whether our behavior was caused by genes or learned from culture. Among scientists, the issue no longer exists in an either/or form. It is fair to say that no serious scientist any longer believes that all human behavior is completely attributable to learning (acculturation).[25] Nor does any scientist today believe (if any ever did) that all human behavior is completely attributable to "hard-wired" genetic programming.[26] Now the issues (largely empirical measuring questions) are the ways and degrees to which genetic programming and culture interact and intermix with respect to particular aspects of our behavior. These issues are being actively pursued[27] and await clarification by social and natural scientists.[28]

What sociobiology offers is a new way of looking at human behavior.[29] It sees behavior varying from society to society as environments and cultures vary, because our behavior results from the interaction of relatively constant genetically programmed predispositions and the more varying natural, social and cultural environments we encounter. We human gene carriers are programmed to head toward a definite ultimate goal— optimum proliferation of our genetic package—and we must bounce off, adjust to and utilize whatever we encounter in our particular environments as we head toward that goal. Let me bring that high-flown metaphor back to a focus on a type of conduct that is of prime importance in this book: aid-giving behavior.

Our aid-giving behavior will vary depending on factors like the weather and the availability of resources (natural environment) and religious dictates (cultural environment). It will also vary depending on the people who surround us (social environment), the degrees of our genetic overlap with them, and such additional factors as their age and health, which bear on their capacity to reproduce and nurture. Let us narrow our focus still further and model another aid-giving dilemma. Assume that the social environment of A (aid-giver) is composed of just two people—A's child and *A*'s mother. Both need aid today, but *A* can aid only one. Which will it be?

Sociobiologists would note that *A* is related to both by one-half, but *A*'s child probably has more remaining lifetime in which to reproduce and nurture. However, if *A*'s child should be chronically ill, and *A*'s parent healthy, weight would be added to the parent's side in the evolutionary calculus. It becomes important, in this calculus, to determine whether *A* is male or female. if *A* is female, the balance may swing a bit toward the child; if male, toward the mother.[30] In part, this is because, although males and females are in the same position regarding identification of their mothers, no ascribed male parent can be as certain a child is his as an ascribed female parent can be that it is hers.[31] The reason for this difference in parenting confidence is quite simple: we apparently have no innate automatic mechanism by which to recognize close biological relatives so as to differentiate between them and comparative strangers.[32] Something in our environment tells us: A female is told that a child is hers by the objective evidence of its emergence from her womb. The evidence a male receives that a child is his is always less certain than that.

Another factor entering into the calculations as to whether aid will go to *A*'s child or *A*'s mother is the status of reciprocal arrangements between the parties. *A* may owe one of the contenders a favor, but not the other.

At this point, it should be obvious that it would be prohibitively difficult, though theoretically possible, to pinpoint individuals randomly in various parts of the world and predict with any certainty whether they will aid their child before their parent today. Their environments and cultures, the personal characteristics of their relatives and the particulars of their interactions with those relative will vary too greatly.[33] However, if all the people of the world are viewed together, sociobiologists feel comfortable in predicting whether the *typical* or *average* person living today would favor the child over the mother in our example. This is so, because with expanded population samples, extremes of environment and culture are neutralized; they counterbalance each other. Similarly, individual variations of aid-givers and contenders for aid counterbalance. For example, half of the aid-givers will be male and half female. Among the contenders in a large population sample, unhealthy children will be counterbalanced by unhealthy parents; for every parent who is owed a favor due to a reciprocal arrangement, there is a child who is similarly owed a favor. With such variables neutralized and thus washed out of the big picture, one is left with *universal constants* on which predictions as to typical behavior can be made with reasonable certainty. Thus, parents and children are *always* related by one-half, but one's child is *always* younger than one's parent. Therefore, in the typical case, one will aid one's child before one's mother because the child has more lifetime remaining in which to reproduce and nurture representatives of one's gene package.[34]

Similar calculations as to typical aid-giving behavior can be done with other sets of contenders for one's aid. let us retain the child in our model and replace the mother with a nephew. One's child is normally of the same generation as one's nephew (and in large population samples their amalgamated typical ages will be virtually identical); thus, the child should have no edge when it comes to life expectancy in the typical case. Nevertheless, the child wins because of its closer genetic relation to the aid-giver (one-half versus one-fourth). In the situation where the contenders are the aid-giver's nephew and aunt, although they are both related by one-fourth to the aid-giver, typically the nephew would win because nephews are normally two generations younger than aunts.

This is a good place to end our first evolutionary biology primer and move to our first illustrative social goal, the accomplishment of which may be furthered by evolutionary learning. I will add specific information on such learning as it becomes relevant in the social goal chapters, and we will eventually have another primer.

As a bridge to the next chapter—on reducing child abuse—let us make a final observation on aid-giving: the reverse of it, its negative mirror image, is failure to aid and, beyond that, actively harming others. If we have evolved to be inclined to aid close relatives before strangers, evolutionary logic holds that we should also be less inclined to harm the relatives.[35] For example, when we are in the presence of a tantrum-throwing two year old for an hour straight, typically we should be less inclined to be uncontrollably annoyed if the child is closely related genetically than if it is a comparative stranger genetically.[36]

Chapter Two

Reducing Child Abuse

Virtually everyone would agree that reducing child abuse is a desirable goal. In this chapter, we will see a science-based suggestion along that line, but first let us get a definition of the problem. What is child abuse?

Child abuse can take many forms depending on whose definition is used. It can range from homicide through lesser physical and psychological impacts and even include benign neglect. It can be debated whether a given form is bad generally, or bad for a given child at a given time. Debate on such issues by the experts has resulted, it has been said, in "definitional confusion."[1]

I suggest we use a sociobiological definition here: conduct toward a child is "abuse" if it is likely to decrease the child's "fitness"—that is, decrease the likelihood of the child's genes being proliferated in future generations. This definition will not, of course, preclude debate (did the spanking make the child more social, thus more likely to attract a mate with which to reproduce in later life?), and it has esoteric sociobiological problems of its own.[2] But it has the virtue of being potentially somewhat measurable: (For example, how many offspring did children thus treated eventually have, on the average, versus those who were not treated in that way?) We can get a handle on it, one might say. Can that be said about any other definition of child abuse? Certainly not if it hinges on things like the child's "endangerment" or "less than optimal development," which are definitions that experts have put forth.[3]

Furthermore, I believe it reasonable to assume that most instances of what popular consensus would consider child abuse would translate into being disadvantageous for the eventual reproduction of a child's genetic package. Let us assume so for our working generalization, while realizing that with any given type or instance of treatment of a child the issue may be debatable.

Now that we have some focus on the target, child abuse, we can look at the question of reducing it. Like most social problems, no single discrete solution of the entire child abuse problem is possible as a practical matter. We cannot send all adults on space ships to Mars, although that would eliminate active abuse (except for children abusing each other). So the realistic search is for partial solutions—steps taken that can each help to reduce the problem somewhat. Many approaches have been suggested and tried. Counseling known offenders may partially reduce the problem as may imprisoning them; providing in-home child-care assistance might also help.[4] These methods have all been suggested and implemented. Here I will make a new suggestion that emerges from recent evolutionary learning. It is directed to single, unmarried adults who have custody of their children, and it involves their choice of people to join in the parenting. A large part of child abuse is connected to the presence in the home of stepparents—people who have married one parent after the other has left the home by death or divorce. Let us look at some recent empirical data on the subject.

STEPPARENT CHILD ABUSE

Several recent studies have been done on child abuse statistics in Canada and the United States.[5] These studies have shown that the risk of child abuse is much greater in families where one of two parents is a stepparent than in those where both are biological parents.[6] One Canadian study showed that the risk in the stepparent home was approximately *seventy times* greater than that in the biological parent home for children 2 years of age or younger.[7] What is going on here? Traditional studies of child abuse do not have an explanation.

Numerous such studies have focused on causation or motivation, but no universal causes or characteristics of abusing parents have been suggested. Instead, the reports speak in terms of tendencies.[8] Thus, from a psychological perspective it has been reported that abusing parents tend to have high expectations of their children[9] and to have experienced abusive treatment from their own parents.[10] There are other reports of abusive tendencies in parents stemming from premarital pregnancies[11] or illness.[12]

Sociologists have emphasized "external factors" such as problem children, large family size and economic hardship.[13]

None of the various factors identified by researchers as correlated with child abuse appears to explain why abuse in stepparent homes should be so comparatively high.[14] It was suggested that poverty in stepparent homes could be the answer.[15] If so, one should find that stepparents tend to be concentrated in lower socioeconomic groups. Researchers Martin Daly and Margo Wilson looked into this matter and found that, within two-parent homes, the proportion of children living with a stepparent is virtually identical on all socioeconomic levels in the United States and Canada.[16] Daly and Wilson also controlled for other factors, such as family size and reporting bias, that might account for the difference in abuse in two-parent homes where one is a stepparent and those where both are biological parents. They concluded that "[s]tepparenthood *per se* remains the most powerful risk factor for child abuse that has yet been identified."[17] If so, why should that be? Sociobiology may have the answer.

OTHER PEOPLE'S CHILDREN

Needless to say, the overwhelming majority of stepparents are not close genetic kin of the stepchildren. Single parents do not often marry close relatives of their own or of their children's other parent. This fact that stepparents are most often "genetic strangers" of their stepchildren sets the foundation for a sociobiological explanation of child abuse. As we saw in Chapter 1, sociobiologists hypothesize that an adult will be more inclined to aid a child identified as a close genetic relative than one who is a comparative stranger, everything else being equal. Not harming is, effectively, a reverse form of aiding; thus, evolutionary principles suggest that adults will be less inclined to abuse a child the closer the genetic relationship thought by the adult to exist between him or her and the child. Considerable empirical research supports this proposition.[18]

Thus, abusive stepparents may simply be reacting as genetic strangers with little biological predisposition to nurture their stepchildren and no early bonding with them.[19] They therefore have less motivation than natural parents to restrain themselves when they encounter pressures such as a child's irritating conduct. Under this explanation, they are motivated no differently than any other genetic stranger. They are merely more often and continuously in the presence of the child, so there is more opportunity for the abuse to occur.[20]

We should note here that couples who adopt a child unrelated to either of them are also genetic strangers to the child. Is there also, then, typically more risk of child abuse in such homes than in two-biological-parent homes? Not necessarily. Statistics on abuse in adoptive parent homes are lacking,[21] but Daly and Wilson suggest that such factors as the mutual desire of adoptive couples to stimulate a natural care-giving situation, as demonstrated by their decision to adopt, could distinguish the adoptive and stepparent situations.[22] And here we must give some attention to another possible distinguishing ground that many find unpleasant to contemplate—the possibility that humans have some genetic programming that *affirmatively* prompts them toward child abuse when they enter the role of stepparent.

"STEPPARENT" INFANTICIDE IN OTHER ANIMALS

During the 1970s and 1980s considerable interest and controversy surrounded reports of infanticide by other animals when in positions similar to that of human stepparents.[23] In independent observational studies, primatologists Sarah Hrdy,[24] first, and, later, Paul Newton[25] both reported patterns among langur monkeys in which males take control over one or more females and their young from another male and then kill the infants. One explanation for this behavior is that the newly arrived male is thereby able to reproduce with the female sooner. This is because milk production in a mother interferes with conception and permanently removing the infant causes milk production to cease. Similar instances of infanticide by males have been reported in other types of monkeys[26] and in lion prides as well.[27] In the case of the lions, the newly arrived male killed the infant cubs and drove off those that were older. This evidence from the animal kingdom has prompted an extension of evolutionary logic to cover the situation—namely, that some animals may be predisposed to behave toward unrelated youngsters in ways that are conducive to liberating the reproductive and/or nurturing capacity of the actor's mate for use in producing and nurturing the actor's own offspring.[28]

This thesis could apply to females as well as males. It is unlikely that the biological programming of female mammals would have evolved so as to create a concern for liberating their mate's reproductive capacity because it would appear that most male mammals are physically capable of inseminating regardless of the existence of young offspring. But female programming could have evolved so as to create a concern for liberating the male's protective and nurturing capacities.[29]

If the above thesis should prove to be valid and applicable to humans, it would, for most of us, be a distasteful revelation of evolutionary biology. In any case, it need not apply in order to explain stepparent child abuse statistics from a sociobiological viewpoint. That might be done sufficiently by the first thesis we noted: that stepparents are normally genetic strangers who are not programmed to be highly solicitous for the children in-volved—as are close genetic relatives. Stepparents choose to marry the natural parents of accompanying children, and, except for the marriage, few stepparents would have independently chosen to live with the children as adopting couples typically do. Nevertheless, while not being pro-grammed to be highly solicitous for the children, the stepparents find themselves in close, prolonged proximity to them and experience the usual irritations that can come from such proximity. With that explanation for stepparent child abuse, we can return to our original question: what does sociobiology in this area suggest to social planners and parents about reducing child abuse?

PROMOTE MARRIAGE OF SINGLE PARENTS TO THEIR IN-LAWS

Biological aunts, uncles, grandparents, parents' cousins and so on, should be less inclined to abuse their young relative given a certain provocation than they would be to abuse a young stranger given the same provocation. This suggests that a single parent looking for a mate would do well, in the child's best interests, to mate with the child's uncle, aunt, etc. The single parent's own siblings and other relatives are the child's uncles, aunts, and so forth. In virtually all societies, however, mating with one's own sibling or other close genetic relative is taboo.[30] Thus, the indicated desirable mates from the child-care standpoint are the siblings, etc., of the other biological parent of the child (the single parent's "in-laws" if the parents were married).

The people who are most closely related genetically to a child after its parents, who are related by one-half, are grandparents and uncles/aunts, who are all related by one-fourth. So the single parent's in-laws related to the child in those categories should typically be the best stepparent candidates. Other in-law adults of an appropriate age to be a mate for the single parent are likely to be much more distantly related to the child (other parent's cousin [one-sixteenth related], for example). Therefore, whatever genetically programmed "give aid/don't harm" inclinations they have toward the child should be comparatively faint in those people. Neverthe-less, on the average, it should be better for children to have even fairly remote biological relatives as stepparents rather than strangers.

Social planners, advisers and single parents themselves should duly note the "on the average" qualification in the last sentence. Put another way, the child's genetic relatedness to stepparent candidates should obviously not be the only consideration if the best interests of the child are the focal consideration. For instance, a single parent may not be well advised to select a stepparent candidate with a quick temper and a history of violence simply because the candidate is a close biological relative of the child. The cautious form of advice here is as follows: *When all other characteristics are essentially equal between desirable stepparent candidates, select the one most closely related genetically to the child if you wish to avoid child abuse.*

If such advice were followed by a substantial number in a society, evolutionary learning predicts that the incidence of child abuse would be correspondingly reduced.

IN-LAW MARRIAGE PRACTICES WORLDWIDE

I will not presume here to tell social planners how they might get the "marry your in-laws for the good of your children" message across to their single-parent populations if the idea should appeal to them. I am confident that modern communications would provide the means.[31] One nationally televised situation comedy in which a particularly loving uncle of the children involved marries the widowed mother—his sister-in-law—would go a long way toward promoting the idea, in the United States at least.

In some countries, however, such as England and the United States, the in-law marriage idea may seem a bit bizarre. As a matter of fact, until recently at least, some forms of in-law marriage were legally prohibited in England and a few U.S. states for reasons that are now obscure.[32] These restrictions have now been essentially eliminated, but the attitudes they reflected could still exist in Anglo-American culture. If so, precedent for in-law marriage may be useful to social planners—and it exists in abundance.

A *preference* for marriage by widows and widowers to brothers-in-law and sisters-in-law has existed and now exists in a large number of world cultures.[33] The reasons that have been given for the practice are several and vary from society to society, but among them is the following observation by Edward Westermarck in his book, *The History of Human Marriage (1925):* "Moreover, as the children of a widow are best cared for if their step-father is their uncle, so the children of a widower are best cared for if their step-mother is their aunt."[34]

With the insights of modern genetics and Hamilton's theorem on kin selection, one might add that the average grandfather and grandmother, being also related by one-fourth, should be equally as desirable stepparents, except perhaps for the fact that they will be older than aunts and uncles and could be less (or more) solicitous caretakers for that reason. Furthermore, everything else being equal, a cousin of the deceased or departed parent (one-sixteenth related to the child) would typically be a better bet as stepparent than a comparative stranger—if reducing the likelihood of child abuse is the focal consideration.

Chapter Three

Reducing Sibling Incest

I believe it fair to say that most, if not all, societies frown on sibling incest—sexual relations between brothers and sisters[1]—and, by consensus, would subscribe to reduction of its incidence as a societal goal. We will take this statement to be true and examine a surprisingly easy adjustment, suggested by evolutionary learning, that many families might make to help achieve this goal.

Incest has been variously defined.[2] Some want to include only sexual relations between two people of opposite sex who are close genetic relatives.[3] In respect to people of the same generational level, this usually means full or half brothers and sisters (siblings). Some commentators, however, might want to include other relatives, like cousins or even relatives through marriage, such as stepsiblings, who are not close genetic relatives. For our purposes we can include all such same-generation categories in the incest definition. Our focus will be on *reduction of the possibility of sexual relations after puberty of any two people of opposite sex who are close to the same age.*

THE SCIENCE OF SEXUAL CONTACT AVERSION
OR NEGATIVE IMPRINTING

Normally, when one reports on human sociobiology, the relevant hypotheses concerning behavior are set out, followed by any empirical evidence that supports or refutes them. This is partly because empirical research is expensive and time consuming, so an infant discipline like sociobiology often has little of it to report. Here, however, I will reverse the order of presentation because the supporting empirical evidence is quite strong. It makes sense to set it out as an apparent, surprising fact of human behavior and then turn to evolutionary biology for a theoretical explanation of why it should be so.

KIBBUTZIM KIDS, CHINESE INFANT BETROTHALS AND OTHER CLOSE YOUNG ASSOCIATIONS

Starting in the middle of the twentieth century, Israelis experimented with communal living in communities called Kibbutzim. The program included raising children, away from their parents, in communal living quarters. Thus, large numbers of unrelated children were sleeping in close proximity, eating, playing and otherwise associating together throughout early childhood. Later studies of these same people as adults showed that they had an aversion to sexual relations with, and did not marry, one another, even though public opinion was not against such unions.[4]

A similar phenomenon has been reported in studies in China where unrelated children were raised together in an institution called the *sim pua* marriage. Here parents chose spouses for their youngsters at an early age, and the pair were raised together in the home of one of the sets of parents. Researchers have reported a clear reluctance of these pairs to consummate the marriage upon reaching adulthood.[5] It is reported that parental pressure generally overcame their reluctance, but the reproductive rate for such couples was one-third below normal.[6] In addition, these couples had a higher rate of extramarital activity and a higher rate of divorce than was normal (24 percent versus 1.2 percent).[7]

The same type of result—lower fertility and higher divorce rates than normal—has more recently been reported for marriages between patrilineal parallel cousins (children of brothers) in a study of a Lebanese village.[8] The interaction of such cousins as youngsters was much the same as that of brothers and sisters in this Lebanese culture.[9]

The mating aversion or disinterest in these three reports, and others,[10] echoes a hypothesis first proposed by Edward Westermarck in 1891.[11] This hypothesis states that aversion to sexual relations will occur between people living in close proximity to one another from early youth.[12] The Westermarck hypothesis has received considerable elaborative explanation during the latter part of this century owing to rapid advances in evolutionary biological learning.[13]

THEORETICAL EXPLANATION FOR THE
REPORTED BEHAVIOR

The starting point is the factual observation that, when close genetic relatives reproduce together, "inbreeding depression" (a lowering of vitality of the offspring) tends to occur.[14] The end result of this negative inheritance process is a high percentage of stillborn or defective offspring.[15] This result is particularly likely with matings between relatives related by one-half, that is, between parents and their children and between brothers and sisters. Vitality of offspring is, of course, a crucial aspect of successful reproduction, so it is assumed that natural selection through the ages would have provided in some way, in the behavior of humans, for the avoidance of inbreeding depression.[16]

Natural selection proceeds economically. If more than one genetic program develops in a species gene pool that perform a needed function for the species, the one that can do the job with the least expenditure of body- and mind-building materials and energy will tend to prevail eventually. For example, a genetic program could have theoretically developed in some individuals whereby they could automatically recognize close kin upon first sight (or smell) for purposes of avoiding sexual relations with them. However, such sophisticated, high-energy consuming "equipment" would lose out in the gene pool to a program that, in effect, told the individual simply to avoid sexual relations with any person to whom it was in regular close proximity during its early youth, if this simpler program did the job in the environments in which humankind existed when the programming developed.[17] Let us look at the hazy picture of humankind social environments ten thousand to one million years ago (the late Pleistocene era) when most of the slow-to-spread genetic programming humans now have is believed to have settled into place in response to the environments then existing.[18]

Some reconstruction of the Pleistocene era has been possible using archaeological evidence.[19] This has indicated, among other things, that in those times hominids (those in the human ancestral line) were hunter-gathers.[20] Moreover, it is assumed that modern-day, isolated hunter-gatherer groups in Africa, South America and elsewhere have been living lives that are not too different from those of the Pleistocene.[21] Furthermore, scientists cautiously look to other primate species—chimpanzees, gorillas and such—for hints as to the lifestyle of remote human ancestors.[22] Studies of these present-day groups suggest that certain social conditions probably existed in the earlier times. For example, we can assume that during the Pleistocene era the people in close, continuous proximity to a growing youth were almost always close relatives (most one-half related, some one-quarter, an occasional one-eighth, etc.)[23]

Thus, genetic programming that, in effect, said "avoid sex with those who were close, continuous associates during your early youth" should have provided a sufficient identification mechanism for avoiding virtually all inbreeding depression. We can assume that occasionally a Pleistocene era infant was, for some reason, displaced from its birth family and raised by another to which it was not closely related and then, by chance, had a mating encounter as an adult with a member of its birth family. However, such occurrences would have had to be very frequent before this identification mechanism would have become so inefficient for its purpose that it would have lost out in evolutionary competition to more foolproof but expensive mechanisms.[24] Therefore, the existence, right up to the present, of a genetically influenced mechanism that makes humans averse to sexual relations with close, continuous childhood associates makes sense from an evolutionary perspective.[25]

That, then, is an evolutionary biological explanation for the fact, as we noted, that in three different world cultures where youngsters other than siblings happened to be raised together in large numbers, generalized sexual aversion later occurred between them. The occurrence of this phenomenon across cultures strongly suggests it is prompted by a universal human characteristic largely independent of culture.[26] Evolutionary learning says it is based in genetic programming, and that is hard to refute.[27]

If this incest avoidance mechanism is universal in humans, it exists in youngsters anywhere in the world where society wants to reduce incest among contemporaries. And where such a goal exists, practical hints for parents are available in evolutionary biological learning. Here they are:

PRE-TEEN COED PROXIMITY

If (1) the focal problem with incest is stillborn or defective offspring, and (2) conception of such offspring cannot occur until the puberty of parents, and (3) negative imprinting regarding sexual relations occurs with close, continuous proximity of youngsters to each other, then a program is suggested for child caretakers who wish to discourage sexual relations between youngsters after puberty: *Take whatever steps are practical to keep the youngsters in close proximity to one another in their prepubertal years*. The ages 2 to 6 appear to be the most sensitive for this purpose.[28]

The sexes should not be segregated and maximum contact should be encouraged. In particular, they should be encouraged to play and study together. They should bathe together. Probably the program with the least supervision necessary compared to the payoff is to have them sleep in the same bed—or room, at least—even when separate facilities are available. Currently this may be done by a majority of families of necessity, but many, for one reason or another, will separate the sexes early.[29] If this is done with the thought of discouraging sexual intimacy,[30] scientific evidence suggests that it is shortsighted and misdirected. Conception (pregnancy) is not possible, of course, until both the male and female involved have reached puberty.[31] And prepubertal intimacy may be the best insurance against postpubertal intimacy with resulting pregnancy.

Again, I will not presume to tell social planners how best to get this message to child caretakers. Undoubtedly, however, the introduction of child-care experts, child psychologists in particular, to the evolutionary biology we have discussed here would be a desirable step. Beyond that, the media might be employed in the same way that it has been used to warn the public of the dangers of certain toys on the market.[32]

Even if a large percentage of caretakers in a population got the message and increased the contact of prepubertal children whom they wished to discourage from sexual relations in later years, it would not, of course, completely prevent postpuberty incest between those children. Little, if any, human behavior is caused by a single factor. Most is caused by a combination of genetic programming, life history and the individual's environment.[33] Although much genetic programming is consistent and universal, life histories and environments are infinitely varied. We know that people will have sexual relations for various reasons with individuals to whom they have slight or no physical attraction. Prostitutes do it routinely for economic reasons. The people who participated in the Chinese *sim pua* marriages, discussed earlier, apparently did it as a result of parental pressure, although with little enthusiasm because of the "negative imprinting" they had received from being raised together.

Nevertheless, if a large segment of a population were to take steps to increase the contact of related or unrelated contemporaries of opposite sex from birth to puberty, scientific evidence points to the probability that an appreciable reduction would occur in the incidence of sexual relations between those contemporaries after puberty.

Chapter Four

Distributing the Property of People Who Die Without a Will

Societies everywhere face the question of what to do with the distributable property of the dead when relatives who survive cannot agree and the deceased did not make her wishes clear in some accepted manner. (For simplicity here we will call the "typical" deceased a "she.") Often the lawmakers' official solution has been to attempt to do what the deceased would have wanted if she had expressed her wishes before dying. Of course, the wishes of any particular individual dying in the future cannot be known at the time laws are written, so lawmakers have enacted schemes that seek to reflect what the "typical" deceased would desire.[1] This has frequently been little more than consensus guesswork or a reflection of the personal preferences of the lawmakers themselves. Presumably they would want to know anything that is relevant to their goal of reflecting what the typical deceased would want. If so, evolutionary learning may have something for them. In previous chapters we have noted what sociobiology has to say about the aid-giving tendencies of all humans living today. When people express their desires regarding how their property is to be distributed after their death, it is a concentrated manifestation of those tendencies.

We have noted that an aid-giving act is in response to (1) genetically programmed inclinations, (2) lifetime experiences, including human contact and (3) the actor's environment at the time of the act, including the social environment—the people around her. Certain aspects of lifetime experience and environment can be vastly different from person to person. But there are some characteristics of the environment—the social, human part of it—that are the same everywhere and have been since time immemorial.

The genetic programming in humans has been reacting and adjusting to these universal human characteristics for all those eons. If we isolate these characteristics, we will have elements from which to sketch "typical" aid-giving behavior. When the differences in life experiences and environment from individual to individual are washed out of the picture, the universals will remain.

Here, then, are relevant human universals that have been obvious for eons: parents are always older than their children; younger people, on average, have longer life expectancies than older people; reproduction is not possible before puberty; reproductive capability gradually diminishes after puberty; and infants and youngsters are not self-sufficient or capable of giving much, if any, aid to others. To these elements we can add facts that people may have vaguely sensed for ages, but science has only recently brought to light: We are genetically related to our parents by one-half, our grandparents by one-fourth, our great-grand parents by one-eighth, etc. Finally, we can add the evolutionary biological propositions that we are genetically programmed to aid those to whom we are most closely related and the younger before the older, all else being equal, if we must choose. Using this list, we can deal with the question of what humans will typically want done with their property after they die.

Probability theory suggests that, *at the moment of their death*, people may typically choose to give all their property to one person—the one who is best able to reproduce the dying person's genes. The applicable principle here is avoidance of the "gambler's fallacy."[2] Experienced gamblers know they should put all their money on the one horse that is most likely to win. I think the reader will join me in finding this notion counterintuitive when applied to leaving our property to relatives who survive us. For example, if we have children, most of us picture giving something to each of them. This could be because we are not contemplating dying soon and (unlike the gambler and the horses) want to maintain good reciprocal relations with each of our children in the meantime. The person who knows she is dying immediately, on the other hand, would have no future in which to have reciprocal relations with anyone. She might, therefore, be inclined to put all her money on one relative. Be that as it may, we cannot know the personal characteristics (reproductive potential) of, for example, the children of our hypothetical typical person dying in the future. All we can know is the universal characteristics that apply to them. Therefore, in our upcoming exercise, we will conclude that relatives with the same universal characteristics would be treated equally by the typical deceased. Presumably, this would mean that each such person should receive an equal share of any dollar amount of distributable property.

When the survivors have different universal characteristics, such as one being a parent and another a child of the deceased, so that the typical deceased could be predicted to be more inclined to aid one than the other, we will conclude just that. Whether that means all distributable property should be given to the preferred person or shared with others in some way, we will have to leave to the lawmakers. Insights from evolutionary biology do not permit us to be more specific.

In trying to draw information from evolutionary biology to assist lawmakers here, we have a time-frame problem. When the lawmakers tell us they are trying "to reflect the normal desire of the owner of wealth as to the disposition of [her] property at death,"[3] on what exact point of time in the life of the deceased do they want to focus? Are they picturing the person, orally or in writing, making her wishes known some appreciable length of time before she dies when she is uncertain as to how long she will live? Or are they picturing the person on her deathbed just before her last breath when she is quite certain she is dying? I know of no indication from lawmakers as to which scenario they have in mind. It makes a difference for our purposes, because when a person is uncertain as to when she will die, she is likely to reflect reciprocal altruistic considerations, explicitly or implicitly, in her pronouncements—such as "I am tentatively giving this to . . . in the expectation that he will . . . for me until I die." She may change her will if the expected return does not materialize. But when a person's death is to be immediate, there is no future in which to get a personal, reciprocal return. This could make a significant difference as to whom she wishes to receive her property. Now she may want someone, like an infant, who is incapable of aiding her, to receive something. Because we do not know whether the pre-death or the moment-of-death time frame is what lawmakers have in mind, we will explore what evolutionary biology suggests as to both scenarios.

In most societies that have spoken to our property distribution problem, the rules dictate that the property go to relatives of the decedent in certain orders of priority. The possible patterns of surviving relatives among any large sampling of dying people are innumerable, but formulas can and have been worked out in most developed countries that cover every contingency.[4] We cannot consider every possible combination here. An earlier book of mine treats several combinations containing various surviving relatives, including parents, spouses, half-"blood" relatives, adop-

Figure 3.
Lineal Descendants

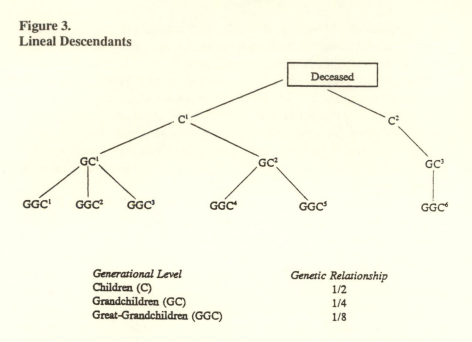

Generational Level	Genetic Relationship
Children (C)	1/2
Grandchildren (GC)	1/4
Great-Grandchildren (GGC)	1/8

tives, "illegitimates," and remote relatives.[5] Here we will explore only those situations that are likely to hold the widest general interest.

These are the cases where biological children, grandchildren, great-grandchildren, and so on down the line, in various combinations, are the only close surviving relatives. I will sketch what evolutionary biology suggests the typical deceased would want in these situations. First, we will consider them in the moment of death scenario, where the possibility of our deceased personally receiving reciprocated aid is out of the picture, and, later, in the pre-death scenario where it could still be a consideration.

THE MOMENT OF DEATH SCENARIO

Assuming that only children, and no more remote descendants (see Figure 3), survive our typical deceased, she should want the children treated almost the same. Since all the children are related to the deceased by one-half, each can produce and nurture offspring related to the deceased by one-fourth. Some difference in treatment could be prompted, however, based on large age differences between the children. From birth to puberty a child's reproductive value, from its parents' genetic standpoint, increases steadily. This is because the child cannot reproduce at birth and may die before it can. The closer it gets to puberty, however, the more likely it is to arrive there and reproduce. It has been observed that this increasing value scale may be offset by a steady decrease in the child's need for parental aid. The newborn, although not as valuable as a reproducer of a parent's genes as is a child at puberty, is less self-sufficient and therefore needs more aid, which must be provided if the parent's initial investment is to come to fruition.[6] If we accept these observations, we can assume that our typical deceased would want to treat two or more prepubertal children essentially the same regardless of their exact ages.

At puberty a child is at peak value as a reproducer of his or her parents' genes.[7] From puberty on, this value decreases steadily because as the child ages its remaining years of life expectancy and, later, its reproductive capacity, decrease. So, assuming there are three children aged 1, 15, and 25, it is likely that our typical deceased would want to aid the first two equally and give the 25 year old less aid than the others.

The same considerations we have been exploring are applicable whenever all surviving descendants are on the same generation level—for example, when no children but several grandchildren survive. Before adding the complication of descendants surviving the deceased on different generation levels, let us clear the board by assuming that our deceased would want to treat equally everyone on the same level. For the purposes of our exercise, this seems sensible as a practical matter, notwithstanding the considerations that could distinguish descendants on the same level. Standards do not exist, at present at least, for quantifying, on a scale of 100, let us say, the reproductive and nurturing value of a 15 year old versus that of a 25 year old. Formulas of this sort would be necessary to write a law distinguishing between descendants on the same generational level.

Now let us assume that there are children and grandchildren surviving. The grandchildren are related to the deceased by one-fourth and can produce offspring related by one-eighth, whereas children (one-half) can produce offspring related by one-fourth. Thus, as reproducers of our deceased's genes, the offspring of a child are twice as valuable as the offspring of a grandchild. This fact, considered alone, would suggest that each child would receive twice as much as each grandchild. However, the grandchildren of our typical deceased are one generation younger than the children and have more reproductive and nurturing time available to them. That would point toward an equalization of the amount of property that our deceased would want to give to children and grandchildren.[8] A grandchild would receive *something more than one-half* of what a child receives. That is as specific as we can be with the tools we have at hand.[9] This same sort of analysis would apply when yet a third generational level is added to the mix. Thus, the analysis would call for giving great-grandchildren something more than one-half of that given to grandchildren who, in turn, would get something more than one-half of that given to children.

Assuming a child and one of its offspring (or a grandchild and its offspring) survive the deceased, how would she want her property distributed as between them? We just saw that the offspring, as well as its parent, would get something. But that need not mean that the offspring's share would be given directly into its hands. Our typical deceased will be aware that the young and immature are unlikely to apply their resources wisely. Because parents are the adults most likely to be solicitous for a child's welfare, they are the most logical people to hold the children's resources and apply them for their benefit as needed. Thus, our typical deceased would want the immature child's share delivered to its parent together with the parent's share. On the other hand, when the child in any parent-child set is mature, the resources should be given directly to the child because the typical mature individual is more likely than any custodian, including its parent, to use resources for its own benefit.

THE PRE-DEATH SCENARIO

Here our typical deceased is making her property distribution wishes known an appreciable time before her death when she is uncertain as to exactly when she will die. There is time for her to receive aid from those around her—and perhaps for many years. All the considerations from the moment of death scenario hold here, with the addition of the possibility that her wishes regarding where her property should go are influenced by reciprocal altruism. She may express her wish that her property be given to certain persons, or to them in larger amounts, than she would if she were, to the best of her knowledge, at death's door because those people are in a position to aid her in return. Conversely, she may designate less or nothing to people who, because of age, illness, or other reasons, are unable to help her.

Two of our universal working elements bear on the ability to reciprocate—age and remaining life expectancy, which are closely tied. At one end of the age spectrum, an infant has virtually no ability to reciprocate but a relatively long life expectancy in which eventually to do it. Toward the other end of the age spectrum, the mature descendant has less life expectancy but is presently capable of reciprocation. Nice quantification of these considerations is not possible. What we can say is that when age and life expectancy alone are considered there is no obvious clear advantage to either young or old descendants on the potential reciprocation question.[10]

When, however, two or more generations are in the picture (for example, children and grandchildren), some weight will be added to the older generation's side of the scale owing to the closer genetic relationship of the older generation to the deceased. Everything else being equal, close genetic relatives are predicted to be more reliable as reciprocators than more distant relatives.[11] So, regarding our typical deceased's pre-death wishes, we can say: as compared to what she would want at the moment of death, here she could be expected to *somewhat* (we cannot be more specific) favor children (one-half related) over grandchildren (one-fourth related) and grandchildren over great-grandchildren (one-eighth related).

MOMENT OF DEATH VERSUS PRE-DEATH SCENARIO

Our theorization has produced an indicated difference in the treatment of descendants between the moment of death and the pre-death scenarios. The difference occurs when the people to whom property is to be distributed are on different generational levels. In the moment of death scenario, recall our conclusion that people on the younger generation levels would get something more than one-half of what those in the generation above them would receive. For all we can say, that something could bring the younger level up to an equal share or a larger share than the older level.

On the other hand, our theorization regarding the pre-death scenario is that *in addition to* the intergenerational considerations of genetic overlap and age that are the same in both scenarios, one must weigh the likelihood of reciprocation factor. And that factor favors the older generation somewhat. So if one chooses the pre-death scenario, there is less reason to treat the younger generation equal or better than the older than if one chooses the moment of death scenario. This distinction could constitute a rule of thumb for lawmakers: the younger generation gets relatively more or less of the estate depending on which scenario the lawmakers choose.

EMPIRICAL TESTING

It would be virtually impossible to interview a sampling of people at the moment of their death in order to test our theorization regarding how the typical person in that situation would want her property distributed. Such people, if their illness or injury permitted them to communicate at all, would likely be too distracted to concentrate on the question. However, empirical testing of theory regarding the pre-death scenario is possible.

One testing ground for the theory regarding what the typical deceased would want in the pre-death scenario would be a study of actual wills that people write and leave intact at their death.[12] However, in the United States at least, actual wills on record in official offices have not provided researchers with sufficient information to cast much light on the surviving descendant questions we have been exploring. Thus, studies of wills that have been done in the United States are of little use to us. However, U.S. interview studies of people (who may or may not have actually made a will) exist that do provide data bearing on our questions, although none, it would appear, was conducted by researchers with sociobiological theory in mind. Nevertheless, their data are consistent with our theoretical pre-death wishes scenario in which reciprocal altruism is in the picture.

Interviewees have overwhelmingly agreed that children should be treated equally regardless of age[13] or sex.[14] They have similarly agreed that grandchildren should be treated equally among themselves regardless of their position in the "branches" of the family tree (see Figure 3).[15] This is consistent with our theoretical pre-death wishes scenario. Recall that we could not see comparative ages affecting the result on the same generation level when likelihood of reciprocation was a consideration.

Some interviewers have asked questions involving survivors on different generation levels. For example, what would be the interviewees' distribution preferences, in terms of percentages of the estate, if only (in Figure 3) child 2 and grandchildren 1 and 2 (nieces/nephews of child 2) survived? The results showed a tendency to give more to the child than to each grandchild.[16] In the most particularized report, 42 percent gave equal share to all three. However, 41 percent gave the child twice or more what they gave each grandchild.[17] Recall our theoretical conclusion that, as between different generation levels, those on the level most closely related to the typical deceased would be better treated in the pre-death scenario than they would be in the moment of death scenario because of the likelihood of reciprocation factor.

When interviewees were asked their preferences when the possible distributees were a child and a grandchild (offspring of the child), they tended to give the child all the property or at least more of it than the grandchild received.[18] With that in view we should remember our hypothesis that our typical deceased would want a grandchild's share to be put into the hands of its parent when the grandchild is immature. The results of the interview studies would be consistent with this hypothesis if the interviewees were visualizing a large percentage of the grandchildren as immature. The studies apparently did not control for this factor.

AID TO LAWMAKERS

Any society seeking to distribute the property of its decedents in the way they typically would want it done should decide whether it wishes to reflect the pre-death or the moment of death scenario. If it is pre-death, well-planned empirical studies in the population could supply helpful information. But many societies are not in a position to have such studies done in the near future. Nevertheless, people will keep dying and schemes will be needed for distributing their property. Until empirical data are available to lawmakers in these circumstances, they would do well to make calculations such as we have illustrated in this chapter, using universal characteristics of humans.

If a society wishes to reflect a moment of death scenario, an empirical study is a practical impossibility. Therefore, calculations of the sort we have illustrated will be the only available supplement or corrective to the lawmaker's intuition or personal preference.

Lawmakers might look to the pre-death scenario based on the rationale that a law reflecting it would provide an effortless, no-expense will for people at the will-making time of their life.[19] By not acting at all, one could consciously choose the property disposition that the law provided in such cases! Even in a developed, highly literate country, however, few people will actually know what the rules for the various survivor patterns are without an expert's advice.[20] (How many of our relatives and friends know the rules that would apply to them?)

I think it fair to say that in the foreseeable future, the vast majority of people around the world who have no will when they die will have been unclear, at best, as to what their law provides in such cases. Under those circumstances, lawmakers may prefer to attempt to reflect the wishes of people who do not know the law and never get around to making a will, for one reason or another, before they die. That brings us up to the moment of death. And if moment of death is the point of focus, then evolutionary biology will be a particular value to lawmakers because, as I have said, empirical data on the typical deceased's wishes at that instant are unobtainable.

Chapter Five

Evolutionary Biology Primer II: Reproductive Strategy

At this point, we need to introduce some more basic evolutionary biology because in our next chapter, on rape reduction, we will be dealing with evolved reproductive strategies and mechanisms about which we have said little thus far.

Much sociobiology is rooted in known biological facts. From these facts, logical hypotheses are derived. A starting point with regard to reproductive strategy is the fact that the ovum, or egg, produced by the human female is many times larger than the sperm produced by the human male.[1] Furthermore, in humans, gestation takes place in the body of the female parent (in some animals it occurs in the male)[2] where she contributes considerable energy and body-building resources. Thus, at the moment of any birth resulting from fertilization of an egg by a sperm, the inevitable investment of energy and resources in the offspring by the female has been vastly larger than that of the male.[3]

Compared to a male, however, a female has a severely limited number of reproductive opportunities. Typically, a human female can contribute only one egg to conception efforts each nine months during her reproductive years. A male, on the other hand, is limited biologically in his reproductive efforts only by the number of sperm-rich ejaculations that he can accomplish in a given period of time. This can be several times in a day and many thousands in a lifetime. Thus, the male is much better equipped by nature to reproduce frequently than is the female.[4]

Another result of gestation of offspring in the female body is that each time a female conceives she can be certain the child is hers. A male can never be as certain that he is the father. Thus, the female has comparatively limited reproductive capacity but assurance of parenthood,[5] while the male has less assurance with much greater capacity.

Given these facts about reproductive apparatus and knowledge of the processes of natural selection, sociobiologists hypothesize that human males will be genetically predisposed to seek to inseminate females as frequently as physical capacity *and environmental and social restraints* permit. In other words, clear biological facts suggest that typically, a male will seek copulations with fertile females as often as permitted by his energy, the females, other males, societal rules, his resources, and other limiting factors in the world around him.

Natural selection principles indicate that males who were genetically programmed for such behavior during evolutionary history would have been the most successful reproducers. Any emerging male genetic programs that encouraged abstinence or limited copulatory activity *regardless of environmental and social restraints* would have been less successful at reproduction in the evolutionary long run and would eventually disappear from the human gene pool. It is assumed, therefore, that all males today will be the descendants of relatively successful reproducers and thus normally be genetically *inclined* to frequent (compared to females) copulatory activity with a variety of partners. That is the hypothesis. Popular impressions and empirical research on human sexual behavior support the hypothesis.[6]

To lay the groundwork for our upcoming chapter on rape reduction, we should continue to focus on male reproductive behavior. First, however, we should glance at the sociobiological view of the female counterpart. The two have evolved together in both concert and conflict throughout evolutionary history, much as the path of a river and its shoreline are formed by each other.

It is assumed that the most reproductively successful females in the evolutionary past (those whose genes are represented in today's human gene pool) were comparatively cautious about mating. If their relatively expensive and few reproductive opportunities were to reach fruition—if their offspring were to reach adulthood in a good position in turn, to, reproduce—it was important that the offspring have good health. The selection of a mate in good health would help to assure that. Resources of various sorts were needed for nurturing and the right mate could help to supply them. The reproductively successful females, therefore, became choosy.[7] Timing was also important. There were good times to have offspring and times when it was best to wait.[8] So females became cautious. Let me take a moment here to emphasize this point about female choice and explain its relevance to rape trauma, the female reaction to the subject we will deal with in the next chapter.

Sociobiology suggests that a human female is programmed to be choosy and cautious in selecting those with whom she will copulate. What happens when that slow, careful approach is overcome by a stronger, aggressive male—particularly one she would not have chosen? Evolutionary biologists would hypothesize that something in her bodily system will react negatively to the thwarting of her strategy. That reaction may be physical, but a psychological reaction also makes evolutionary sense.

Although evolutionary psychologists are just beginning to understand the function of psychological pleasure, pain and emotions,[9] they are fairly clear on the general principle that psychological pleasure and pain serve to heighten our awareness of events happening to us that are of particular significance for the reproduction of our genes. Through that heightened awareness, we are better prepared to pursue or avoid such events as they occur and to seek out or avoid similar experiences in the future.[10] For example, if today something happens to our body (weight loss?) that makes us more attractive to a mate or we gain resources with the same result, it makes us happy. We like the happy feeling, so we are motivated to continue to do the things that brought it about. If our mate pays attention to some attractive person of our sex, it makes us jealous—the jealousy is uncomfortable. The discomfort alerts us thereafter to avoid whatever made our mate's attention stray.

Similarly, returning to our rape subject, when a female is thwarted in her programmed, careful mate selection process by an overwhelming male, she receives a psychological wallop[11]—call it anger, frustration, insult—whatever. It hurts. It hurts a lot, because this is not a relatively insignificant event like a mate being temporarily attracted to another woman. This event strikes at the very core of the genetic reproductive process. The trauma she experiences has evolved to alert her to the fact that what has happened may have serious consequences for her reproductive process, so avoidance action is desirable. It also sensitizes her to avoid, in the future, the circumstances that brought it about.

The important insight here is the sociobiological prediction that women who are forced to copulate against their wishes will be seriously hurt psychologically. Now, presumably that is not news to any woman who has ever been raped, but it may be to some other women and it will most certainly be to many—or maybe most—men. I am thinking particularly of the male public figures who in the past have been quoted as advising women to the effect that "if rape is inevitable, sit back and enjoy it."[12] To be sure, these males have never been in that female position and never will be, and neither their biology nor their life experiences can help them to project themselves into the female experience. But evolutionary learning, if social planners can find a way generally to expose men to it, could provide them with intellectual insights as to why women don't just relax and enjoy it.[13]

However, let us return to male reproductive strategy so as to lay the groundwork for the next chapter. We have said that females evolved to become choosy, and males competed with each other to be chosen. Viewed from the male's biological standpoint, the objects for which they were competing were female eggs in a position to be fertilized. This was the "limiting resource" in the reproduction game; compared to sperm, eggs were scarce. The males in our evolutionary past who were most successful, in terms of being represented in today's human gene pool, were those in whom genetic behavioral programming developed that led them to accomplish more lifetime egg fertilizations than their competitors. Other behavior undoubtedly became important for successful reproduction by males as well, such as nurturing activity toward at least some of the children born of the fertilizations. But programmed activity aimed at a large number of lifetime fertilizations is assumed to have been passed down from generation to generation in males as the initial element in successful reproductive strategy.

To this point in our narrative of evolved reproductive behavior, we can find authoritative support from all corners of the evolutionist community.[14] It is at the most solid part of the core of evolutionary biological learning. In the last decade specialists have added detail to the theoretical male reproductive strategy picture by continuing to build on factual bases and natural selection principles.

Lea and William Shields and Nancy and Randy Thornhill have postulated that human males will have evolved to have a flexible reproductive strategy.[15] It is assumed that all males will be programmed to adjust their strategy to achieve egg fertilizations in order to accommodate to the circumstances they may find themselves in from time to time during their reproductive lifetime.

The best reproductive position for a male to be in regarding a given female is to have exclusive copulatory access to her during the fertile period of her ovulatory cycle. That period, the estrus period, is not externally obvious in human females as it is in many mammals. From observing females, the male cannot detect when they are in condition to become pregnant. As a practical matter, this means the male should have exclusive *continuous* access to the female when she is not pregnant. Given the desirability, from the females' standpoint, of having a male's attention and resources to apply during gestation and nurturing, few males are able to maintain such access to more than one female at a time.

Thus, monogamy—one spouse at a time— has been the most common mating pattern among humans during recorded history.[16] Getting and staying in that position has been one of the behavioral options in the evolved male reproductive repertoire. However, natural selection logic suggests the unlikelihood that genetically influenced behavioral mechanisms equipping males *exclusively* for monogamous pair-bonding type behavior would have had much success in the reproductive sweepstakes of evolutionary history. This is because any competing program that popped up in the gene pool, which prompted its carriers to strive for or take advantage of insemination opportunities outside of pair-bonds while not rupturing the bonds, would win in the long run. It would be responsible for more offspring in each generation and thereby slowly push the "faithful, monogamy only" program out of the human gene pool.

Similarly, in the long run any genetic program that equipped its carriers for only pair-bonding, plus other, less committed consensual courting behavior, would have lost out to a genetically programmed repertoire that, in the appropriate circumstances, also included nonconsensual copulatory behavior—that is, rape.

Thus, it is assumed that, with the possible exception of isolated programs that are the result of recent genetic developments (mutations and such) that have not had time to be tested in natural selection competition, all of today's human males reflect ancestral reproductive programming that allows for the rape option under certain circumstances.[17] The environmental circumstances that can call up the genetically programmed rape option will be discussed in the next chapter.

Chapter Six

Reducing Rape

Here we have another universally approved goal. I suspect that even most convicted rapists would vote for a program likely to reduce rape—by others, at least.[1] Sociobiological learning suggests such a program,[2] but the idea did not originate with sociobiologists. It has been around for a long time, and in simple form it makes intuitive sense to the layperson: To decrease rape, increase the penalty for it.[3] What recent sociobiological thinking has done is to endorse a more complicated form of this old idea. Now, some readers may wonder why any endorsement is needed for the concept that increasing punishment for a crime in a society can eventually result in reducing its incidence. It turns out, however, that the issue is quite complicated, and this has left the door open for ideologists who oppose punishment, for whatever reason, to argue that it does not result in deterring potential criminals.[4]

There has been a long-standing debate over whether and when penalizing convicted criminals can deter, and thus reduce, crime in general.[5] Attempts to research the question in regard to rape have been difficult to conduct and have yielded hazy results. Studies have been limited by the type of data available. For instance, in looking for a possible deterrence connection, comparisons have been made between the number of executions for rape in given geographical areas and the rape rate in those areas over a period of time. (How many potential rapists actually learned of the executions, one wonders.)[6] Moreover, confounding factors complicate the researcher's task. For instance, a drop in unemployment leaving many more men idle could pressure the rape rate upward while an increase in punishment could simultaneously pressure it downward and so result in essentially no change.[7]

Various suggested approaches to reducing rape other than through punishment often seem to be improbable or visionary. For example, raising the socioeconomic level of unemployed young males is a popular notion that sociobiologists, incidentally, might also endorse.[8] But adding the voices of anti-rape advocates to the large numbers that are already wishing for such a difficult socioeconomic improvement is not likely to further its accomplishment appreciably.[9]

If other suggestions for reducing rape seem unpromising, and if it is considered to be a serious enough social problem, the time may be right for social planners to bet on a scientifically based prediction. Sociobiologists predict that a properly sculpted increase in the penalty for rape will reduce its incidence.

Let us be more specific about what is meant by an increase in the penalty for rape. About two hundred years before sociobiology arose from various branches of natural and social science, European social philosophers Cesare Beccaria and Jeremy Bentham were theorizing that, if punishment were to act as a general deterrent to crime in a society, a proper mixture of three elements must exist: (1) the prescribed penalty, (2) the likelihood of its imposition on the offender and (3) publicity regarding (1) and (2) conveyed to potential offenders.[10]

In other words, a severe penalty provided on the statute books will be ineffective if too few offenders are actually made to suffer the full penalty. On the other hand, even if the enforcement system should be very efficient in making offenders suffer the prescribed punishment, there can be little deterrent effect if the prescribed punishment is insufficient. In any case, a deterrence program will not work unless the likelihood that an offender will suffer the indicated penalty is communicated to potential offenders.

My reading of sociobiologists, in particular Lea and William Shields who have spoken most directly to the rape deterrence issue, is that they would agree with this formulation of classical deterrence theory as applied to rape.[11]

The sociobiological prediction on rape reduction through a punishment system, then, is the following: *If adequate publicity is given to an appreciable increase in likelihood and/or severity of punishment for rape, and all other factors affecting the rape rate remain stable, the rate will decrease.* Let us look at the reasons why this prediction is made—at the mechanism in men that can be influenced in this way.

THE HIDDEN COST-BENEFIT CALCULATOR

As we have noted, evolutionary biologists generally agree that behavior is a product of (1) genetically programmed predispositions, (2) which are slanted in various directions by life histories, and (3) played against the environment the actor is in at the time of the behavior. But evolutionary biology, indeed natural science in general, is not yet developed enough to know much about the particular mechanisms involved. Little is known thus far about how genetic material, sensory intake (seeing, hearing, etc.), memory, brain circuitry, nerves, muscles, and so on, work together to create behavior.[12] Evolutionary biologists are fairly confident about one point, however. From the observation of various natural phenomena, it is assumed that, whatever behavioral mechanisms exist in humans, they will normally operate according to economic logic.

Evolutionary psychologists describe mechanisms quite similar to computers that exist in each of us, although we are only vaguely conscious, if at all, of their operation.[13] These "computers" make calculations as to whether various courses of action are worthwhile, given the costs and benefits involved, in view of an overall goal.

Let us draw an analogy to a business calculator that we might program to maximize profits.[14] Profits are our goal, and we want the computer to tell us whether a business venture we are contemplating will produce them. We put in information regarding current facts, past experience and "best guess" projections. Our business computer then estimates the costs and benefits and only signals "go" if the benefits appear to outweigh the costs sufficiently.

The human calculators do much the same thing while replacing the goal of monetary gain with that of maximizing what biologists call "inclusive fitness."[15] This translates into the goal of getting copies of the human actor's genes into succeeding generations through personal reproduction (having children) and otherwise (actor's genetic relatives having children). Let us look now at decisional situations that confront human male calculators—mechanisms—in the area of personal reproductive behavior aimed at maximizing inclusive fitness.

We ended the last chapter on evolved reproductive strategy with the observation that all men have genetic programming that allows for rape under certain circumstances. Indeed, all men are assumed to be capable of a variety of strategies for achieving inseminations that can be practiced concurrently if circumstances permit. The strategies blend into one another on a range of options. Lea and William Shields have labeled three prominent points on the range. At the top is *honest courtship* (cooperative pair bonding with a female); in the middle, *deceitful or manipulative courtship* (pseudo-commitment); and at the bottom, *forcible rape.*[16]

When men confront a reproductive strategy option, their behavioral mechanisms will "compute" whether the possible benefits of pursuing the option are worth the costs of it in view of other options that may or may not be available for pursuing the "inclusive fitness" goal.[17]

At the top end of the male reproductive strategy range—honest courtship—benefits are likely to be comparatively great because of continual access to the pair bond female during her entire ovulatory cycle and the relative exclusiveness of that access.[18] So a male with resources sufficient to satisfy the female's requirements for them is likely to choose an honest courtship option—perhaps exclusively. Deceitful or manipulative courtship involves less commitment in terms of time and resources (costs), but benefits are correspondingly less likely because of a more limited access to the female. At the bottom on the option range, rape usually entails virtually no time and resource commitments (costs), but benefits are again correspondingly unlikely because, among other things, one copulation—blindly random in terms of the female's fertile period—has a low likelihood of resulting in conception.[19]

At this point it is useful to emphasize that no one is likely to be conscious of such detailed considerations as a prelude to their actions. But evolutionary biologists believe that millions of years of natural selection have resulted in such considerations being reflected in behavioral mechanisms.

Although a male is highly unlikely to gain reproductive benefits from a rape, his calculator may tentatively signal "go" on occasion either because he lacks status and material resources,[20] or is expending them elsewhere in higher potential benefit arrangements, or has unattractive physical or other personal characteristics—or some combination of the above—and is therefore unable to gain consensual access to the female on whom he is focusing. A prime example is the teenager who has not had the time or opportunity to gather resources but has a fully developed reproductive apparatus[21] (normally at its lifetime peak capacity, as a matter of fact).[22] What has he got to lose? The costs in terms of required expenditures in time and resources are essentially zero for rape, but another possible cost is virtually always in the picture—retribution.

Death, bodily injury, imprisonment (away from females) and even loss of reputation through actions of the female, her mate, her kin or society as a whole are possible costs to the rapist. Such costs could not only counterbalance any slight potential benefit realizable from rape but also subtract from any positive "inclusive fitness" results the male has achieved, or might achieve, by pursuing other reproductive options. If the possibility of retribution is serious enough, it should result in a "bottom line" on the normal (nonpathological)[23] internal calculating male mechanism that says "stop—don't rape."

CAN THE MALE MECHANISM COMPUTE TODAY'S CULTURAL CONDITIONS?

It takes ages for an advantageous genetic change that emerges by mutation or otherwise in some individuals in a species gene pool to spread, generation to generation, throughout the pool. The exact time frame is still subject to dispute among scientists,[24] but they generally agree that whatever genetic programming for behavior humans have today was set in place thousands of years ago and reflects the environments of those times. As we have noted earlier, the primary focal point is thought to be the late Pleistocene era when our ancestors were hunter-gatherers.[25] Since those times, exponential, exploding changes have taken place in the human cultural environment.

So the question arises whether a given behavioral mechanism adapted to conditions of the Pleistocene can process and compute data put into it by today's environments. If not, the resulting behavioral pattern could not be predicted from sociobiological principles.[26] For our present purposes, this means that a type of punishment given rapists in modern times that was too radically different from anything human males experienced during the Pleistocene might not compute and result in deterrence.

Reconstruction of Pleistocene social life, from studies of present-day human hunter-gatherer groups and other closely related primate species, provides reasonable assurance that the retributive possibilities for what we now call rape were similar in some respects then to what they are today. Most probably, males forced themselves on females, on occasion at least, and the females would resist, on occasion at least. It is likely that consorts and allies of the females would sometimes intervene. Furthermore, we can assume that the females, their consorts and allies had the capacity to inflict bodily harm on the male with teeth, fists and perhaps sticks and stones. Banishment from the group may also have been a possibility. Such conservative, minimal reconstruction provides the grounds for assuming that the male mechanism computing the rape option will respond to at least some of today's retributive possibilities, much as it did to those during the Pleistocene. Recall from Chapter 1 that sociobiologists expect a certain evolved flexibility in the information processing capacities of behavior mechanisms.[27]

Let us first assume that the information processing capacity of the relevant male mechanism is very flexible or open, so that it can compute virtually any retributive possibility that would have a negative effect on his "fitness" (ability to survive and reproduce) even if the particular means by which he suffered that effect were a modern novelty such as the electric chair. Perhaps only the *likelihood* of suffering fitness damage as a result of the contemplated aggressive act and the possible *severity* of that damage are weighed by the mechanism while the means is incidental. Any means, ranging from the victim's fists (immediate result) to an electric chair ten years after the event (long-range result) may be computable. That sort of open, flexible mechanism might have evolved because during the Pleistocene, it was advantageous for males contemplating forced copulations to be able to appreciate the different severities of retribution that could be anticipated from potential victims and their kin based on how *physically strong* those people were—regardless of the particular means they might use (teeth, fists, sticks, stones). If so, then, today essentially the same mechanism may be able to compute an announced increase in severity of penalty (damage to fitness) for rape that can be anticipated from a group of people (now, the modern state), regardless of the particular means to be used—such as modern novelties like life imprisonment, the gas chamber, or the electric chair.

But let us now assume that the male mechanism is not so flexible—that the means of retribution used today must be quite similar to those used during the Pleistocene before the mechanism can process them. In that case, a death sentence might not be easily computable, if at all. At least it seems clear that it was more difficult for one human or even a group to put an individual to death during the Pleistocene than it is today. Knives, guns, the noose, and lethal injections have facilitated the job considerably.

On the other hand, the difference between being banished by a group (kept away) for some length of time and being imprisoned (kept in, and thus away) for some length of time seems negligible, does it not? So the possibility of imprisonment may be more easily processed by the mechanism than the possibility of death.[28]

Furthermore (and this will be significant for later observations in this chapter), the possibility of physical injury at the hands of the modern state (a group of people) seems virtually indistinguishable from the same possibility at the hands of a Pleistocene groups (the victim and her consort or allies). If any modern retributive possibility is computable by the Pleistocene-constructed male mechanism, one that entails physical injury should be.

Does the foregoing provide reasonable assurance that human males today contain a behavioral mechanism that is sensitive to at least some types of increases in the possibility of retribution for rape? I think so. Let us assume that it does and go on to talk about how one might bring about increases in retribution. Nothing said hereafter is meant to be a prescription for what ought to be done about rape. I intend merely to survey feasible means of increasing male perceptions of retribution on the assumption that a society might want to pursue that route. We should keep in mind that a penalty may be quite effective as a deterrent, yet undesirable.[29]

INCREASING RETRIBUTION POSSIBILITIES: THE LAW-ENFORCED PENALTY ROUTE

Retribution can come from several directions, and it can be quick or delayed.[30] A way of increasing quick retribution for rape is to teach self-defense tactics to women. Sociobiologists would also endorse this method.[31] If, as we speculated above, the possibility of physical injury to a man attempting rape has particular deterrence value, then a well-publicized increase in the ability of women to physically injure their assailants could be a particularly effective way to reduce the rape rate. But here we are focusing on more delayed retribution—penalties imposed by governmental legal schemes. I will not attempt to detail a sociolegal plan for decreasing the incidence of rape following sociobiological insights. However, I think it appropriate to say more here about implementing the idea that sociobiology provides than I did in the chapters on child abuse and incest. There the indicated social plans, though not without problems, would be fairly straightforward—publicize. Here the indicated plan is much more complicated because it involves publicity as only one of three interdependent courses of action.

Recall that classical deterrence theory identifies three interrelated factors in deterring crime through punishment: (1) *publicity* given to (2) the *likelihood* that an offender will suffer (3) a prescribed *penalty*. Theoretically, the largest decrease in the crime rate should come from a maximum increase in all three factors, but a rate decrease could also occur from a significant increase in any one of the factors while the other two remain constant.

Let us deal with the factors one at a time. I will start with publicity because I suspect it would be easiest for social planners to achieve an increase there. The primary hurdle would be finding funds for the publicity.

THE NEW COMMUNICATION LINE TO THE POTENTIAL RAPIST

If lower socioeconomic groups produce a disproportionate percentage of rapists (more on that presently) and if, as appears to be the case, members of those groups read less but watch television more than or as much as the average citizen,[32] then a large window has opened up in recent decades through which rape may be decreased—the television set.

Available data suggest that the vast majority of rapists are relatively poor young men. There is some question as to whether these data present an accurate picture.[33] Whatever the percentage of potential rapists in the lower socioeconomic and age groups, society, with its television technology, is better able to reach those men than it was before 1950 when reliance would have had to have been put on the print media, education in schools and word of mouth. Although these other avenues should not be neglected, in any future effort to reach potential rapists at all socioeconomic and age levels, television seems to offer the most fruitful publicity route.

PUBLICIZING THE STATUS QUO

Recent law enforcement history can be used to predict the likelihood that a rapist, in the immediate future, will suffer a punishment and the range of the possible penalty. One could pick any jurisdiction and roughly calculate, for example, that in the last year, the typical sentence of those convicted of rape was nine years imprisonment of which only four were normally served before release. However, only 10 percent of rapists were caught, convicted and received any sentence at all.[34]

Very few men know such details. What they do know will vary along a spectrum. Most presumably will know that a penalty is possible for rape, and many will have an impression of its severity ("could be locked up"—"might get the chair"). At a minimum virtually all men, in North America today at least, will learn before or soon after puberty that raping is considered "bad"conduct in their society.

Increased publicity can increase the degree of awareness in the male population all along the spectrum. If, pursuant to a concerted effort, television stations around the entire country over a period of months were to show several recently convicted rapists being arrested or behind bars, a great many men would see it, some would hear about it, and others would simply receive the general impression from the community that rape conduct is even "worse" than it used to be. If sociobiological theory is correct, this should have a downward effect on the rape rate in the country.

A publicity campaign of this sort featuring atypical results of rape (it is probable that most men who are technically guilty of rape are never caught and punished in any way) would involve some distortion of the true picture.[35] This could be troublesome to social planners. Furthermore, any decrease in the rape rate that occurred as a result of such a campaign could be short-lived. The fact that it does not reflect reality (the fate of the average rapist) might eventually filter into public consciousness and negate any initial decrease in the rape rate. For these reasons, an increase in one or both of the other two operative factors in deterrence through governmentally imposed punishment may be called for in addition to or instead of an increase in publicity. Let us consider the second of those factors.

INCREASE IN PRESCRIBED PENALTY

An increase in the severity of the prescribed penalty—that which the written law allows a court to impose on an offender—is probably the second easiest of the three intersecting avenues for social planners seeking a decrease in rape using governmentally imposed punishment. The principal hurdle here lies in the political process, which could block legislation that would prescribe the penalty increase.[36] Some element of the population is almost certain to oppose any legislation dealing with an emotionally charged issue like rape.

Once enacted into law, however, an increase in the severity of the prescribed penalty will filter into popular knowledge, to some minimal degree at least, without any special publicity effort.[37] This should be especially true to the extent that the penalty increase is actually imposed on some offenders.

DEATH OR CASTRATION FOR RAPISTS

The type of penalty the law prescribes for rape often depends on the circumstances surrounding it. In some countries of the world and a few states in the United States, a death sentence is prescribed for rape of the most aggravated types—such as rape of a child.[38] Generally speaking, however, imprisonment for some specified maximum time, less than life, appears to be the prevalent prescribed penalty around the world for even the most aggravated rape.[39]

Thus, in many jurisdictions—any place where the maximum sentence is currently less than life—there is room to increase the prescribed penalty by expanding on the years in prison option. In those circumstances where life imprisonment is already a specified possibility, any increase in the severity of prescribed penalties for aggravated types of rape would seem to call for the death penalty. In the present atmosphere of the United States, at least, such a step is not likely to be taken.[40] The prevailing opinion in the United States seems to be that, if the death penalty is to be imposed for crime at all, it should be given for crimes such as murder or treason, which the public appears to consider more serious than rape.[41] It might therefore seem that there is little room for increasing the severity of the prescribed rape penalty in places like the United States.

Another type of penalty—peculiarly related to the rape offense—might be considered in those jurisdictions where increased severity is desired but the death penalty is not feasible: castration. Surgical castration of the male represents, in sociobiological terms, personal reproductive death, although the individual's life continues.

Castration of certain categories of convicted sex offenders is currently performed in five Western European nations: Denmark, Germany, Norway, Sweden, and Switzerland.[42] In the last decade it has also been advocated in other countries.[43] Although the current practice in Europe is to perform the procedure when the offender chooses it in lieu, or reduction, of another penalty,[44] Denmark's law permitted involuntary castration from 1929 through 1967.[45]

A large majority of men who undergo surgical castration are, as a result, completely unable to engage in sexual activity.[46] Furthermore, the effects on the libido and physical capacity are such that Danish authorities were able to report that "we have not seen a rapist [rape again] after castration."[47]

DETERRENT AND OTHER RAPE REDUCTION EFFECTS OF CASTRATION

Recall our earlier discussion of the types of retribution for rape that are most likely to be picked up and "computed" by the relevant male behavioral mechanism. Based on presumed parallels between modern and late Pleistocene conditions, it seemed that the possibility of bodily injury inflicted by the victim, or a group of people, was most likely to be readily processed by the mechanism. If that is right, then castration may have particularly high potential as a route to rape deterrence.

From what we know of today's human males, we can speculate that most men would not consider surgical castration to be "worse than death." Some, however, may consider it worse than life in prison (the possibility of release from there always exists), and many may consider it worse than long prison terms less than life. It seems likely that men's attitudes toward castration for rape would be reflected in their behavior when threatened by blows to their pubic region. We have all observed or experienced that quick protective reflex.

Castration of convicted rapists could operate to reduce the rape rate on three levels. Primary or immediate level reduction would occur when potential rapists were deterred by the threat of castration as punishment. A secondary (probably lesser) reduction effect would occur when repeat offenses were prevented by castration of convicted offenders.[48] Beyond that, Shields and Shields have suggested that a third, long-range reduction effect might occur when, owing to castration, some men proven to have raped are unable thereafter to have offspring who could otherwise have inherited characteristics that might have contributed to the rapists' actions.[49] As we have noted, social environmental conditions are partly responsible for rape behavior, but inherited characteristics may also contribute to the behavior.[50]

Here the commentators are not referring to the behavioral mechanism we have discussed which predisposes men to rape under certain cost-benefit conditions. All men are presumed to inherit that predisposition. The speculation is that other inheritable characteristics or traits that some men have and others do not may be contributing to raping behavior.[51] Whether those characteristics are physical, attitudinal or whatever, if they exist at all, are issues that the scientists need to develop.

CASTRATION PUBLICITY

In North America at least, any legislation establishing castration as punishment would likely receive wide publicity. Indeed, even when castration for rape is merely proposed by authorities in a state, there is national press coverage.[52] Thus, with no special effort by social planners, considerable news media coverage would likely occur when a state enacted castration as a penalty. According to sociobiologists who have focused on the question, a corresponding reduction in the rape rate could follow where castration represented an increase in the severity of punishment over that previously specified. If the penalty were limited to rapes of only the most aggravated types, such as the rape of a child, a more generalized deterrent effect might nevertheless occur. By the time the message filtered through the population, a large proportion of those receiving it would likely hear only that castration was possible for "rape" (without qualification). However, for reasons discussed earlier, any resulting reduction in the rape rate could be short-lived if the enforcement system failed to impose the new penalty.

CASTRATION AND THE PENALTY/ENFORCEMENT SEESAW

Commentators have suggested that jurors may be inclined not to convict those accused of crime when the penalty that can be imposed on the accused is too severe.[53] In criminal jury trials, the jury is normally charged with deciding whether or not the accused committed the offense. If the answer is yes, judges then impose the penalty from a range provided by statute. When jurors are deciding whether the accused committed the offense, they are not supposed to consider the penalty that might be imposed. However, there is little doubt that some jurors, consciously or unconsciously, do consider the penalty. It is hard to gauge the extent of that phenomenon.

Commentators who have observed that severe penalties might have a backlash effect were usually focusing on death and life imprisonment.[54] It is anyone's guess at this point whether jurors would consider castration too severe a punishment for rape of even the most aggravated types. It is possible that male jurors would react differently than female jurors.

INCREASE IN IMPOSITION OF PRESCRIBED PENALTY

Let us now turn to the last of the three interrelated factors that are said to operate in crime deterrence through governmentally imposed punishment. This factor, the imposition of prescribed penalties, is probably the most intractable of the three. For this reason, and because voluminous literature exists on the subject,[55] I will simply outline the breadth of problems that social planners seeking to reduce the rape rate will face when they approach this route.

In a Western-style legal system, before a person who is technically guilty of rape can suffer a prescribed penalty, several conditions must occur: (1) The rape must come to the attention of authorities who 2) must apprehend the offender; 3) authorities must exercise their discretion to prosecute; (4) the jury or judge hearing the case must adjudge guilt; (5) a penalty must be pronounced; and (6) the penalty must be imposed.

At any point along this line, the enforcement system might be perfected or tightened, resulting in a prescribed penalty actually being suffered by more offenders. Each point has its own set of major problems for anyone seeking to increase deterrence of crime. How, for example, does one go about encouraging more victims to report rapes in the first place?

At the other end of the line, what does society do when a rapist, who is sent to prison, reforms and becomes a "model prisoner?"[56] If he is released before serving his sentence, how does this translate, for deterrence purposes, to the mind of the potential rapist who is paying attention to the penalty imposed for committing rape?

Although increased enforcement of already prescribed penalties for rape may be comparatively difficult to accomplish, studies have suggested that enforcement increase is a more fertile field for rape deterrence than is prescribed penalty increase.[57] These studies did not attempt to control for the third operative factor, however—the publicity (or knowledge received by the public) concerning increases in one area or the other. It may be that publicity is *the* most sensitive of the three operative factors we have been discussing.

SUMMARY

We have surveyed problems involved in any attempt to reduce rape through generally increasing male perception of societal retribution as a cost of doing it. These problems are not new to criminologists. What is new is the sociobiological endorsement of the proposition that if such male perception *is* increased, a reduction of the rape rate will actually result. There has been confused debate among authorities regarding this proposition, and empirical evidence addressing it has been unclear. The sociobiological endorsement of the proposition may encourage social planners, who are intent on decreasing the rape rate, to embark on efforts generally to increase the male perception of societal retribution as a cost of rape by using one or more of the three interrelated routes we have explored.

Chapter Seven

Reducing Street Crime

Perhaps some criminals would oppose reducing street crime, but virtually everyone else would support it as a societal goal. Numerous means to that end have been tried in modern societies, and more have been suggested. In this chapter we will see that evolutionary biology may provide support for a controversial means of reducing the rate of such crime.

We will use the term *street crime* as shorthand for a grouping of antisocial behavior that occurs on or just off streets and other public areas.[1] It includes armed robberies, purse snatchings, home and store burglaries, rapes and other assaults on the person, even murder. To somewhat anticipate our conclusion: It includes any antisocial behavior that might be detected by police patrolling in public areas. "Inside" crimes such as theft and violence between members of a household, income tax evasion, embezzlement and treason are not within our definition.

When a street-crime wave occurs in a community, the public often calls for an increase in the police force. The popular impression is that an increase in police will result in a decrease in crime.[2] Thus, it may come as a surprise to many of us that empirical studies on the effects of increasing police patrols have produced confusing results. Some studies have reported subsequent decreases in the crime rate,[3] while others have not.[4] Those studies that have reported no decrease in crime after increases in a community's police patrol capacity have prompted opponents "to dismiss foot patrol programs as more cosmetic than substance"[5] and to flatly claim that "police activities cannot suppress crime."[6] Proponents of police patrolling have responded that such reactions "ignore the known limitations of crime statistics."[7] The problem is that numerous variables can confound such statistics. For example, when people who officially report crime in an area are aware that it is undergoing intense study by researchers, they tend to report more conscientiously. The result is a paper increase in crime regardless of the actual incidence.[8] A paper increase could thus mask an actual decrease. Another confounding factor is the possibility that criminal activity will be displaced by increasing police patrols in a given area. Some potential criminals may simply transfer their activities to neighboring areas. Thus, the crime rate may go down in the target area, but if one could know where the criminals transferred their activities, one might find no overall decrease when those places were surveyed together with the target area.[9]

PSYCHOLOGICAL DETERRENCE FROM POLICE PATROLLING?

The main dispute concerning the effectiveness of police patrolling is over whether it can result in *psychological* deterrence.[10] Does the mere presence of police make criminals back off from the crime they had intended to commit? There seems to be no question that increased patrolling can result in crime rate reduction in a way that has little to do with psychological deterrence.[11] The more police there are, the more likely they are to apprehend offenders. To the extent that some proportion of those who commit crime today are *physically* taken off the street, the rate should decline tomorrow and stay down until replacements grow or move into the criminal ranks. This likelihood is yet another factor that clouds the issue of whether increased police patrolling can result in psychological deterrence of street crime. When a decrease in the crime rate subsequently occurs in the target area, to what extent is it due merely to an increased physical removal of potential criminals from the scene?

The problems that muddy the research waters have created a situation where empirical resolution of the issue of whether police patrolling psychologically deters crime is elusive at best. Under these circumstances, evolutionary biology may be of particular value. It could provide support for the proponents of police patrolling. Theory and initial evidence suggest that, up to a saturation point, if the capacity for efficiently used[12] police patrolling is increased in an area, there should be a psychological deterring effect on those potential criminals who remain in the area. Let us look at the relevant evolutionary science concerning the behavioral programming in question.

IDENTIFYING A BEHAVIORAL MECHANISM

In order to be able to predict, from biological insights, a typical reaction in a population to something like an increase in police patrolling, we need to establish that everyone in the population has an underlying genetically influenced mechanism that is sensitive to the occurrence. If there were no such mechanism, then biology would have nothing to add to the debate over whether a police patrol increase can reduce street crime.

We should not expect to find a "hard-wired," automatic mechanism of the sort that makes us pull away when we touch a hot stove—that is, see a policeman and back off. It is true that natural selection proceeds economically, and such a reflexive system would be relatively inexpensive in terms of body-building materials and neural energy needed for its operation. However, such a quick, automatic mechanism would probably have us recoiling whenever we saw something we wanted or needed while there was an authority figure in the vicinity (mother, when we are infants). Not obtaining what we need to survive, we would not last beyond infancy if that were the case. So, we are looking for a more flexible mechanism that can handle variations in the environment and learn to respond appropriately. Such a mechanism exists: it is the cost-benefit calculating mechanism we discussed in the last chapter.

A flexible, cost-benefit calculating capacity exists not only in humans, but throughout the animal kingdom. Whether or not nonhuman animals can "think," whatever that means in human terms, they clearly "compute," consciously or unconsciously, the costs and benefits in terms of survival (and reproductive fitness) of various courses of action. We can see that happening in virtually any species in situations quite similar to our human street-crime scene: an animal comes upon a resource or territory that it and another animal wants, or it is faced with an aggressive conflict with the other. It proceeds when the other is smaller and backs off when it is larger.[13] Every animal probably makes such calculations, although some, like the relatively immobile, entrenched corals in the sea, do it over such a long time frame that it is difficult for us to observe.

Thus, I believe we can be comfortable in assuming that humans,[14] together with the other animals, have a cost-benefit calculating mechanism. The harder question is whether that mechanism will operate in our street scene to deter criminal conduct—whether it is sensitive to police presence. It is one thing to establish a general capacity like computing cost-benefits and quite another to establish that particular aspects of costs and benefits are computable. Even when we assume that it is a programmed *learning* capacity that is involved,[15] an organism simply cannot learn certain things at a given point in its evolutionary history. This will be the case when the phenomena that pop up in its environment are too different from anything its information processing mechanisms have previously experienced. Until the genetic programming underlying the mechanisms has had time, through slow-moving natural selection, to adjust to the phenomena, those phenomena simply will not register on the computing mechanism.

Now, the new phenomenon we are talking about here is police presence. Is the presumably very ancient mechanism in the human animal that says back off from a resource or a conflict when the other individual is larger or stronger, sensitive to a *third* party on the scene who is ready to take action that can cost the actor? In order to answer that question, again we need to reconstruct relevant environmental conditions that existed during the Pleistocene. If the human social environment contained the substantial equivalent of police during the Pleistocene era, we can be fairly comfortable that today's human cost-benefit mechanisms will be sensitive to their presence.

RECONSTRUCTING ANCESTRAL SOCIAL ENVIRONMENTS

One method of reconstructing ancestral experiences that is apt for our purposes is called phylogenetic comparison.[16] In this approach, one lines up today's primate species that are closely related genetically—such as chimpanzees, gorillas, orangutans, bonobos (a type of chimpanzee) and humans—in order to determine common characteristics between them. When these characteristics have been identified, the assumption is made (with good evolutionary logic) that their common ancestor (see Figure 4) had these same characteristics. Then, it is assumed that the ancestors of any of today's species that were intermediate between the common ancestor and today's species representatives also had the characteristics. Primatologist Richard Wrangham has observed that this method "offers a logical starting point for behavioral reconstruction at any time during human evolution. It will not always be correct, but it is unlikely to be far wrong."[17] One might add that the more conservative one is in identifying common characteristics among today's primates, the more accurate will be the inference that their intermediate ancestors had the same characteristics. For example, from looking at today's primate species, it just must be that their intermediate ancestors (1) were alive, (2) ate food, (3) did other things daily that promoted their personal survival and (3) had close relatives living during the same time they did. That's a conservative set of inferences.

Figure 4.
Reconstruction of Human Ancestral Conditions Using Phylogeny of the Homanoids

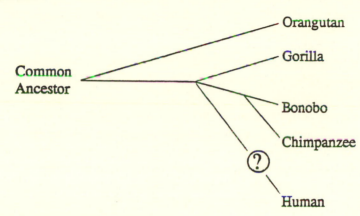

Adapted from Wrangham, "The Significance of African Apes," 56.

We can be almost as conservative in proposing a set of characteristics and circumstances that will give us a working model from which to assess the probable effectiveness of police patrolling as a psychological means of deterring crime in today's world.[18] I propose that the following social characteristics and circumstances typified the life of primate (including human) ancestors during the Pleistocene: (1) Individuals had a need for resources, such as food, and would regularly come into possession of such resources[19]; (2) individuals had other members of their species around them who also had a need for the same types of resources[20]; (3) an individual would, on occasion, attempt to obtain a resource from another who was first in possession while the latter was in the immediate presence of the resource or some distance away[21]; (4) individuals would occasionally be bodily aggressive toward another during attempts to gain a resource, or otherwise[22]; (5) allies of a first possessor or an individual being attacked would often come to the aid of that individual in order to protect her or him or the resource with resulting adverse consequences for the aggressor[23]; and (6) when the threat of bodily harm or other adverse consequences owing to the presence of allies was sufficient to outweigh an aggressor's need for the resource or motivation for attacking, he or she would back off.[24]

Reports on present-day chimpanzees, bonobos, gorillas, and orang-utans, to the extent that they contain relevant information, support the propositions I have listed (see notes).[25] Present-day humans, of course, fit all the propositions.[26] I expect that further relevant information on primates will soon be forthcoming, for the observation and reporting of nonhuman primate behavior is a rapidly growing enterprise.

ALLIES ON THE STREET TODAY

So, let us assume that our proposed characteristics and circumstances reflect the ancestral social conditions of the Pleistocene era—a fairly comfortable assumption. If so, then genetic programming set in place by natural selection has had time to adjust to the conditions. Humans should be genetically prepared to learn the significance of a third-party presence. If the people around a targeted individual's person or property are unknown to us, it is sensible to conclude that they are more likely allied with the targeted individual than with us. This was surely the case during the Pleistocene when individuals tended to move in closely related bands.[27] If the unknown party has a look about him indicating alliance with the targeted one (similar amount of body hair during the Pleistocene?—a uniform today), the same conclusion will follow. At this juncture, I think we can put police on patrol in the position of Pleistocene apparent allies of the targeted individual and conclude that they should be the functional equivalent. We should be genetically prepared to learn the same respect for police that our ancestors had for allies of another individual and be sensitive to police presence when contemplating an attack on a person or theft of a possession.[28]

INCREASING CHANCE OF COSTS

To this point in our assessment of whether increasing police patrols should increase the psychological deterrence of crime in a community, we have proposed behavioral mechanisms (or one mechanism with several aspects) that promote (1) cost-benefit calculations and (2) sensitivity to police presence. What response should the mechanism(s) have to an increase in police presence? I suggest that, within limits, an increase in police presence will increase the *chance* of costs and thus increase the psychological deterrence value. There should be a "saturation point," however, for police patrolling in any community area, beyond which any further additions to the force would be fruitless. When I can expect the police to come by every hour or so, I am more likely to do the deed than when I can expect them every half hour. When I can expect them about every ten minutes, I might well desist. If the city were to go further and add enough police so that one could pass about every minute, that might be more than what is needed to deter me. Establishing the area's saturation point will involve complex calculations, and it will vary from community to community. But I would guess that few large communities have ever reached it except for limited periods of time in limited areas.

ILLEGAL PARKING PLACE

Let me end this exercise with a do-it-yourself test that we can apply to the questions of whether the mechanisms I have proposed exist in us today and are likely to psychologically deter street crime when police patrolling is increased. When we are greatly in need of a parking space (territory) and see one in front of a fire hydrant, we are inclined, you will agree, to be deterred by the possibility of receiving a parking ticket and fine. But are we not more inclined to be deterred by the presence of uniformed police within view of the hydrant? And if we were aware that police would pass the hydrant every ten minutes rather than every hour or so, would we not be more likely to drive right by? If the answers are yes, evolutionary biology suggests that this reflects the operation in us of anciently fixed psychological mechanisms that have sensitized us to avoid doing something when the costs to us outweigh the benefits. In the hydrant situation the additional cost—beyond the fine—brought about by the police presence could be a tongue lashing and public embarrassment.

I know of no reason why the behavioral mechanisms of people in the street contemplating serious crimes would not make similar cost-benefit calculations. The costs to them of police presence (ultimately in terms of reproductive fitness) might be much more serious—rough physical treatment, perhaps, and physical detention for a time, if not the death penalty.

Of course, not every potential criminal is going to be deterred by police presence even in a fully saturated area. For various reasons, some may want to be apprehended,[29] and some psychologists believe that certain personality types cannot be deterred from offending behavior by any means.[30] However, sociobiology suggests that an increase in police patrolling to the saturation point in a community will result in an increase in psychological deterrence of the typical or average person who is contemplating a street crime.

Chapter Eight

Help on the Horizon

We are exploring help that evolutionary biology can provide to social planners. Those insights discussed thus far are comparatively well grounded in developed theory and empirical support. In this chapter we will deal with "softer" information regarding the existence and operation of human behavioral mechanisms. It awaits further development and testing before it will be as solid as the body of information we have explored to this point. Developments in evolutionary biology have occurred very rapidly in the last two decades, however, so it is likely that further research concerning the mechanisms discussed in this chapter will take place between the time I am writing and the day you are reading. Anyone interested in applying the behavioral insights we will explore in this chapter should consult with evolutionary biologists concerning recent developments.

The chapter will begin with a reprise of the first social goal we treated in this book—the reduction of child abuse—and look at a mechanism involving maternal instinct, which might be influenced in that direction. Next, we will explore a way to increase the male parent's voluntary compliance with court orders calling for child support payments. Then I will introduce some general sociobiological learning on recognizing relatives as a prelude to the rest of the chapter where that subject will be relevant to the social goals addressed there. The first of those goals will be decreasing or increasing patriotism. (Take your choice—natural science is ideologically neutral as we have noted.) That will lead us into concluding observations on the reduction of international wars.

REDUCING CHILD ABUSE THROUGH
POSTPONEMENT OF CHILDBEARING

Many reasons have been offered as to why it is better for women to have children late rather than early in their childbearing years.[1] Sociobiology has recently come up with an additional possible effect of postponement—the reduction of child abuse. This possibility was highlighted by psychologists Martin Daly and Margo Wilson who have used North American records on child abuse and homicide to test sociobiological hypotheses. One such hypothesis is that human mothers tend to be more solicitous for the welfare of offspring born late than for those born early in the mothers' lives.[2] This may happen with fathers as well, but that possibility is less distinct for reasons we will presently note.

This proposition on parental solicitude derives from the more basic sociobiological proposition that the processes of natural selection have genetically programmed humans to behave in ways that optimize the reproduction of the particular gene package they carry. The most obvious means of doing this is *personal* reproductive activity—being directly responsible for the birth of children. Since a child carries one-half of each parent's genes, after the birth of a child its welfare is of particular importance to a parent because the child can, in turn, reproduce about one-fourth of the parent's genes. However, no one is in a better position to reproduce one's genes than oneself. Thus, as long as a person is able to personally reproduce, it can be expected, typically, that considerable time, attention and resources will go into putting and keeping oneself in a position to do so. These activities will sometimes conflict with the needs of one's existing children and call for compromises. ("I should see my doctor today so we'll have to postpone Bobby's dental appointment.") To the extent that existing children are "shortchanged," it can be counterbalanced in the parent's calculation mechanism by the possibility that other offspring will be forthcoming.

However, as a person reaches the end of his or her reproductive years, the principal means of genetic reproduction shifts from self to relatives. The last years of reproductive capacity represent last chances for a full generative effort in behalf of one's gene package. Thus, evolutionary learning predicts that there will be a tendency to give more time, attention and resources to children born late in a person's reproductive years.

Now some readers may be thinking of family situations that do not seem to agree with this hypothesis (i.e., the older child was mother's "favorite"), but remember from our earlier chapters that sociobiology deals in tendencies, predispositions and typical, average "everything else being equal," behavior. Furthermore, general behavior rules can be modified by equally valid competing rules. Let me illustrate how the general rule in question here can be invaded by competing rules and still hold true in the end.

Let us begin with the general hypothesis (prediction) about increased solicitude for children born in a person's later years. A good setting for observing and testing the operation of that general proposition would be the following: A first, healthy child is born to a woman when she is 15 years of age. The child grows up and leaves the home. A second, healthy child is then born to the woman when she is 45. The prediction here would be that the woman is *highly* likely to be more solicitous of the second child than she was of the first.

Now, assume a woman has a first child at 27 and her second at 33. The first is healthy, the second is sickly. Here, if one were to measure solicitude at any given time when the children were subteens together in the home, the general hypothesis that a person will be more solicitous of children born later in life is less likely to hold true. That is because it has been invaded and modified by two other sociobiological hypotheses or rules of thumb. They are (1) parents will tend to be more solicitous of children who are healthy (likely to reproduce one day)[3] and (2) parents will tend to be more solicitous of children the closer the children get to puberty (capable of reproduction).[4] In our example, the general prediction that parents will be more solicitous of children born later in the parents' lives was not too strong to begin with because the difference in the mother's ages at birth of the two children was not too great (six years). Furthermore, whatever strength the prediction had may well be overcome by (1) the ill-health of the last-born child and (2) the fact that the first-born was closer to puberty.

Nevertheless when considering all parents at all times when their children are young, sociobiologists believe that one could rely on the general proposition that the parents will tend to be more solicitous of children born late in their lives than of those born early.

Everything said thus far may apply to some extent to male as well as female parents, but the picture regarding males is hazier because their reproductive capacity has a less clear cutoff time than that of females.[5] Physically, the male's reproductive capacity declines gradually from the late teens but can apparently extend at least into his nineties.[6] Thus, the end of the reproductive years when "last chances" occur is less apparent for males than females. For this reason hereafter we will focus on females, with the realization that everything said may apply in some form and degree to male parents as well.

The heightened solicitude that sociobiology hypothesizes for children born late in parents' lives includes not only beneficial actions, but also the avoidance of harmful actions toward the children. In other words, to the extent that you are solicitous toward a child you will be inclined to feed, clothe, love and keep it happy, but you will also be less inclined to neglect it when other matters distract you and less inclined to actively abuse it when some annoyance like persistent crying occurs. And you should be less inclined to kill it—which is extreme abuse! The last point is important because two studies on infanticide, one in South America[7] and one in North America,[8] have shown that the older the mother, the lower the risk of infanticide. When the data were put on a chart, a steady decrease in risk was shown as the mother's age increased.

We must note, however, that both studies showed a slight increase in infanticide when the mothers were around 40 years of age. This inconsistency may have simply reflected two other sociobiological predictions: (1) The likelihood of abuse (lack of solicitude) will be higher when impaired or chronically unhealthy children are involved,[9] and (2) in families with large numbers of children the parents' finite (limited) capacity for solicitude will be spread out over the brood.[10] Both predictions are consistent with a slight blip up at around 40 in an otherwise steady decline in infanticides by mothers as their ages progress because (1) older women tend to give birth to children with certain types of defects and (2) older women tend to have larger numbers of minor children in the home. (The peak for that may well be around 40 years.)

In any case, I would await further developments in this area of sociobiology before relying on the assumption that a generalized postponement of childbirth will reduce child abuse in a population. Further research needs to be done on the following sociobiological prediction: As the age of mothers at birth increases, they will be less and less likely to abuse a healthy, first-born child.[11] If this prediction proves to be right, then postponement of childbirth in a population could result in a corresponding reduction in child abuse because the average age of mothers with young children in the home would increase. This result would be most likely if the fertility rate (the number of children the average woman bears) were reduced by the increase in late starts because there would then be fewer large broods in the population; smaller broods mean more solicitude for each child. Even if the population should act to "catch up" in view of the late starts, however, and have the same number of children, on average, as at present, a reduction in the incidence and/or severity of child abuse may occur due solely to the increased average age of mothers.

INCREASING CHILD SUPPORT PAYMENT

A substantial element of the U.S. population subscribes to the goal of increasing the collection of court-ordered monetary child support, as evidenced by the state and federal government's elaborate administrative machinery to collect delinquent payments from those who have been ordered to make them.

The collection apparatus serves to force payments from those who do not make them voluntarily. However, a large percentage of ordered payments are never received in spite of the apparatus.[12] Furthermore, it appears questionable whether the collection apparatus is cost-effective—whether it collects more than it costs to operate it.[13] Thus, any cost-effective idea for increasing *voluntary* compliance with child support orders should be of interest to social planners, in the United States at least.

Sociobiological insights together with a recent breakthrough in tissue testing, called DNA fingerprinting, may provide that idea. This test could increase the confidence of ascribed fathers in their paternity and thereby increase their voluntary compliance with child support orders. Before concentrating on the test and how it might be utilized in this way, we need to introduce some additional evolutionary biology.

Evolved Parental Aid-Giving Mechanisms

We have enough background from earlier chapters to start here with the proposition that natural selection has inclined people to render aid to close genetic relatives before strangers—and the closer the genetic relative the stronger that predisposition to aid will be. Furthermore, people are inclined to aid younger before older relatives, everything else being equal.[14] These two inclinations should reach their highest manifestation in the case of parental solicitude for their young children. If the parent's generative investment is to reach fruition, the children must reach maturity when they can, in turn, reproduce.

In order to be in a position to exercise the particular solicitude for close genetic relatives that sociobiology hypothesizes, people need to be able to identify the relatives. Later in this chapter we will explore what mechanisms may exist in all people for this purpose. But those in the role of parent are special cases that we must concentrate on here. Human female parents have had relatively little identification difficulty throughout history. The child emerges from the mother's womb. And, incidentally, there is some evidence that this solid initial identification is thereafter supplemented by olfactory and auditory learning devices. Human mothers may be able to identify their infants by their personal odors[15] and voices.[16]

Human males have been in a different position with regard to identifying their offspring.[17] The information available to the male is more indirect and circumstantial than that available to the female. Historically, males have never been able to be as certain of their parenthood as females (at least until now—more on that later).

The degree of doubt about paternity that males have should, in theory, affect their conduct toward the children in question.[18] Empirical studies tend to confirm this theory. In societies where promiscuity ("sleeping around") is comparatively high, anthropologists report a common pattern of males having more solicitude for their sisters' children than for those ascribed to themselves. (Confidence can be high that a man's sister and he share the same mother and that his sister's children are actually hers.)[19]

Therefore, sociobiologists are comfortable that paternity confidence— or lack thereof—affects male solicitude for the children involved. However, scientists have said little about the relevant behavioral mechanism.[20] Sociobiology is a young discipline with a huge agenda, and the issue of the behavioral mechanism behind paternity confidence is one that has not been developed very far.

Initially, what is needed is a reconstruction of the relevant "environment of evolutionary adaptiveness," as scientists call it.[21] As we have seen, the focus for that purpose is generally on the late Pleistocene era of ancient history. We can further narrow that focus for present purposes and ask what the human living conditions relevant to paternity identification were like many thousands of years ago. To date, the study of nonhuman primates has yielded virtually no information that can help us answer that question. And when modeling is attempted using conservative propositions applicable to living primate species closely related to humans, we run up against the question of whether a subjective (in the "mind") issue like who is the male parent of a child is even important in the social life of nonhuman primates. Recall from the last chapter that one can have some confidence that the possibility of being injured as the result of an aggressive or acquisitive act toward another is an issue in nonhuman primate life. The same cannot be said on the paternity issue.

However, even if primates closely related to humans do not effectively treat paternity as an issue in the way that humans presently do, so that the common ancestor cannot be assumed to have done so (see Figure 4 in Chapter 7), the possibility still exists that the issue became important to the *intermediate* ancestors of humans during the Pleistocene. The social environment, including mating and nurturing patterns, may have been such, even at that early time, as to create a behavioral mechanism that was sensitive to the issue.

On the more specific question of *how* paternity identification was accomplished *when* it became an issue, we have a suggestion from evolutionary biologists that phenotypic matching—that is, matching the facial and other characteristics of the body of the child to those of the father—should have been important.[22] That is the best we have on the question from scientists to date. Beyond that, evolutionary psychologist John Tooby and primatologist Irven De Vore have observed that, with increasing intelligence and ability to communicate, the time came during human evolution when a male was able to receive information from his relatives concerning his mate's infidelity in his absence.[23]

Common observation in today's world tells us that males everywhere are sensitive to second-hand information received from mothers and third parties bearing on the paternity question.[24] Is there a man anywhere who would not think twice, at least, when told that his mate was seen alone somewhere with an attractive male stranger? If this is due in some substantial part to genetic programming that has evolved out of the remote past and settled into today's humans, then we can expect males everywhere to have the sensitivity. Social planners could count on it in their populations when trying to increase confidence about paternity. To be cautious, however, in view of the hazy picture just surveyed, we must allow the possibility that men are learning to react this way solely because of their cultures—that all the world cultures where this sensitivity exists have developed it in their males independently, with no genetic programming traceable to the phenomenon.[25] If so, then it could be that some men in some cultures have not been paying attention, or whatever, so they have not *learned* to be sensitive to information received from others regarding paternity. But it seems at least equally as likely that a genetically influenced kin recognition mechanism exists in all of today's human males, wherever they may be, that has *some* degree of receptivity to such information. If so, the next question is: Is the second-hand information processing capacity sufficiently flexible to receive and process information from others that says "you can be certain the child is yours because (a late twentieth century scientific advancement) DNA testing says so"?[26] If the answer to that question is yes, then evolutionary biology has important information for those who would like to find ways to increase child support payments.

DNA "Fingerprinting" to Increase Child Support Payments

DNA tissue testing as a means of identifying humans was first announced in the press in the late 1980s. It can be used to identify the parents of a child. Its degree of identification specificity is similar to fingerprinting. When properly conducted, the test is, with infinitesimal reservations, "certain." In technical English-language usage there can be no qualification to certainty, but we commonly speak about degrees of certainty. The experts say that, properly conducted, thorough DNA fingerprinting showing a man to be the genetic father of a child is about as certain as it can be without being absolutely certain. The tests are complex, and the reports on how certain they are vary owning to a variety of factors.[27] However, even conservative estimates of the degree of certainty the tests can produce are staggering.[28] For example, one cautious court directed an expert witness to *reduce* to one chance in 84 million his estimate of the probability of two American male Caucasians (anywhere in the United States) having the same DNA "fingerprint" that a given child's father must have.[29] In other words, if a man is identified as a child's father through properly conducted DNA testing, he can comfortably "know," like mothers can, that he is the child's biological parent. With time, the tests should become yet more "certain."[30] Steps are being taken to reduce the cost of the testing procedure and to make it generally available. When that happens it could be regularly used in lawsuits that lead to child support orders.

Paternity Actions Calling for Child Support

Court-ordered child support can result from paternity actions—which normally involve out-of-wedlock births—and from divorce proceedings. DNA testing to eliminate or reduce doubt as to paternity could be useful in both types of proceedings, but its use in divorce proceedings would be difficult to institute. To begin with, the issue of children's paternity does not usually arise in divorce proceedings; it is just the opposite with paternity actions, however.[31] The paternity of the child involved is usually the center of attention, and evidence of various sorts is admitted on that issue. Thus, DNA testing could be used for two purposes in paternity actions: first, to help resolve the central issue in the lawsuit of whether the defendant was the child's genetic father[32] and, second, to give the defendant confidence that he was the child's father in order to promote voluntary compliance with any child support order issued by the court.

If positive DNA paternity tests can help dispel doubt about paternity and bring about increased solicitude for the child involved, they should be particularly effective when shown and explained to men who have child support orders entered against them after a paternity action. Probably no other group of fathers have more collective paternity doubt than defendants in paternity actions, first, because of the often casual nature of their relationship with the mothers and, second, because of the very fact that they are in the courtroom. They are normally there denying paternity.

Some men will, of course, fail to comply with child support orders for reasons other than lack of confidence in their paternity of the children. However, when a significant number of out-of-wedlock fathers nationwide are told that tests indicate the virtual certainty of their being the fathers of the dependent children, then sociobiology will predict a corresponding increase in voluntary compliance with child support orders[33]—provided, as we said earlier, that an evolved mechanism exists in human males that is flexible enough to receive and process an affirmation of paternity provided by a modern scientific test. Whether or not it exists is a question awaiting the closer attention of scientists.

KIN IDENTIFICATION IN GENERAL

In the last section we saw that, if genetically programmed inclinations of aid-giving by parents to their children are to be well understood, the mechanism by which the parent identifies the child needs to be explained. Now let us turn the picture around and expand it by looking at how a child (or an adult) identifies its parents, as well as its siblings, cousins and all other close relations. We need to do this because the underlying theme in the rest of this chapter is why and how *anyone* decides to aid, cooperate and reciprocate with and not harm other humans. So here we need to discuss the basic mechanisms by which people outside the parenting status identify those with whom they thus interact. The starting point for most such interaction, scientists believe, is *kin recognition*. Not only are we predisposed to give aid first and foremost to those we identify as close genetic kin, but we are also inclined to hook up with those people in long-term, reciprocal give-and-take relationships.[34] Furthermore, on the reverse side of that, we are especially careful not to harm such people.

Experiments have been done in recent years with a large variety of animals in order to discover their kin identification mechanisms.[35] Bees, birds, mice and squirrels are fairly easy to manipulate and study in this regard. Humans, to say the least, are difficult subjects for behavioral researchers. You cannot readily put them in boxes with relatives they have never seen before, as you can with mice, and observe their reactions. As a consequence, behavioral scientists have, to a large extent, built their hypotheses on how humans recognize kin from what has been observed in other animals, supplemented by what is known and surmised about human social environments throughout the ages during which human behavioral mechanisms were being set in place. The result is that at the moment the human kin identification mechanism picture is best described as quite speculative and undeveloped. For this reason, and because the social goal facilitation suggestions that will be discussed in the remainder of this chapter would hinge largely on specifics of the mechanism, one must say that such goal facilitations are away in the future. With the current burgeoning development of the scientific disciplines feeding into evolutionary biology, however, they might be only ten or twenty years away.

Having said that, let us set the stage for a squint into the future by looking at four kin recognition devices that scientists discuss.[36] The actual human mechanism may well be some combination of the four—or at least three of them. We will start with the theoretically possible device that is thought least likely to exist in humans[37] or in any other living thing.

Recognition "Alleles" (Genes) Device

This device would involve genes that gave their carriers the capacity to "automatically" recognize a relative at first meeting. No comparisons to self and no prior learning of family characteristics would be required. This would be an error-free device but would have to be extremely complex in its operation.[38] Scientists have found no clear evidence in animal studies that such a device exists anywhere in nature.[39]

Location (or Spatial Distribution) Device

This kin recognition method relies on a correlation between genetic relatedness and the geographic distance between the actor and others. If another individual is found in the actor's nest or home vicinity, it is treated as a close relative. There is clear evidence that this device exists in other animals, especially in birds.[40] It is subject to mistakes or "invasions" when nonrelatives and relatives can wander in and out of the location in question.

What is the likelihood that humans employ this device in kin recognition? Recall again that scientists focus on the late Pleistocene as the primary molding period of today's human behavioral mechanisms. Our ancestors then were nomadic.[41] Nomads, by definition, have a fluid or shifting home base. Nomadic groups can be thought of as carrying their home base with them, so one can roughly locate it at any given moment, just as one can with sedentary groups. However, continued movement of the nomadic group would increase the likelihood of group members being lost or left behind while encountered strangers were mixing into it. To the extent that this was true, a locational device would have lost efficiency as a kin recognition mechanism for the nomads. Furthermore, to the extent that one's home is not on any particular piece of land, but rather wherever one's group and its personal effects are at any given time, the focus is less on geographic location and more on association with other people as the reference for kin identification. This suggests that the locational kin recognition device may be less strongly represented in today's humans than are the two remaining nominees.[42]

Association (or Familiarity) Device

Scientists believe that an association device is the most prevalent kin recognition method in nature generally.[43] In respect to humans, it is beyond doubt that our remote ancestors were born and grew up with others of their species around them. It is almost as certain that those others were typically close genetic relatives.[44] Thus, natural selection would likely have favored a genetically influenced mechanism that said, "treat like close genetic relatives those with whom you have most often associated since birth."

"Invasions" that bring about mistakes are possible here just as with the locational device, discussed above, and occasions for such mistakes most probably existed somewhat during the Pleistocene. Surely at least some individuals for one reason or another happened to associate closely in their early years with other individuals to whom they were not closely related genetically. However, the hominid (humankind and its ancestors) populations of the Pleistocene probably lived and foraged in fairly cohesive bands of thirty or so individuals.[45] To the extent that this arrangement tended to keep relatives together, it would have favored a representation of associational cues in any kin recognition mechanism that developed in humans.[46]

Phenotypic Matching Device

This device is based on the assumption that the various types of gene "packages" (genotypes) that animal bodies (phenotypes) carry are manifested in characteristic appearances, odors, behaviors and the like, of the carriers. An individual learns its own characteristics (and perhaps those of others identified as genetic kin through locational or associational devices)[47] and then matches them against the corresponding characteristics of an unfamiliar individual. Scientists have demonstrated that this phenotypic matching exists in various animals.[48]

With regard to humans, there are no comparable studies as yet, but during the time periods when today's human behavioral mechanisms were set in place, bodily characteristics probably differed from individual to individual and some such characteristics "ran in the family"—just as they do today.[49] Thus, natural selection principles suggest that phenotypic matching is represented to some extent in whatever kin recognition mechanism exists in today's humans.[50] Like the locational and associational devices, this one is subject to mistakes. If a stranger can have a nose like that of one's mother today, the same was probably true during the Pleistocene. Thus, any phenotypic matching element in today's human kin recognition mechanism is likely to be subject to such "invasion" errors.

We can speculate that the susceptibility to error of the three indirect kin recognition devices—locational, associational and phenotypic matching—could be checked and balanced if they worked in combination. It would be much like three people—one blind, one deaf and one with no sense of smell—joining forces to identify objects around them. It might well be that a combination of indirect kin recognition devices could in this way evolve to become essentially as efficient in kin recognition as a direct, foolproof device such as recognition alleles.

But we do not know. All scientists have told us at the moment is that the three indirect kin recognition devices exist in many animals. Common sense as well as sociobiological theory tells us that some mechanism exists in humans.[51] Until we know more details, the following discussions of mechanism manipulation to decrease or increase patriotism and decrease war can only be tentatively suggestive.

DECREASING OR INCREASING PATRIOTISM

It is easy to identify organized groups for whom a social goal is the increase of patriotism within a population—the Veterans of Foreign Wars in the United States, for example. It is not as easy to pinpoint groups that want to decrease patriotism because it is at the root of war—but that sentiment is pervasive.[52] Therefore, patriotism is a good subject with which to illustrate the value-free nature of scientific learning in general and sociobiology in particular. We will treat the two opposing goals of decreasing and increasing patriotism together in this section and illustrate how the achievement of either might soon be facilitated by evolutionary learning. Many authorities have contributed to the insights we will report,[53] but Gary R. Johnson,[54] Pierre van den Berghe,[55] R. Paul Shaw and Yuwa Wong[56] are most prominently represented.

As in our previous discussions of goal facilitation through evolutionary learning, we need to do our best, with the information available, to identify and understand the relevant behavioral mechanisms—in this case the mechanisms that influence the human behavior we call patriotism. Let us begin by once more going back to the early times when the genetically programmed input to human behavior is assumed to have been set in place. During those early eras, scientists believe that hominids moved in small dispersed bands consisting mostly of close family members.[57] We can reasonably speculate that a large part of their waking hours were spent in resource gathering, which was more successful when individuals cooperated with one another. A band would often conflict with another group over scarce resources.[58] Individuals from different bands would, on occasion, find themselves working the same territory, and skirmishes would sometimes result.

Imagine yourself as a band member in this setting and assume you could understand the "inclusive fitness"[59] effects of your behavior. You would know that when mutually beneficial cooperation was called for it would make sense to do it with close genetic relatives, and the closer the better because the benefit to your gene package would be supplemented by the benefit to them to the extent your package overlapped with theirs. And any aid given to others was like helping yourself to the extent that your genes overlapped with their genes.[60] Thus, it would often make sense to risk your own well-being in conflicts with other bands when in so doing you could benefit one or several close relatives.[61]

Here is another important scene in this ancient picture we have put you in. When you come upon a skirmish on the savanna, with which side do you align? Why, the side to which you are most closely related, of course.[62] Let's see now, which side is that?

It is fairly clear that our ancestors did not think explicit thoughts like this in their daily affairs; certainly we ourselves don't. But sociobiology holds that mechanisms developed in them through natural selection that prompted them to act *as if* they were consciously making such calculations. One mechanism behind the sort of actions we just outlined was the kin recognition mechanism. As we have seen, the details of that perceptual/psychological mechanism are still hazy today, but it probably consisted of some mixture of locational, associational and/or phenotypic matching elements.

As the ages progressed, our ancestors formed larger and larger groupings—tribes, chiefdoms and eventually cities and nations. The members
of these larger groupings carried with them, largely intact, the kin identification mechanism that was adapted to band living in the Pleistocene.[63]
Natural selection works too slowly for much change to have occurred in
the mechanism. The social environment typically has changed drastically
for the human species in the last several thousand years, but the mechanism
remains much the same. To use a computer analogy, the "hardware" and
"software" in today's mechanism were constructed to reflect conditions
of the Pleistocene. The mechanism takes "input" from today's social
environments which can prompt "output" (behavior) that would have
served an individual's inclusive fitness interests during that early time, if
not today.[64] Thus, the unrelated person who lives immediately next door
today may be treated like the cousin who lived in the next dwelling during
the Pleistocene (location cue); live-in nannies and adopted siblings may
be treated as genetic nuclear family were (associational cues);[65] and
unrelated people in the neighborhood, city or nation who look, sound and
behave like the actor may also be treated with special deference (phenotypic matching cues).

Let us call this ability of the kin recognition mechanism to include
people it was not specifically designed to include its *flexibility feature*. The
ability for which it was specifically designed—to identify close genetic
relatives—we will call its *core function*. The flexibility feature, if and as
it exists (remember we have made certain assumptions as we've progressed in this narrative), should be subject to influence by social planning
to increase or decrease patriotism.

Influencing the Mechanism

The types of social programs we are about to illustrate could never
guarantee an increase or decrease in patriotism. Factors other than kin
recognition influence patriotism and its manifestations. For example, if by
manipulating kin recognition cues someone convinced a subgroup in a
nation that they were more like "family" than the rest so they should
secede, a threat of attack on the whole nation by outsiders could act as
patriotic *counter pressure* in the subgroup and prolong the national allegiance of its members.[66] But to the extent that the following social program
suggestions turn out to touch sensitive cues in the human kin recognition
mechanism, such programs should pressure patriotism in the desired
direction.

We will concentrate on nations as the units within which patriotism is to be increased or decreased, although the programs outlined could be applied to virtually any grouping of people such as a village or the whole international community.

Let us take "national patriotism" to mean an attitude that prompts an individual to work with others in the nation toward common goals and make personal sacrifices in the collective interest. We will assume that social planners want to increase or decrease that attitude generally throughout the population.

As we have noted, the human kin recognition mechanism may be made up of one, two or perhaps all three of the following devices or aspects: associational, locational and phenotypic matching. We will take them in that order and see how they might be influenced to increase and decrease patriotism if they exist in the human kin recognition mechanism.

Using Associational Cues

The idea here is to mix together (to increase patriotism) or separate (to decrease patriotism) people of different genotypes (gene packages) within the nation. This wholesale mixing or separating would take advantage of the flexibility feature of the kin recognition mechanism. You will recall that its core function was to identify close genetic relatives with which to interact in cooperative and aid-giving exchanges. But it is expandable and contractable; it can be "tricked."

Mixing the population would trick any associational aspect of the mechanism by increasing the likelihood that any given individual would interact closely, face to face, at an early age and thereafter with people of very different genotypes. If this were done, those involved should come to think of and treat the people of various diverse genotypes with which they associate much like "family." This, in turn, should make them responsive to appeals for cooperation and sacrifice in the national inter-est—the collective interest of the diverse national population.

Conversely, separating the population so that any individual is more likely to interact early in life, and therefore only with people of very similar genotypes, should eventually result in a decrease in responsiveness to national patriotic appeals because individuals cannot identify those citizens of different genotypes as being like "family."

Probably the best practical indicator of genotype for these purposes would be ethnicity.[67] Thus, one should either mix or separate the ethnic groups that are represented in the nation. To illustrate: to increase patriotism, one might mix the population by enacting laws that encourage

racially integrated housing in a community, as was done in the United
States for other purposes. Conversely, to decrease patriotism, one might
create separate residential regions within a nation for the different ethnic
groups, as was done in South Africa for other purposes. If such programs
are well designed and carried out, the typical citizen should eventually be
more inclined or less inclined, respectively, to cooperate with other
citizens toward common goals and make sacrifices for the "nation," as
represented by its total inhabitants—to the extent that the associational
device exists in the human kin recognition mechanism.

Using Locational Cues

Here the idea is again to mix or separate people of the various genotypes
that exist in the nation—"genotypes" being translated, for practical use,
into ethnic groups. But now the reason for such population shifts would
be to increase the likelihood of any individual citizen being born and raised
in close *spatial* proximity to people of quite different genotypes than its
own (to increase patriotism) or of much the same genotype (to decrease
patriotism).

The program for increasing patriotism would be aimed at tricking the
mechanism into signaling an individual to think of and treat the people in
the next dwelling (whoever they might be) like the uncles, aunts and
cousins who usually lived there during the Pleistocene. The program for
decreasing patriotism would aim at presenting the mechanism with ap-
proximations of the conditions that existed in the Pleistocene where
genetic "strangers" lived at a distance and were seldom seen.

The same ethnic regional grouping and community housing programs
we outlined above, in connection with associational cues, could be employed. For illustration purposes, let us turn them around here. One might
encourage ethnic groups that are bunched in certain regions or cities to
disperse and mix with other elements of the national population (to
increase patriotism) or, conversely, enact laws encouraging community
housing segregation along ethnic lines (to decrease patriotism). Completed
programs of those sorts should prompt an increase or decrease, respectively, in national patriotism—to the extent, now, that the locational device
exists in the human kin recognition mechanism.

Using Phenotypic Matching Cues

Recall that when an individual animal uses phenotypic matching as a
kin recognition device, it apparently matches its own characteristics (and

perhaps those of close genetic relatives identified through associational and/or locational cues) against those of the individual whose identity is in question. What characteristics are being compared? They need not be limited to visually perceptible things like facial features. Body odor may be included,[68] and scientists suspect that the vocal sounds an individual makes are also matched.[69] Let us speculate on how vocal sound as a kin recognition cue might have operated in human ancestors during the Pleistocene.

Assume that an individual, whom we will call Ego, had an older brother who left the band before Ego was born. Now Ego runs into this stranger for the first time while gathering roots on the savanna. Ego's kin recognition mechanism takes in various cues through her senses. Included are the sounds the stranger makes—grunts and screeches in various patterns. Ego's mechanism "computes" these as being very similar to her own and to those of her family. They are similar because the stranger (brother) *inherited* his tone and pitch[70] and/or because he spent his early years in Ego's family area where the inhabitants *learned* sound patterns from each other.[71]

That scenario illustrates the possible core function of a sound cue in a phenotypic matching device—the recognition of close genetic kin. Today's social planners could take advantage of the device's flexibility feature. Current human languages are present-day manifestations of the vocal sound cue.[72] Therefore, to the extent that the device is represented in the human kin recognition mechanism, social planners might use the language within a nation to increase or decrease patriotism.

The fullest use of this kin recognition cue to encourage patriotism would have everyone speaking the same language with the same accent, vocabulary, etc.[73] Various steps toward this end can be envisioned, such as laws mandating that instruction in schools everywhere be in the same language. Conversely, a law calling for all road signs and public notices to be in more than one language should tend to decrease patriotism—to the extent, as I said, that phenotypic matching is in the kin recognition mechanism.

The foregoing descriptions of how social planners might use associational, locational and phenotypic kin recognition cues to increase or decrease patriotism in a nation are only illustrative. I have been brief because social planning using evolutionary information is not as imminent in this area as it is in the areas we have explored earlier in the book. Interested readers should be able to extract many similar program ideas for the increase or decrease of patriotism from the relevant sociobiological literature.[74]

Echoes from Around the World

Before leaving the subject of patriotism, we should glance at current events. I expect the reader has already thought about relevant happenings around the world. I will just mention three of the most obvious. While not directly confirming the sociobiological theory we have explored, they create echoes that are consistent with it.

Language differences in the former Soviet Union were very prominent in the independence moves of various regions of that nation.[75] Similarly, the French language in Quebec, Canada, is at the root of that province's agitation for independence from the largely English-speaking remainder of the Canadian nation.[76] An editorial in a major U.S. newspaper noted that the Canadian "conflict underlines the crucial importance of language in tying a country together. . . . Language is central to identity, and competing languages erode the national identity required to sustain a nation."[77]

On our residential theme, the enforced separation of ethnic groups that occurred in South Africa under the apartheid system may well have kept that country from a more cohesive, cooperative national ethic.[78] A test of that proposition would occur if the government were to attempt to mobilize all its inhabitants for war against another nation while the effects of the enforced separation are still being felt in the country.

REDUCING WAR

Reducing the risk of war must be one of humankind's most significant and universally subscribed goals. And the reader may be way ahead of me here. Did the often-heard pacifist phrase "universal brotherhood" occur to you while reading the last section? Let us hold that for a moment.

Shaw and Wong recently surveyed the myriad approaches to resolving conflicts between nations[79] and concluded that an approach that takes human biology into account is more likely to succeed than an approach using some other starting point.[80] David Barash[81] has pointed out that many analyses of aggression in humans have had direct or indirect biological references.[82] These have included various studies concentrating on territoriality[83] (the need and struggle for living space) and approaches that described human social organization goals and the frustration that "triggers" aggression when such goals are not achieved.[84] In older studies, the biological discussion was often little more than vague speculation on how much of human aggression, conflict and war was attributable to "innate" (genetic in today's terms) human tendencies. Of course, writers in the past have been limited by the amount of biological information available at the time they were writing.

What Shaw and Wong and others influenced by sociobiology are suggesting is that efforts to understand the human conflict puzzle and come up with ways to solve it can be more successful now that modern biology has exposed genetic programming with its overall goal of "inclusive fitness" maximization. Human conflicts ranging from a fistfight to nuclear international war can be better understood and controlled, sociobiologists believe, when it is realized that the basic unconscious motivation behind most (nonpathological) human behavior is the proliferation of each individual's genes, whether through personal reproduction or reproduction by close genetic relatives (inclusive fitness).

As we have seen, in order for a gene carrier to be able to further the inclusive fitness goal in day-to-day affairs, it must be able to identify genetic relatives to whom it will be particularly solicitous—solicitous in terms of aiding and cooperating with them *and not harming them*. Kin recognition mechanisms are the means for such identification in other animals, and such a mechanism presumably exists in humans. Thus, sociobiologists suggest that social planners who want to reduce war might well begin by focusing on the human kin recognition mechanism.[85]

The suggestion, of course, is not that war can be completely eliminated by simply learning about the kin recognition mechanism and manipulating it. Too many other variables are involved for that. Even on the two-person conflict level one cannot assuredly stop fights and killings by pointing out to the participants that they are close genetic relatives. On that level, however, a recent empirical study of homicide statistics has indicated that humans are *much less* inclined to kill others when they have identified them as close genetic kin.[86] This agrees with sociobiological theory. The inclusive fitness concept holds that a person will harm a stranger before a close genetic relative, given the same provocation or incentive. Thus, to the extent that an individual can be led to identify another person, rightly or wrongly, as a close genetic relative, the chances of that individual harming the other should be reduced.

It is one thing to predict human conduct on a person-to-person level in a narrow setting like a family home—it is quite another to attempt to predict it in a group-to-group setting in large arenas like an international battleground. Sociobiology, at least in its present stage of development, provides its best insights in situations where just a few—preferably only two—people are interacting in the narrow setting.[87] In such situations, the genetic part of the genes/learning/culture mix of human behavior can be kept in clearer focus. Nevertheless, one senses the likelihood that if citizens can be convinced that people whom their government would label as "enemy" are much like "family," actual hostilities—trigger pulling and bomb dropping—can be reduced. Such reduction could occur because of the attitudes of those at the battlefront or among citizens at home, or indeed, among a country's leaders themselves, whose reasons for encouraging hostilities may be tempered by the perception that another nation's citizens are like "family."

So what is new about this? Pacifists have been preaching "universal brotherhood" and similar slogans for centuries. What is new is that sociobiology may be able (1) to provide scientific evidence that such preachments are not as naive and impractical as cynics might claim and (2) to uncover the mechanisms in humans that can be influenced toward such a goal.[88]

One such mechanism, we have seen, is the kin recognition mechanism. As we noted, this mechanism is likely to have a flexibility feature that can be "tricked" or influenced in various ways. Let us continue here to follow the two social planning themes we outlined by way of illustration in the last section on patriotism—the mixing of ethnic groups and the promotion of language similarity—this time in order to decrease international war. These two themes will illustrate, respectively, (1) how sociobiological learning can lose its potency as the social arenas to which it is applied increase in scope and (2) how it may nevertheless still provide insights on the larger scenes.

With regard to loss of potency, recall that the mixing of ethnic groups within a nation to promote patriotism (national "brotherhood") keyed into locational and associational devices that may exist in the human kin recognition mechanism. This is well and good *within* a nation, but how does one mix ethnic groups *between* nations that are separated by oceans, or even by artificial borders, in order to promote global "brotherhood" and reduce international wars? With difficulty, to say the least. (Note, however, that civil wars are another matter.)

More promise exists in social planning measures keyed to taking advantage of a possible phenotypic matching device in the human mechanism that has language (sound) as a kin recognition cue. The idea here is to get everyone on both sides of a potential conflict speaking the same language with the same accent. One universal language is no longer farfetched. The twentieth-century explosion in international travel and communications has provided the routes by which it could occur. Without conscious design, one language, English, has already made particularly impressive inroads in all corners of the world. With conscious social planning effort, the whole world could conceivably be speaking one language with the same accent some day.

Let us be more specific and speak directly to the practical-minded social planner who wants to reduce international warfare. Keep your eye on sociobiology. It may eventually provide you with concrete insights as to how you might utilize a human kin recognition mechanism to reduce the risk of wars. It may show you several routes, which could have a cumulative effect if more than one were employed.[89] One such route could be to get everyone speaking the same language around the world. One of the ways that you might promote that sub-goal is to raise funds to send teachers of the target language to various parts of the world. Such teaching programs are clearly feasible, for they have been done before on limited scales.[90] So keep your eye on sociobiology.

Chapter Nine

When Evolutionary Learning Will Not Help

Some readers may wish to skip this chapter and turn to the Epilogue. The chapter is aimed primarily at social planners and others who might wish to develop ideas for using evolutionary biology to address social problems other than those we have explored. I hope it will help them to craft solutions and avoid pitfalls. I will illustrate situations where the science may at first appear to offer help toward accomplishing a social goal but, on deeper analysis, does not.

There will be social goals that sociobiological learning cannot help achieve. Of course, in some cases assistance that is lacking at present may develop in the future. The discipline is still young, and, as it matures, more and more solid insights of practical use to society should emerge. Much of sociobiology's limitations at the moment derive from lack of technical knowledge of the details of human behavioral mechanisms. We saw that in the last chapter. Even when the gaps in technical knowledge are filled, however, there will be many social goals in respect to which science will not be able to provide practical advice to social planners. One reason for this will be that behavioral mechanisms developed in earlier human environments are not attuned to some aspects of the current world—or to use another metaphor, something in modern environments has short-circuited the ancient mechanisms.[1] Let me give a graphic illustration involving contraceptive devices.

Unimpeded insemination through copulation leads quite directly to childbirth whether or not the parents are aware of the connection or consciously plan on it. Thus, we can speculate that, during the Pleistocene, conscious planning for childbearing, if it even existed, was less important to successful reproduction by humans than was the urge to copulate.[2]

Now, along comes modern culture with its invention of contraceptive devices which create roadblocks in that direct route leading from copulation to childbirth. Not enough time has elapsed for natural selection to generally reflect these devices in human genetic behavioral programming.[3] So the human behavioral mechanisms are still intent on copulation while something more is now needed for successful reproduction in those societies where contraceptives are readily available—namely, a way to deal with these cultural devices that have short-circuited the ancient mechanisms' process. Existing genetic programming in humans that permits conscious planning may or may not be up to the task.[4] In any event, today's social planners who want to raise or lower the birthrate in a population[5] would be well advised not to concentrate solely on raising or lowering people's desires to copulate. Manipulations of that sort might have been sufficient had they been applied in the environments of ten thousand years ago, but not today, now that contraceptive devices have gotten in the way. Today, attention must also be paid to the conscious thought processes of prospective parents about the desirability of having children.

In this state of affairs, what can sociobiology tell planned parenthood advocates? Well, to the extent that *conscious* parenthood planning is involved it may not be much. That is because, as suggested above, a mechanism involving such conscious planning may have developed poorly, if at all, in our human ancestors during the Pleistocene. And we have the same basic equipment they had.

On the other hand, harking back to one of our earlier chapters, those who would reduce the incidence of rape in today's world should be able to accomplish much by concentrating solely on the mechanism behind the male drive to copulate. It does not appear that anything has popped up in modern environments to short-circuit the connection between the drive to copulate and the amount of rape that occurs in the way that contraception has interfered with the number of births that occur.

If we set aside things like contraceptive devices—modern cultural innovations that could thwart those looking to science for social planning assistance—we can generalize as to the types of social goals that are most likely to be realized through evolutionary learning. Those are goals closely tied to facts of our lives that should not have changed much since the Pleistocene: for instance, the fact that we are related to people around us in varying degrees; the fact that we give aid to and take it from others; and the fact that we reproduce and need to mate with someone in order to do so.[6] Thus, the accomplishment of goals like getting people to stop abusing their children or to give food to the needy[7] or to refrain from forcing themselves on those of the opposite sex is apt to be facilitated by influencing relevant evolved mechanisms that are essentially as well designed for today's circumstances as they were for those of the Pleistocene.

On the other hand, getting the stock market to go up or persuading people to buy a newly published book are goals far from the core experiences of Pleistocene life. This distance decreases the likelihood that ways can be found to influence human behavioral mechanisms in order to further those goals.

So evolutionary learning is not a cure-all or a do-all for the social planner. And those of us who are trying to build bridges between evolutionary scientists and social planners must be careful not to send the planners up "blind alleys." Futile advice of that sort can come in at least three different varieties: the advice is unclear; the advice is clear but it is impractical; or the advice is clear and practical, but the goal in view is not one to which a significant segment of society would subscribe. Let me illustrate these three varieties. It should be helpful to readers who are wrestling with ideas of their own on facilitating social goal attainment through sociobiological insights. It may save some wasted effort.

Numerous authors have briefly speculated on the use of evolutionary learning to change society.[8] Jerome Barkow and Austin Hughes have done so quite recently.[9] I am going to use one of Hughes's ideas, one of mine, and one suggested by other writers, as illustrations of thoughts that lead up blind alleys. These three illustrations are academically interesting and instructive for those who seek to advise social planners, but I do not believe that any of them would lead to feasible social planning projects, for reasons I will explain.

UNCLEAR ADVICE

Austin Hughes focuses on the goal of minimizing violence in slum and working-class neighborhoods in industrialized nations. He draws on kin selection theory to suggest that one element behind violence in such communities is the fact that inhabitants are living among nonkin. To alleviate this situation, he contemplates social policies that would "lower the long-term mobility of urban populations." He apparently visualizes policies that would keep old residents from leaving and new residents from entering target communities. As a specific measure along these lines, Hughes suggests keeping industries from leaving the target communities and thereby dislocating parts of the population. Eventually, such measures

would lead to an increase in the interrelatedness of the community members (people have children, who stay and have children, etc.). This would
increase cooperative, aid-giving tendencies among the population, while
correspondingly reducing xenophobic (us versus them) hostilities within
the community.

Hughes is quite clear so far. The social planner looking to reduce urban
violence could get a tentative sense of direction from this approach. But
then Hughes cautions the reader that kinship can also be a *disruptive* force
in a community. Concentrations of highly interrelated people who control
wealth collectively and between whom there are "abrupt discontinuities"
can lead to feuding kin groups. (Feuding Mafia families come to mind.)
So, Hughes suggests, to minimize violence in an urban community, social
planners should seek to avoid those negative aspects of interrelatedness
while "avoiding the situation in which kin ties are absent altogether."[10]
That is the end of Hughes's advice. He must be visualizing some optimum
degree of interrelatedness that will minimize violence in a community.

So what can a social planner do with Hughes's counsel? Urban slum
and working-class communities are already interrelated somewhat. What
is the optimum degree? If we could learn the answer to that question, we
could (with a lot of effort) measure existing interrelatedness in a sizable
community so as to know whether to invest time and resources toward
changing the degree of it. Anthropologists recently made such measurements in South American tribal communities.[11] But, again, what is the
optimum degree of interrelatedness in a community if minimizing violence in it is the goal? I wonder if scientists will ever be able to answer
that question. Certainly they cannot today. Hughes may well be right in
his theoretical speculations, but if I were a social planner, I believe I would
turn away here and put my energies into projects where the recommended
means for facilitating the goal can be made clearer.

ADVICE CLEAR BUT IMPRACTICAL

Let me give an example from my own ruminations where the advice
may be clearer but still of no practical use to the social planner.

Recall in Chapter 8 where we discussed the use of DNA fingerprint
testing in paternity actions to serve the goal of increasing voluntary child
support payments by fathers of children born out of wedlock. The test
would dispel doubts in the fathers' minds that the children in question were
theirs. As we noted, sociobiologists may soon be able to predict that
large-scale use of DNA testing would significantly increase voluntary
child support payments by out-of-wedlock fathers.

Well, what about divorced fathers who default on orders to pay child support? DNA testing might also increase their child support payments. Furthermore, the test might serve other goals, such as reducing child abuse by fathers[12] and the incidence of incest between them and their daughters. All fathers are on a continuum concerning paternity doubt. Those at the lower end of the continuum—those with the most doubt—should be more likely to commit child abuse and incest than those at the higher end.[13] If biological fathers in the population were generally to be given increased paternity confidence through DNA testing, the incidence of father-imposed child abuse and father-daughter incest might decrease significantly in the population.

But how would one institute such general testing programs? With out-of-wedlock fathers subject to child support orders, the framework is already in place for the test. When a man is brought to court to have him declared the father of a child and ordered to pay support, his paternity is the central issue and all reliable evidence on the question is called in. Note that all immediately interested parties—woman, man and child—have implicitly submitted themselves to the test if it will resolve the issue. No such framework is in place for testing paternity in general in the population or even the paternity of children whose parents are in divorce court. I visualize extreme reactions to any steps taken in those directions on the basis of invasion of privacy.[14] So, although abstract advice to social planners could soon be available here, it may be of no practical use under current social conditions.

CLEAR PRACTICAL ADVICE BUT NO SUBSCRIBERS FOR THE GOAL

The following illustration is for the business community. Writers have recently suggested that sociobiological insights surrounding genetically programmed "inclusive fitness" behavior in humans may be applied to business organizations. The prediction is that if aid-giving, reciprocal relations and cooperation are more prevalent the closer people are genetically related to each other, then business organizations will tend to display those characteristics in proportion to the degree of interrelatedness of the personnel.[15] In other words, a family-owned and -operated business should typically be more cohesive and cooperative internally than an organization made up of nonkin.

I know of no empirical data from the business community that directly supports this sociobiological hypothesis. Cohesiveness and cooperativeness are difficult to measure even when just two people are involved in simple tasks and relationships.[16] However, popular impressions support the hypothesis. *Newsweek* magazine, in a recent feature article on family business enterprises, reported that "[t]he roots run deep, embedded in family values. The flash of the fast buck is replaced with long-term plans. Tradition counts."[17] A professor at the Wharton School of Business was reported as saying that "they stick to their knitting."[18]

Let us assume that the hypothesis is correct—that family businesses do tend to be more cohesive and cooperative internally. The advice from behavioral science would then be that businesses should be formed of family members whenever possible to promote cohesiveness and cooperation internally. But, is cohesive cooperation, in and of itself, a goal of anyone in the business community?[19] Perhaps it is in a socialist ideal of socioeconomic development. However, in the free enterprise, capitalistic climate prevailing worldwide at the end of the twentieth century, most business and community leaders subscribe to other goals. Efficient production of high-quality goods and services, resulting in maximum profits and, ultimately, an increased gross national product, are the goals most often expressed.

Those goals may even be hindered by kinship-influenced behavior. For example, will maximization of profits, either immediately or in the long run, be served when the company president keeps his son in the firm (a cohesive move), even though he would have discharged him for inefficiency had he been nonkin? Perhaps, but presumably not in most cases.

If sociobiology or some other discipline can show that internal cohesiveness and cooperation in family businesses result in increased efficiency and profit maximization, the business community should be interested to learn that. Without such a showing, however, the mere fact, if it is true, that business organizations tend to be more cohesive and cooperative internally when there is a high degree of genetic interrelatedness among the personnel is not likely to interest many people in today's business world.

What have we learned in this chapter and those preceding it? I think it is this: The application of evolutionary learning to social issues is a complex enterprise, and every new insight into human behavior offered up by a discipline like sociobiology cannot be easily translated into a practical benefit for humankind. Nonetheless, every new insight can turn the lights up further for those looking to achieve goals involving the interactions of human beings. Sometimes, when conditions are right, the proper insights may help accomplish the goals.

Epilogue: What Are We Likely to Do Tomorrow?

First know thyself, someone said. Surely good advice. And the social problems we have are created by us—if not by you and me, then by humans like us. So we ought to do everything we can to inform ourselves about human behavior before tackling the social problems it created.[1] Sociobiology offers a way to identify universal characteristics—behavioral tendencies shared by everyone in our society. These universal tendencies prompt typical behavior, and the best-guess assessment of typical behavior is at the core of much of the planning by social reformers, private and public administrators and lawmakers. Indeed, it is implicitly behind much of the planning you and I do regarding our daily contact with other people.

Evolutionary biology is still developing as a field of learning. It has been developing since at least Charles Darwin's time, and the development may go on indefinitely. New discoveries and theories in science tend to feed on each other and to grow exponentially. When can or should we use the information sociobiology provides in planning our social affairs? I suggest the answer is today or never or sometime in between depending on (1) how solid is the aspect of sociobiology that we want to use, (2) how reliable is the other information we have on the typical behavior that is relevant to our social problems and (3) how pressing is the need for solution of the problems.

Evolutionary scientists are taking their time and going slowly in their academic laboratories. That is good. They should proceed deliberately. Outside the ivory towers, however, social problems do not wait. They continually call for attempts at solution. The people embroiled in the problems must make plans to solve them with whatever information is available. Up to this point in history, decisions calling for predictions of typical human behavior—what it will be like tomorrow and into the future—have had little to inform them other than the intuition of the decision makers (slanted by personal biases perhaps). Controlled observation of actual human behavior is difficult to achieve. Even when such observation can be accomplished, the results, together with opinion surveys and such, as to what typical behavior was yesterday, may be invalid tomorrow when cultural winds have shifted. Fortunately, as it turns out, one element in the mix that makes up our behavior is relatively stable: Genetic programming that influences our behavior will be the same in all of us tomorrow as it was yesterday, and it provides an element of reliability for predictions of future typical behavior. That is perhaps the most useful message that sociobiology has given us. This book has been an attempt to convey that message to those with social problems that need solution.

Notes

INTRODUCTION: PROPER USE OF
EVOLUTIONARY SCIENCE IN SOCIAL AFFAIRS

1. See Flew, "From Is to Ought," 146–47. See also Ruse, *Taking Darwin Seriously*, 78; Lopreato, *Human Nature and Biocultural Evolution*, 8; Hofstadter, *Social Darwinism in American Thought* (generally).

2. Social Darwinism, it has been said, was behind the U.S. Supreme Court's overturn as unconstitutional, early in the twentieth century, of legislation that had provided for minimum wages and restrictions on child labor and hours of work. Morris, *Evolution and Human Nature*, 50–51. See also Edel, "Attempts to Derive Definitive Moral Patterns from Biology," 111, 112.

3.

The more extreme form of social Darwinism argued for a policy of complete laissez-faire in order to give free rein to economic competition. The state must withdraw from all efforts to limit the freedom of individual action, leaving everyone to rise or fall according to their ability. Progress would only occur if the fittest were allowed to fight their way to a dominant position in the economy, while those unfit to work would have to take the consequences.

[Herbert] Spencer insisted that the state should concern itself solely with external affairs; internally, it had no business trying to regulate the lives and activities of the people. There should be no state control of health care, education, or relief for the poor (Bowler, *Evolution*, 286–87).

We should note that socialists have also attempted to support their ideology by reference to evolutionary learning. "A. R. Wallace [claimed that] [i]f all differentials of wealth are removed . . . husbands and wives will choose one another for their biological qualities, just as nature intended" (ibid., 288).

4. "When we examine the history of humankind, whether from paleontological evidence, documentary accounts such as the Bible, or culture specific myths and fairytales, we see evidence of children being mistreated, abandoned, sacrificed or eaten . . . in culture after culture." (Burgess and Garbarino, "Doing What Comes Naturally? An Evolutionary Perspective on Child Abuse," 90; see also Jay and Doganis, *Battered*, 2).

5. For a discussion of the "naturalistic fallacy," as viewed by David Hume, Edward Moore and other philosophers, see Ruse, *Taking Darwin Seriously*, 86–93, and Maxwell, *Human Evolution*, 232–33. Albert Einstein said that "[a]s long as we remain within the realm of science proper, we can never meet with a sentence of the type 'thou shalt not'. . . . Scientific statements of facts and relations . . . cannot produce ethical directives" (Einstein, *Out of My Later Years*, 114).

6. See Flew, "From Is to Ought."

7. In Edel, "Attempts to Derive Definitive Moral Patterns from Biology," 112–13, the author expands on this point:

To be rooted in a biological mechanism serving a biological need seems to many to be all the justification human behavior can ultimately ask. In this way, eating and drinking and sexual activity are obviously sanctioned. A variety of other invariant human tendencies, it is believed, will be found similarly to rest on more subtle biological mechanisms and thus prove their merit. Any institution claiming roots in instinct can thus clothe itself with the moral authority of absolute fixity. The history of social psychology is strewn with the wreckage of instincts intended to support prevalent institutional forms, such as pugnacity instincts to prop up war, acquisitive instincts to reinforce private property, and a variety of specific instincts to support the family.

8. See Lumsden and Wilson, *Promethean Fire*, 182–83: "[T]he philosophers and theologians have not yet shown us how the final truths will be recognized as things apart from the idiosyncratic development of the human mind. In the meantime, by appealing to the core principles of neurobiology, evolutionary theory, and cognitive science, practitioners of a new human science can reach a deeper understanding of why we feel certain courses of action to be intrinsically correct."

9. Alexander, *Darwinism and Human Affairs*, 220; Hinde, *Individuals, Relationships and Culture*, 137 ("Knowledge of nature must be used to aid us to achieve rather than to set our goals"). Philip Kitcher, in an in-depth critique of sociobiology enthusiastically endorsed on the dustcover by the two best known early critics of the discipline, Richard Lewontin and Stephen Jay Gould, gives his approval to this use of the natural sciences. Kitcher, *Vaulting Ambition*, 420.

10. Alexander, *Darwinism and Human Affairs*, 278; Lumsden and Wilson, *Promethean Fire*, 183. See also Konner, *The Tangled Wing*, 180; Barash, *The Whisperings Within*; Breuer, *Sociobiology and the Human Dimension*, 73–76; and Murphy, *Evolution, Morality, and the Meaning of Life*, 100.

11. See Murphy, *Evolution, Morality, and the Meaning of Life*, 100:

The view of reason adopted by sociobiology is Humean [reflecting the philosophy of David Hume] and regards it as an instrument that allows us to calculate the best means to the attainment of our ends. It can even evaluate ends where these are seen as subordinate to even higher or more important ends. What it cannot do, however, is evaluate the ends finally accepted as ultimate, for these are given by the passions and, at this level, reason is the slave of the passions. And where do these basic passions come from? Evolutionary biology surely has, at least, part of the answer to this question.

12. This is Masters's statement:

Respect for individual and cultural differences follows necessarily from the discovery of the natural causes of variation in human behavior. If the phenotype [the living body] is merely the 'vehicle' by which genes replicate themselves (Dawkins 1982), human beings are equal in a more profound sense than would appear from the conventional view of civil rights. Difference is not inferiority. Who can know which of us carries a mutant gene that is a valuable adaptation to a future environment and will someday spread throughout the human gene pool?

Nothing is more threatening to the survival of a species than the disappearance of naturally occurring genetic variation. Evolutionary principles teach us to expect humans to overvalue their own interests, perspectives, and importance. These principles also teach us, however, that this selfishness of the human phenotype is an evolved behavioral strategy that is naturally balanced by social cooperation, without which we could not have evolved or survived. Because human societies, particularly modern civilizations, depend on reciprocity, respect for others is enjoined on us by the very nature of the social system in which we live (Italics in original; Masters, "Evolutionary Biology and Political Theory," 205).

13. This is Wilson's statement:

I believe that a correct application of evolutionary theory also favors diversity in the gene pool as a cardinal value. If variation in mental and athletic ability is influenced to a moderate degree by heredity, as the evidence suggests, we should expect individuals of truly extraordinary capacity to emerge unexpectedly in otherwise undistinguished families, and then fail to transmit these qualities to their children. The biologist George C. Williams has written of such productions in plants and animals as Sisyphean geno-types; his reasoning is based on the following argument from elementary genetics. Almost all capacities are prescribed by combinations of genes at many sites on the chromosomes. Truly exceptional individuals, weak or strong, are, by definition, to be found at the extremes of statistical curves, and the hereditary substrate of their traits come together in rare combinations that arise from random processes in the formation of new sex cells and the fusion of sex cells to create new organisms. Since each individual produced by the sexual process contains a unique set of genes, very exceptional combi-nations of genes are unlikely to appear twice even within the same family. So if genius is to any extent hereditary, it winks on and off through the gene pool in a way that would be difficult to measure or predict. Like Sisyphus rolling his boulder up and over to the top of the hill only to have it tumble down again, the human gene pool creates hereditary genius in many ways in many places only to have it come apart the next generation. The genes of the Sisyphean combinations are probably spread throughout populations. For this reason alone, we are justified in considering the preservation of the entire gene pool as a contingent primary value until such time as an almost unimaginably greater knowledge of human heredity provides us with the option of a democratically contrived eugenics (Wilson, *On Human Nature*, 198).

14. Compare the following quotation from *Biophilia*, in which Wilson clearly took off his scientist's cap and expressed his personal feelings as a conservationist: "For if the whole process of our life is directed toward preserving our species and personal genes, preparing for future generations is an expression of the highest morality of which human beings are capable" (121). Also see Barkow, *Darwin, Sex and Status*, 389: "We must engage in social engineering [because of] the unthinkable reality of thousands of nuclear weapons awaiting a word to destroy our civilization and perhaps even our species."

15. I am more confident that the indicated response would be given by Wilson than by Masters. Several years after writing the passage in question, Wilson said: "For now . . . scientists can offer no guidance on whether we are really *correct* in making certain decisions, because no way is known to define what is correct without total reference to the moral feelings under scrutiny" (Italics in the original; Lumsden and Wilson, *Promethean Fire*, 183). Later yet, Wilson said, "[t]he naturalistic fallacy has not been erased by improved biological knowledge, which still describes the 'is' of life but cannot prescribe the 'ought' or moral action" (Wilson, "The Relation of Science to Theology," 430–31). Compare Alexander, *The Biology of Moral Systems*, 220: "[N]o [solution of moral issues arises] out of evolutionary understanding. But perhaps our view of the issues can be clarified and our collective response as a result altered—and perhaps in a direction likely to be judged by those concerned as positive."

Masters, on the other hand, may have been suggesting that one can derive "ought" from evolutionary learning when, before writing the passage in question he said:

Contrary to the nihilism or relativism that have predominated in the West over the last century, an evolutionary approach to human social behavior offers a reasonable basis for judging the rightness or justice of political institutions. More specifically, the new naturalism provides objective criteria for preferring a constitutional regime in which citizens are subject to the law and play a legitimate role in political life. While evolutionary biology also explains the necessities that challenge civilization and its institutions, this naturalistic approach can help us to formulate more decent and humane standards of social life (Masters, "Evolutionary Biology and Political Theory," 204–5).

16. "Since the term *program* was taken over from the field of informatics, it is sometimes rejected as an anthropomorphism. Yet, the use of the term in biology is fully justified [citation]. Even though the mechanism by which the DNA stores and codifies information is of course different from that of a computer, the basic principle is remarkably similar, as demonstrated by the researches of molecular biology" (Emphasis in original; Mayr, *Toward a New Philosophy of Biology*, 4).

17. An excellent objective summary and critique of a major part of the relevant theory and empirical studies, as of the early 1980s, can be found in Gray, *Primate Sociobiology*.

18. If sociobiology were to be used normatively, Roger Masters has illustrated how it "could well challenge existing sociopolitical beliefs and institutions rather than support them" (Masters, "Is Sociobiology Reactionary? The Political Implications of Inclusive-Fitness Theory," 275). Donald Campbell, a psychologist, has said that theorization by him and others has laid the grounds "for an explanation of the human predilections for an ideology of equality . . . as well as the liberty and fraternity that standard sociobiology may also explain" (Campbell, "The Two Distinct Routes beyond Kin Selection to Ultrasociality: Implications for the Humanities and Social Sciences," 36).

19. See Hinde, *Individuals, Relationships and Culture*, 107: "Cultural influences can magnify, distort and redirect the behavioral tendencies of individuals, even to their own disadvantage." More specifically, see Barkow, *Darwin, Sex and Status*, 382: "[T]he goals of those using [advertising] may be different from our own. It is salutary to recall that the seminal genius in the field of advertising/propaganda was named Goebbels. Politicans and political propagandists are often expert in inputting information into what might be termed the 'ethnocentrism' module, triggering a behavior pattern that causes us to rally around a leader in defense of our group and against an external enemy. Would-be leaders are adept at inventing external enemies."

20. It has been suggested that one reason why sociobiology has generated controversy and debate is that "[h]umans are already excellent psychologists who sometimes resent people with PhDs telling them how their minds work" (R. Thornhill and N. Thornhill, "The Evolution of Psychological Pain," 94).

21. See Chapter 6.

22. On the provability of scientific hypotheses, Jerome Barkow has made the observation that those who want "Absolute Truth" should look to religion, not science. See Barkow, *Darwin, Sex and Status*, 10. The reasoning behind Barkow's observation is given by Peter Bowler as follows: "If a hypothesis is successful in passing the tests to which we subject it, we might be tempted to regard it as an established truth about how nature works, but this is *not* a proper interpretation of knowledge gained in this way. It is always possible that a false hypothesis was lucky enough to pass the first, less rigorous tests and may then reveal its weakness by failing new tests in the future. It thus is necessary to regard all scientific knowledge as provisional in nature, accepted as a useful guide for the time being but open to potential falsification by further research" (Italics in original; Bowler, *Evolution: The History of an Idea*, 15).

23. In 1991 the U.S. Food and Drug Administration urged that a drug to treat Alzheimer's disease be made available "under a program that allows people with life threatening diseases and no alternatives to receive experimental drugs." Only AIDS drugs had previously been approved under the program. "Opponents of approval [for the Alzheimer drug] said that experimental evidence of the drug's effectiveness was weak and that it could cause liver damage. Supporters

of the drug said even the small improvement it might offer was important, given the lack of other treatments" (*New York Times*, March 23, 1991, Section 1, p. 8, col. 1). See also "AIDS Shot Proven Safe for Humans," *Chicago Tribune*, January 15, 1991, Section 1, p. 3, col. 1: "The first AIDS vaccine to be tested on humans is safe, but its effectiveness has not been proven, researchers reported. . . . In an accompanying editorial, a researcher said that because of the pressing need for preventive measures against the AIDS virus, scientists may have to go ahead with development of vaccines that may not block infection but could prevent or delay the onset of the disease."

24. For ease of entry, an initial reading program might start with Dawkins, *The Selfish Gene*, then move to Trivers, *Social Evolution* and Gray, *Primate Sociobiology* and end with a recent, as of this writing, comprehensive treatment of the subject, Barkow, *Darwin, Sex and Status*, which reflects developments up to the late 1980s. These books will lead the reader to other sources.

CHAPTER 1: EVOLUTIONARY BIOLOGY PRIMER
I: AID-GIVING BEHAVIOR

1. Here are three reasons why, initially at least, some people may be unreceptive to sociobiology: (1) Its basis in evolutionary explanations for the development of the earth and all living things on it is contrary to the fundamental teaching of many religions; (2) it modifies somewhat the concept that humans are born with "clean slates" (no behavioral inclinations or limitations), so that culture and learning are solely responsible for all human behavior—a concept that is ideologically appealing to many people in all walks of life; (3) many articulate, influential academics have built careers on the "clean slate" assumption that acculturation and learning are solely responsible for all human behavior, and it would be difficult for them to learn biology and amend their positions accordingly.

Given these reasons to be put off by basic Darwinism and its modern development, sociobiology, the tenacity of the evolutionary perspective is remarkable (see Degler, *In Search of Human Nature. The Decline and Revival of Darwinism in American Social Thought*). This alone may be testimony to a (sometimes grudging) intuitive sense in humans that the evolutionary perspective rings true in its core insights about human behavior.

2. See note 24 to the Introduction for a suggested initial reading list. For those ready to tackle heavy reading, Richard Alexander's two volumes, *Darwinism and Human Affairs* and *The Biology of Moral Systems*, are very advanced texts.

3. This is an altered, elaborated version of a chapter appearing in Beckstrom, *Evolutionary Jurisprudence*.

4. The list of disciplines that have produced input to sociobiology and in which those working in the field have academic appointments includes anthropology, biology, population genetics, psychology, psychiatry, sociology,

ethology, political science, economics and law. Maxwell, ed., *The Sociobiological Imagination*, is an anthology with contributions from authors in eighteen different disciplines or subdisciplines. The Human Behavior and Evolution Society, located in North America, the European Sociobiological Society, and the Association for Politics and the Life Sciences (North America) all have been formed to deal with the subject.

5. For an advanced treatment of the subject, see Williams, *Adaptation and Natural Selection*. An excellent introduction for the layperson is in Daly and Wilson, *Sex, Evolution and Behavior*, 1–19.

6. See Wilson, *Biophilia*, 46:

Darwin was a great expansionist. He shocked the world by arguing convincingly that life is the creation of an autonomous process so simple that it can be understood with just a moment of reflection. . . . It can all be summarized in a couple of lines: new variations in the hereditary material arise continuously, some survive and reproduce better than others, and as a result organic evolution occurs. And even more briefly as follows: natural selection acting on mutations produces evolution.

7. A mutation is a

sudden random change in the genetic material of a cell that may cause it and all cells derived from it to differ in appearance or behavior from the normal type. . . . Mutations occur naturally at a low rate but this may be increased by radiation and by some chemicals. . . . The majority of mutations are harmful, but a very small proportion may increase an organism's fitness; these spread through the population over successive generations by natural selection. Mutation is therefore essential for evolution, being the ultimate source of genetic variation (*Dictionary of Biology* [Warner Books], 155).

8. See Dawkins, *The Extended Phenotype*, 81–96, regarding what it is, in particular, on the chromosomes of living cells that is operative in the natural selection process.

9. To a population geneticist "[a]n individual is fit if its adaptations [to its environment] are such as to make it likely to contribute a more than average number of genes to future generations" (Williams, *Adaptation and Natural Selection*, 158).

10. Although evolutionary biologists think that natural selection operates, at least primarily, at the level of the genes or genetic material that an individual (and that individual's close kin) contains, there is some room in current thinking for the operation of selection at the level of whole, extended populations of individuals. See Williams, *Adaptation and Natural Selection*, 92–250; Williams, ed., *Group Selection*; Barash, *Sociobiology and Behavior*, 107–15, 125–28, 137–38; and Brandon and Burian, eds., *Genes, Organisms, Populations*.

11. The phrase "survival of the fittest" was apparently coined by Herbert Spencer. It does not appear in the first edition of Darwin's *Origin of the Species*, but it was incorporated into later editions to replace the term "natural selection." Strahlendorf, "Evolutionary Jurisprudence," Chapter Three, p. 13.

12. "Darwin was ignorant of the principles of Mendelian genetics, even though Gregor Mendel's path breaking work had been published soon after the

appearance of [Darwin's] *The Origin of the Species.* (Darwin's own copy of Mendel's paper on the basic principles of genetics was found in his library after his death with its pages still uncut.)" Degler, *In Search of Human Nature,* 20–21.

"[T]he 'Modern Synthesis,' by which is meant the bringing together of Darwinian evolution and its theory of natural selection with the science of genetics [was not completed] until roughly the time of the Second World War.... principally through the work of Sewell Wright in the United States and R. A. Fisher in the United Kingdom" (ibid., 230).

13. See Williams, *Adaptation and Natural Selection,* 83: "Complex systems of behavior, such as the more elaborate reproductive patterns, will usually be a blend of learned and instinctive elements. There are things that have to be learned, such as the individual characteristics of a particular mate or the location of a nest site. All elements that can be instinctive, however, will be instinctive. Instinct costs less than learned behavior in the currency of genetic information."

14. "[W]hen we talk about complex and flexible behavior, in terms of the [body characteristics] involved, we mean the evolution of the brain into an extremely efficient computer for simulations.... [T]hese attributes reside in the individual and are the focus for selection. The conditions ... that promote them are a complex social and ecological environment in which an organism needs to make rapid and flexible responses to the problems it faces" (Foley, "How Useful Is the Culture Concept in Early Hominid Studies?," 31).

15. See, generally, Barkow, ed., *Evolved Constraints of Cultural Evolution.*

16. See Trivers, *Social Evolution,* 90, 94.

17. See Gould, *Ever Since Darwin,* 260–67, and Kitcher, *Vaulting Ambition,* 79. Kitcher's book was endorsed on the dustcover by the two best known critics of sociobiology, Stephen Jay Gould and Richard Lewontin, as "the best dissection of" and "the last word" on sociobiology. Kitcher later appeared to alter considerably his position on sociobiology in Sterelny and Kitcher, "The Return of the Gene."

18. "Estimates of [gene] mutation rates range from about 10^{-4} to 10^{-10} per generation" (Williams, *Adaptation and Natural Selection,* 24). See also Trivers, *Social Evolution,* 91.

19. Hamilton, "The Evolution of Altruistic Behavior," 354; Hamilton, "The Genetical Evolution of Social Behavior," 1, 16.

20. See, generally, Dawkins, *The Selfish Gene,* 97–100; Alexander, *Darwinism and Human Affairs,* 44, 45, 130; and Breuer, *Sociobiology and the Human Dimension,* 11–16.

21. Barkow describes an aid-giving situation involving relatives which is of more likely occurrence and then makes the following observation:

Presumably, there is a particular subsystem of the brain making use of a complex algorithm to generate this kind of nepotistic altruism behavior. We can predict that this algorithm will produce behavior somewhat similar to that predicted by the arithmetic above, simply because the closer the behavior produced by an algorithm comes to the arithmetic model, the stronger the selection pressure in favor of it will be. Since we are not perfect biological machines, however, actual behavior will be only a rough approximation of the arithmetic (Barkow, *Darwin, Sex and Status, 51*).

22. See, for example, Trivers, *Social Evolution*, 169–79. Hamilton's exposition of kin selection is a cornerstone of sociobiology. His rule states that selection will favor an action by one animal that causes a loss to itself of c offspring and a gain of b offspring to another animal to which it is related by r provided $rb - c > 0$. For a nonmathematical discussion of some of the intricate implications of Hamilton's rule, see Grafen, "A Geometric View of Relatedness," 28.

23. See, generally, Trivers, "The Evolution of Reciprocal Altruism;" Trivers, *Social Evolution*, 361–94; Taylor and McGuire, eds., *Reciprocal Altruism: 15 Years Later*.

24. See Alexander, *The Biology of Moral Systems*, 82, 94, 97, regarding the complexities of direct and indirect reciprocity.

25. See Degler, *In Search of Human Nature, 215–44*.

26. "It is extremely difficult to identify the degree to which human behaviors are genetically influenced, and to some extent, it is foolish to try. If a person stands five feet tall, it is meaningless to suggest that three feet and four inches of her stature is due to her genetic makeup, with the remaining one foot and eight inches due to her environment (nutrition, health, etc.). Rather, every inch of her height results from the interaction of her genes and her experiences" (Barash, *The Hare and the Tortoise*, 51–52). "Attributing behavior to either learning or instinct is about as fruitful as arguing whether it is the sugar or the flour that make the cake" (Goldsmith, *The Biological Roots of Human Nature*, 87).

27. Regarding the developing science aimed at measuring the relative contributions of environmental and genetic influences on behavior, see Plomin, DeFries and McClearn, *Behavioral Genetics*.

Extensive studies at the University of Minnesota, in Rome and elsewhere have been done in recent years on identical twins (each pair share the same genes) who have been raised and lived apart—in different environments. The similarities in their personalities and behaviors as adults are beyond coincidence and often startling. See "All About Twins," *Newsweek*, November 23, 1987, p. 58. See also Emmerman, "Alter Egos," *Chicago Tribune Magazine*, July 1, 1990, p. 10, 12: "Preliminary results from the 10 year study being conducted by the Minnesota Center challenge the popular notion that all of us—not just twins—begin life as formless lumps of clay that are then molded by our circumstances. The study instead strongly suggests that we're all born with a genetic blueprint that influences such traits as leadership, sociability and even the tendency to be moved to tears by a sad movie.

[Psychology professor Thomas Bouchard of the University of Minnesota says] "[w]hat we inherit are propensities toward personality traits. . . . Just because something's genetically influenced doesn't mean it's chiseled in stone. Environment also plays a key role" (ibid).

28. For an overview of the variation in the approaches of certain social and natural scientists to the interaction of genetic programming and culture, see Boyd and Richerson, *Culture and the Evolutionary Process*, 12–14, and Barkow, *Darwin, Sex and Status*, 231–92. See also Ball, "Memes as Replicators," 145, 157. Examples can be found in Alexander, "Evolution and Culture," 50; Cavilli-Sforza and Feldman, *Cultural Transmission and Evolution*; Dawkins, *The Selfish Gene*, 203; Durham, "Toward a Coevolutionary Theory of Human Biology and Culture," 39; and Lumsden and Wilson, *Promethean Fire*.

29. See Dunbar, "Darwinizing Man: A Commentary."

30. For other reasons why human females are posited, typically, to be somewhat more solicitous of ascribed offspring than are males, including the factual premise that females invest more of their reproductive potential in any particular offspring than do males, see Daly and Wilson, *Sex Evolution, and Behavior*, and Symons, *The Evolution of Human Sexuality*. See also Beckstrom, *Sociobiology and the Law*, 81–88.

31. See Gaulin and Schlegel, "Paternal Confidence and Paternal Investment: A Cross-Cultural Test of a Sociobiological Hypothesis;" Russell and Wells, "Estimating Paternity Confidence;" and Kurland, "Paternity, Mother's Brother, and Human Sociality."

Compare the following statement by the Wyoming Supreme Court in an action to determine the paternity of a child: "The woman carries the child through pregnancy. When born of her, the fact of motherhood is obvious. Not so the man. The proof of fatherhood, or the proof of the lack thereof, must come from an external source" (A v. X, Y, and Z, 641 P.2d 1222, 1225. [Wyo. 1982]). Recent developments in embryo transplants would call for a slight modification of the Wyoming court's statement. The possibility of mixups in community maternity wards can also affect maternity confidence to some degree.

32. Barash, *The Whisperings Within*, 106, and Irons, "Kinship," 80–81. "[I]t has recently been shown that some insects have a specific chemical signal enabling them to recognize their kin by smell. It is not beyond the realm of possibility that a simple signal exists in humans as well—although I would be inclined to expect a visual rather than an olfactory signal" (Konner, *The Tangled Wing*, 321–22). See also Daly and Wilson, *Sex, Evolution, and Behavior*, 53. Compare Alexander, *The Biology of Moral Systems*, 100: "[T]here is yet no undisputed evidence of unlearned recognition of relatives in any species." For more detailed discussion of human kin recognition mechanisms, see the section titled "Kin Recognition in General" in Chapter 8 of the present work, and authorities cited there.

33. See Irons, "Investment and Primary Social Dyads," for a discussion of how basic genetic programming that furthers the reproduction of genetic materials contained in us humans and our close genetic relatives ("inclusive fitness") may produce varying treatment of others depending on differences in environment. See also Kurland, "Paternity, Mother's Brother, and Human Sociality," 164–66, for a discussion of various idiosyncratic characteristics that can influence whether one invests in sons and daughters or nieces and nephews.

34. See Fisher, *The Genetical Theory of Natural Selection*, 27–30, on the possibility of quantifying the reproductive and nurturing value of a generational age difference.

35. Daly and Wilson, in *Homicide*, analyze data from around the world indicating quite clearly that people are much more likely to be killed by strangers than by close genetic kin, everything else being equal (pp. 20–30). For example, homicide data from Detroit showed that people living in the same household "who are not blood relatives of the killer are more than eleven times as likely to be murdered as cohabitant kin [even when] spouses are removed from the analysis" (p. 23).

36. "If child abuse is a behavioral response shaped by natural selection, then it is more likely to occur when the situation is one of reduced inclusive fitness payoffs due to uncertain relatedness or low benefit-cost ratios" (Burgess and Garbarino, "Doing What Comes Naturally? An Evolutionary Perspective on Child Abuse," 98).

CHAPTER 2: REDUCING CHILD ABUSE

1. Geovannoni, "Definitional Issues in Child Maltreatment," 34. See also Dome, *Crimes Against Children*, 5–6.

2. If a poor parent kills a severely demented child, does this increase the likelihood of the child's genes being proliferated through healthy siblings who, because of the death, will be able to get a larger share of the family's scarce resources? See, generally, Hausfater and Hrdy, eds., *Infanticide*; Daly and Wilson, *Homicide*, 37–93.

3. Geovannoni, "Definitional Issues in Child Maltreatment," 34–35.

4. See Olds and Henderson, "The Prevention of Maltreatment," 725.

5. Daly and Wilson, "Abuse and Neglect of Children in Evolutionary Perspective," 405, 408–9 (analysis of abuse data reported in 1976 to American Humane Association from areas comprising 44.6 percent of U.S. population); Lenington, "Child Abuse: The Limits of Sociobiology," 17, 23 (author combined broad-based abuse data from studies by Gil and Johnson with conservative estimates of national household type frequencies); Lightcap, Kurland and Burgess, "Child Abuse: A Test of Some Predictions from Evolutionary Theory," 61, 64 (analysis of relative occurrence of abuse within sample of twenty-four two-parent households in rural Pennsylvania); Daly and Wilson, "Child Abuse and Other Risks of Not Living with Both Parents" (analysis of Hamilton, Ontario, child abuse data for 1983). For a survey of all the above and other data, see Daly and Wilson, *Homicide*, 85–90.

6. Some data do not specifically identify the abuser in the stepparent homes, but some studies have been able to report that children are more likely to be abused by stepparents than by biological parents. Lenington, "Child Abuse: The Limits of Sociobiology," 24 (author's analysis of data in studies by Gil and Johnson); Lightcap, Kurland and Burgess, "Child Abuse: A Test of Some Predictions from

Evolutionary Theory," 64 ("Given the choice between abusing a stepchild or a biological offspring, these individuals never abused their own kin").

7. Daly and Wilson, *Homicide*, 89.

8. Thomas, "Child Abuse and Neglect Part I: Historical Overview, Legal Matrix, and Social Perspective," 336.

9. Lenington, "Child Abuse: The Limits of Sociobiology," 25.

10. Steele and Pollock, "A Psychiatric Study of Parents Who Abuse Infants and Small Children," 97.

11. Nomura, "The Battered Child 'Syndrome,' A Review," 389.

12. Lynche, "Ill Health and Child Abuse," 319.

13. Lenington, "Child Abuse: The Limits of Sociobiology," 25.

14. See Ibid., 26.

15. See Ibid., 23.

16. Daly and Wilson, *Homicide*, 87.

17. Ibid., 87–88.

18. "[T]he absence of perfect genetic identify *guarantees* that individual fitness interests will not be perfectly congruent and hence that relatives will experience conflict [citation]. What the adaptationist analysis suggests is that blood relationship mitigates conflict, other things being equal, and there is plenty of evidence in the homicide literature that this is indeed the case [citation]" (Italics in original; Daly and Wilson, "Evolutionary Psychology and Family Violence," 303).

19. "[W]e assume that several proximal mechanisms encourage discriminative parental care, such as the attachment bond resulting from early parent-infant contact. It follows that surrogate parents will find it more difficult to develop deep affection for their charges" (Burgess and Garbarino, "Doing What Comes Naturally? An Evolutionary Perspective on Child Abuse," 93).

20. Many of my observations in this area were earlier expressed in Beckstrom, *Sociobiology and the Law*, 127–32.

21. Based on U.S. cases of reported child abuse and neglect in adoptive and foster homes (undifferentiated) in 1976, Daly and Wilson show that the percentage of abuse cases in the aggregate was approximately the same for adoptive and foster homes as for stepparent homes. Daly and Wilson, "Abuse and Neglect of Children in Evolutionary Perspective," 409–10. This, of course, is not the same as a report that adoptive and foster homes, combined, have as high an incidence of abuse and/or neglect as stepparent homes. Furthermore, the combination of adoptive and foster homes in the statistics makes them of little value in assessing conditions in either. Foster homes differ from adoptive homes in that foster homes usually involve temporary boarding of children for hire.

22. Daly and Wilson, *Homicide*, 84–85.

23. "Nature's Baby Killers," *Newsweek*, September 6, 1982, p. 78.

24. Hrdy, *The Woman That Never Evolved*, 72–95.

25. Newton, "Infanticide in an Undisturbed Forest Population of Hanuman Langurs."

26. For a summary, see Hrdy, "Assumptions and Evidence Regarding the Sexual Selection Hypothesis: A Reply to Boggess," 317–19. See also Pereira, "Abortion Following the Immigration of an Adult Male Baboon" (abortions apparently resulted from blows inflicted by the newly arrived male.)

27. Bertram, "Social Factors Influencing Reproduction in Wild Lions," 473.

28. See Boggess, "Infant Killing and Male Reproductive Strategies in Langurs," 286.

29. This could have occurred if, during the stages of development of the species in question when its behavioral predispositions were being set in place, "widowed" or deserted males entered into new pair bonds with offspring from former unions in tow.

30. In Massachusetts, for example, sexual intercourse between specified close relatives is a crime (Mass. Ann. Laws ch. 272, Section 17, Law. Co-op., 1990) and marriages between them are invalid (ibid., ch. 207, Sections 1,2).

31. "Perhaps research specifically into the structure of human attention and of how information is 'inputted' into the intra-individual system will eventually provide more effective means of altering behavior, but commercial advertising shows at least one way of doing this" (Barkow, *Darwin, Sex and Status*, 382).

32. Marriages to siblings-in-law—uncles and aunts of the children—were prohibited in England until the early twentieth century. Marriages to parents-in-law—grandparents of the children—are still arguably illegal in some U.S. states. The reasons for these apparent restrictions are obscure, and they are being gradually eliminated. See Beckstrom, *Sociobiology and the Law*, 117–25. Sociobiologists have suggested that social rules against "incest" between people, including affines (in-laws), are often established and continued by socially powerful people in order to protect their economic and, ultimately, their reproductive (inclusive fitness) interests. N. Thornhill and R. Thornhill, "Evolutionary Theory and Rules of Mating and Marriage Pertaining to Relatives," 377, 387; N. Thornhill, "The Evolutionary Significance of Incest Rules," 115.

33. See Stephens, *The Family in Cross-Cultural Perspective*, 194, where he summarizes G. P. Murdock's analysis of the Human Relations Area File: "Both the sororate [marrying a sister-in-law] and levirate [marrying a brother-in-law] are common in primitive societies. In Murdock's sample (1949), the levirate was reported present for 127 societies and absent for 58 societies; the sororate was reported present for 100 cases and absent for 59." See also footnote lists in Westermarck, *3 The History of Human Marriage*, 94–95, 208–10.

34. Westermarck, *3 The History of Human Marriage*, 264. By way of illustration, the author notes that "[t]he Omaha custom of marrying sisters was explained as serving the purpose of 'holding the family intact for should the children be bereft of their own mother they would come under the care of her close kindred and not fall into the hands of a stranger.' " (ibid., 96).

CHAPTER 3: REDUCING SIBLING INCEST

1. For example, the state of Massachusetts treats sexual intercourse between a brother and sister as a criminal felony (Mass. Ann. Laws ch. 273, Section 17 [Law. Co-op., 1990]).

"No known society unreservedly approves of sex between relatives sharing one-half or more of their genes by common descent. However, there are 40-odd societies with preferential marriage between close kin . . . usually limited to members of royal families or high aristocrats [citing authorities]." (van den Berghe, "Human Inbreeding Avoidance: Culture in Nature," 92). See also Shepher, *Incest*, 2 *("Incest prohibition is . . . a universal")*.

2. See Shepher, *Incest*, 170–71; van den Berghe, "Human Inbreeding Avoidance: Culture in Nature," 94.

3. "Sibling incest has been estimated to be least 5 times as frequent as parent-child incest (25 cases per 1000) [citations omitted]. Lindzay (1967) sees this as a substantial underestimate" (Banks and Kahn, *The Sibling Bond*, 170). To the same effect, see Trepper and Barrett, *Systematic Treatment of Incest*, 2.

4. Shepher, "Mate Selection among Second-Generation Kibbutz Adolescents and Adults: Incest Avoidance and Negative Imprinting." See also Spiro, *Children of the Kibbutz*.

5. Wolf and Huang, *Marriage and Adoption in China, 1845–1945*, 82.

6. Shepher, *Incest*, 65.

7. Ibid.

8. McCabe, "FBD Marriage: Further Support for the Westermarck Hypothesis of the Incest Taboo?," 50.

9. Gray, *Primate Sociobiology*, 215–16.

10. See van den Berghe, "Human Inbreeding Avoidance: Culture in Nature," 96, for a summary of similar, but more fragmentary, evidence.

11. Westermarck, *3 The History of Human Marriage*, 353.

12. For a general acceptance and restatement of the hypothesis, see, in addition to others cited in these notes, Bishof, "Comparative Ethology of Incest Avoidance;" Bixler, "Incest Avoidance as a Function of Environment *and* Heredity;" Parker, "The Preculture Basis of the Incest Taboo: Toward a Biosocial Theory;" Fox, ed., *The Red Lamp of Incest*.

13. See, for example, Pastner's recent study of cousin marriages in a village in Pakistan with which she demonstrates "how the effects of childhood familiarity, if such exists, can be modified by cultural practices. In other words, biopsychological influences may be present, but they can be either reinforced or [as in the case she presents] attenuated by institutionalized social arrangements" (Pastner, "The Westermarck Hypothesis and First Cousin Marriage: The Cultural Modification of Negative Sexual Imprinting," 573).

Unelaborated reports exist of brothers and sisters marrying frequently among various royal families and among common people in ancient Egypt. See summary in Degler, *In Search of Human Nature*, 267. It has been suggested that the royal family unions may have been "symbolic" in nature. See Irons, "Incest: Can

It Be a Viable Strategy for Humans?," paper presented at the 1989 meeting of the Animal Behavior Society at Northern Kentucky University. Sociobiologists would undoubtedly like to know whether the brothers and sisters in such unions were raised in close proximity to one another and, if so, the degree of reproductive success of such unions. As noted in the text, in recent societal settings when contemporaries were customarily raised together and later married, their reproductive rates were considerably below normal.

14. Shepher, *Incest*, 89–93; van den Berghe, "Human Inbreeding Avoidance: Culture in Nature," 9.

On a genetic level, populations of inbreeding organisms incur increasing rates of homozygosity [having two identical genes controlling a particular feature] at all gene loci. Thus, a deleterious recessive allele [gene] that two relatives receive from a common ancestor will, on average, be passed on to three of every four progeny resulting from mating of these kin. Twenty-five percent of the offspring produced by these parents would be homozygous for the recessive allele in question and therefore phenotypically [in their body characteristics] manifest the deleterious trait. An additional 50% of the inbred offspring would be heterozygous [having two different genes capable of controlling a particular feature—but one dominates and suppresses the other] for the recessive allele. The accumulation of deleterious recessive alleles in this manner is believed to be the primary basis for reduced fitness of inbred offspring (i.e., inbreeding depression) (Porter, "Kin Recognition: Functions and Mediating Mechanisms," 181).

15. See the authorities cited in note 12, above, and the striking empirical evidence produced at the Institute of Child Development Research in Czechoslovakia: Seemanova. "A Study of Children of Incestuous Matings" and *Chicago Sun-Times*, September 12, 1972, p. 21, col. 1.

16. The evolutionary logic of this assumption is outlined by William Irons as follows:

Incest reduces the viability and probable reproductive success of offspring [citation]. This is a characteristic which human beings share with a wide range of habitually outbreeding species [citations]. Given this, natural selection should favor any behavioral mechanism which would cause human beings to avoid incest and such mechanisms should have been consistently favored in populations ancestral to the hominid line long before the emergence of symbolic forms of communication and the propensity to be highly cultural [citation]. Selection for mechanisms preventing incest would be especially strong in species in which social groups are based principally on kinship, since in the absence of such mechanisms a high proportion of matings would be with close kin. Given what we know about living primates and about living and historically described human populations [citations], the most probable situation is that ancestral human populations were characterized by sociality based on kinship and that selection for mechanisms preventing incest was strong. One psychological mechanism, the Westermarck effect, has been identified which appears to have been designed by natural selection to prevent incest in human populations [citations] (Irons, "Incest: Can It Be a Viable Strategy for Humans?," paper presented at the 1989 meeting of the Animal Behavior Society at Northern Kentucky University).

Traditional sociological and psychological explanations for incest avoidance behavior with little or no biological orientation are reviewed by Shepher, *Incest*, 125–73. See also Lopreato, *Human Nature and Biocultural Evolution*, 315–16.

17. See the section on "Kin Recognition in General" in Chapter 8 for a more detailed discussion of the human kin recognition mechanism possibilities.

18. Today's human genetic programming can be likened to a state of the art computer, which is an accumulation of ideas and inventions, each building on those that came before. The human genetic programming has built up for several billion years starting at about the time life first appeared on earth. But there is always a certain amount of lag time between the point when a beneficial genetic change appears in some members of a species and the point when it has spread to everyone in the species by the generation-to-generation reproductive process. Some believe this can happen, if conditions are ideal, in as little as one thousand years (see Lumsden and Wilson, *Promethean Fire*, 152; Kitcher, *Vaulting Ambition*, 390). But normally many thousands of years would be required. Scientists vary somewhat in estimating the time range in the past when most present human programming was "put together," so to speak. (They call it the Environment of Evolutionary Adaptedness or EEA.) Most estimates, however, fall within the late Pleistocene era, ten thousand to one million years ago. See Irons, "Let's Make Our Perspective Broader Rather Than Narrower," 368–70; Tooby and Cosmides, "The Past Explains the Present: Emotional Adaptations and the Structure of Ancestral Environments," 386–88; Barkow, *Darwin, Sex and Status*, 42, 259–60.

19. See, for example, Potts, "Home Base and Early Hominids." See, generally, Bishop and Clark, eds., *Background to Evolution in Africa*. For a reconstruction relating to human aggression by an evolutionary biologist, see Alexander, *Darwinism and Human Affairs*, 227–28.

20. Eibl-Eibesfeldt, *Human Ethology*, 616; 14 *Encyclopedia Britannica*, Populations, Human, 839. See, generally, Fisher, *The Sex Contract*; Pfeiffer, *The Emergence of Society*.

21. See Pfeiffer, *The Emergence of Society*, 33; and Barkow, *Darwin, Sex and Status*, 198. For a perceptive study comparing what is known of early hominids from archaeological evidence to the lifestyles of modern hunter-gatherer groups, see Potts, "Home Base and Early Hominids." For a conservative appraisal, see Binford's judgment that, from present archaeological evidence, "[t]he imposition of the assumption of culture on the pre-modern ancestors of man is no more justified than the imposition of a [modern day] Bushman way of life on the early hominids" (Binford, "Human Ancestors: Changing Views of Their Behavior," 322).

22. See Wrangham, "The Significance of African Apes for Reconstructing Human Social Evolution."

23. "The Westermarck Effect is an excellent example of an evolutionary rule of thumb, since for most of human history, people who were raised in the same household would have been genetic relatives" (Gray, *Primate Sociobiology*, 215). "Our best guess is that early humans might have lived in relatively small

groups made up of about a dozen to a hundred or so individuals, in which biological relatives would have been fairly closely associated and the average coefficient of relatedness would have been relatively high (say approximately 0.3 to 0.4)" (Wells, "Kin Recognition in Humans," 410).

24. See Tooby and Cosmides, "The Past Explains the Present: Emotional Adaptations and the Structure of Ancestral Environments," 385, for a tentative outline of the mechanism's development and operation.

25. Some sociobiologists believe that negative imprinting also occurs between youngsters and older people who are in close, continuous proximity to them during their early years. For example, see N. Thornhill and R. Thornhill, "Evolutionary Theory and Rules of Mating and Marriage Pertaining to Relatives," 377. However, no supportive empirical evidence exists for that proposition comparable to that for negative imprinting between contemporary youngsters.

26. Incest "taboos" are considered to be a cultural phenomenon, and their relationship to "negative imprinting" has been explained by van den Bergh as follows (restating Fox's suggestions):

If it is indeed close early childhood association . . . which breeds non-breeding, and if the "immunizing" kind of contact includes nudity, intimate touching, cuddling and exploratory sex play, as seems to be the case, then those cultures that are sexually permissive (especially of prepubertal sexuality), and that rear boys and girls together are the ones that least *need* incest taboos. Conversely, sexually repressive, sex-segregated societies which thereby interfere with "natural immunity" to incest will resort to strong taboos backed up by punitive sanctions to prevent incest (Italics in the original; van den Berghe, "Incest Taboos and Avoidance: Some African Applications," 354). The author then goes on to examine some cultures that illustrate this thesis (ibid.).

27. Gregory Leavitt, lamenting the absence of a thorough examination of the evidence for the Westermarck Hypothesis, has recently done a critical review. He concluded that the "evidence is far from conclusive and does not negate or displace cultural/environmental theory" (Leavitt, "Sociobiological Explanations of Incest Avoidance: A Critical Review of Evidential Claims," 983). One might note, in response to Leavitt, that conclusive confirming evidence of any human behavioral hypothesis may be unachievable (see note 22 to the Introduction of this book). Furthermore, sociobiology, in general, including that relating to incest avoidance, holds that evolved, genetically influenced inclinations mix with and are modified by culture/environment in producing human behavior, a fact that is often lost sight of or not appreciated by critics who confuse genetic influence with genetic "determination" of behavior. See, for example, Lewontin, Rose and Kamin, *Not in Our Genes*, 6, 9, 223–64, and Sociobiology Study Group of Science for the People, "Sociobiology—Another Biological Determinism," 280–90. Compare note 13, above.

28. Van den Berghe, "Incest Taboos and Avoidance: Some African Applications," 354.

29. For differing attitudes on rooming youngsters together, compare the two following quotations. In introducing a chapter titled "Making One Room Big Enough for Two," Weiss, in *Your Second Child*, 217, wrote, in 1981: "This chapter would have never been included in a book written a generation or two ago. Then, everyone did it. Well, almost everybody, except for the chosen few with the maids and mansions. No one thought twice above two or more children sharing a bedroom, or even a bed. Sometimes whole families slept in one of those sofas that even a four-year-old could transform into a bed. Most children took it for granted. They knew no other way." Three years later in 1984, Lansky noted in *Welcoming Your Second Baby*, 39: "Many American parents feels their children should have their own rooms, separate from their siblings, that each child needs the privacy and space offered by a room of one's own. In fact, a lot of parents here are now opting to put their kids together even if they don't have to. And kids don't always object. As one child put it, 'You and Dad share a room. Why do I have to be alone?' "

30. Because children of opposite sex become more conscious of their sexual differences around school age, many psychologists have recommended separate bedrooms at that time. See Weiss, *Your Second Child*, 220.

31. A female normally cannot conceive until a matter of days before her first menstrual experience. See Dwyer, *Human Reproduction*, 9, 11. A male cannot fertilize before he begins to produce sperm. That happens at puberty (Gondos, "Oogenesis and Spermatogenesis," 147) and is accompanied by an increase in testicle size together with a wrinkling and reddening of the scrotal skin. Shortly after the onset of puberty, the penis size increases and public hair starts to grow. The mean age at which puberty occurs varies but is around eleven years. See Aafjes and Vreeburg, "The Hypothalmo-Pituitary-Testis Axis at Puberty," 285.

32. See Barkow, *Darwin, Sex and Status*, 382.

33. Compare Pastner, "The Westermarck Hypothesis and First Cousin Marriage: The Cultural Modification of Negative Sexual Imprinting," 575: " 'Innate tendencies may be present, but they can be ignored, amplified, or countered by cultural arrangements."

CHAPTER 4: DISTRIBUTING THE PROPERTY OF PEOPLE WHO DIE WITHOUT A WILL

1. For example, until 1990 the Official Comments to the Uniform Probate Code, which was recommended for adoption in all U.S. states, stated that its "intestate succession" provisions attempted "to reflect the normal desire of the owner of wealth as to the disposition of his property at death." *Uniform Probate Code*, 24 (6th ed. Official 1982 Text). As to the purpose of the 1991 edition of the Code, see note 19, ahead.

2. See Altmann, "Altruistic Behavior: The Fallacy of Kin Deployment," 958: "To maximize your long-term gain, bet only on that outcome with the highest expected value. The proof of that dictum can be found in almost any

textbook on probability theory." See also Bertram, "Problems with Altruism," 251.

3. See note 1, above.

4. See, for example, *Uniform Probate Code* (U.S.), 43–57 (10th ed. Official 1991 Text).

5. Beckstrom, *Sociobiology and the Law*, 17–59.

6. Daly and Wilson, *Sex, Evolution and Behavior*, 136–37 (1978).

7. Irons, "Natural Selection, Adaptation and Human Social Behavior," 29–31.

8. Mathematicians and biologists have not yet developed a method of precisely quantifying the reproductive and nurturing value of a one-generation advantage. But see Fisher, *The Genetical Theory of Natural Selection*, 27–30.

9. Ibid. Freedman, *Human Sociobiology*, 115, refers to an empirical study (apparently never published) by Ginsburg which asked grandparents to choose between "saving" children and grandchildren. The children prevailed except when they were beyond the reproductive years. Then there was a tendency to "save" the potentially reproductive grandchildren.

10. Older people will have had more time to accumulate resources with which to reciprocate than will the young. But during the period in evolutionary history when present human genetic programming was largely set in place, and in some present-day societies, resource accumulation will have been relatively insignificant. In any case, any tendency of the old to have accumulated more may well be offset by the tendency of the young to have more physical vigor with which to reciprocate.

11. Alexander, "The Evolution of Social Behavior," 356; Flinn, "Resources, Mating and Kinship," 36–37; and West Eberhard, "The Evolution of Social Behavior by Kin Selection," 19. See also Hames, "Relatedness and Interaction among the Ye'kwana: A Preliminary Analysis," 247. (Patterns of reciprocity vary according to the amount of "relatedness" shared by the individuals interacting.)

12. A serious drawback to the study of wills is the clear danger that legal advisors who draft wills for clients have influenced them to adopt the advisors' personal preferences of "culturally and socially approved" choices. Young, "Meaning of 'Issue' and 'Descendants' ", 226. See also Sussman, Cates and Smith, *The Family and Inheritance*, 54 n. 42; Fellows, Simon and Rau, "Public Attitudes about Property Distribution at Death and Intestate Succession Laws in the United States," 325.

13. Sussman, Cates and Smith, *The Family and Inheritance*, 101; Fellows, Simon and Rau, "Public Attitudies about Property Distribution at Death," 369; Fellows, Simon, Snapp and Snapp, "An Empirical Study of the Illinois Statutory Estate Plan," 737; and Contemporary Studies Project, "A Comparison of Iowans' Dispositive Preferences with Selected Provision of the Iowa and Uniform Probate Codes," 1102.

14. Fellows, Simon and Rau, "Public Attitudes about Property Distribution at Death," 369; Fellows, Simon, Snapp and Snapp, "An Empirical Study of the Illinois Statutory Estate Plan," 737.

15. Fellows, Simon and Rau, "Public Attitudes about Property Distribution at Death," 382–83; Fellows, Simon, Snapp and Snapp, "An Empirical Study of the Illinois Statutory Estate Plan," 740–41; Contemporary Studies Project, "A Comparison of Iowans' Dispositive Preferences," 1111; and Young, "Meaning of 'Issue' and 'Descendants' ", 225.

16. Fellows, Simon and Rau, "Public Attitudes about Property Distribution at Death," 738; Contemporary Studies Project, "A Comparison of Iowans' Dispositive Preferences," 1106.

17. Fellows, Simon and Rau, "Public Attitudes about Property Distribution at Death," 384. Eight percent split the estate between the grandchildren, giving the child nothing, and another 8 percent were reported as "other." (Two percentage points were lost in my rounding up and down.) Ibid.

Some grandchildren in a representative sample will be so young that they will not be able to reciprocate for awhile, whereas relatively fewer children will be thus incapacitated by infancy and immaturity. This could partially account for some "pre-death" interviewers in a sample greatly favoring a child over grandchildren. Idiosyncratic needs and deserts would undoubtedly also be involved.

18. Fellows, Simon and Rau, "Public Attitudes about Property Distribution at Death," 374; Contemporary Studies Project, "A Comparison of Iowans' Dispositive Preferences," 1106.

19. In the 1991 edition of the Uniform Probate Code, draftsmen have added the comment, missing in earlier versions, that the "intestate succession" part "was designed to provide suitable rules for the person of modest means who relies on the estate plan provided by law." *Uniform Probate Code* (U.S.) 43 (10th ed. Official 1991 Text).

20. In a 749-person interview sample randomly selected in various U.S. states, "[O]ver 70 percent of the respondents indicated they know who would inherit their estates if they died without wills. But when asked in the succeeding question to name the heirs and the proportion of the estate received by each heir, only 44.6 percent responded correctly or nearly so. These findings are consistent with prior studies and clearly demonstrate that most citizens do not know who will inherit their property and are not relying on existing intestacy statutes" (Fellows, Simon and Rau, "Public Attitudes about Property Distribution at Death," 339).

CHAPTER 5: EVOLUTIONARY BIOLOGY PRIMER II: REPRODUCTIVE STRATEGY

1. "In many creatures the male is bigger than the female, but in others, like hamsters and hawks, the female is larger. But the sperm is always smaller than

the ovum, and that defines the sexes. From this trite distinction, much follows" (Daly and Wilson, *Sex, Evolution and Behavior*, 82).

2. Symons, *The Evolution of Human Sexuality*, 24 ("pipefish and seahorses, for example").

3. "The female provides the raw materials for the early differentiation and growth of their progeny. Here, at the very fundament of sexuality, is love's labor divided, and it is the female who contributes the most" (Daly and Wilson, *Sex, Evolution and Behavior*, 82). See also Symons, *The Evolution of Human Sexuality*, 23. But see *New York Times*, National Edition, December 3, 1992, p. A1, col. 6: "Production of Sperm Is Found to Cut Life, in a Worm, at Least." (Production of sperm may be more expensive than biologists had previously thought.)

4. Betzig, "Mating and Parenting in Darwinian Perspective," 4.

The male's reproductive output is limited by his access to fertile females, whereas access to all the males in the world would not elevate the female's capacity. In our own species, for example, it is hardly possible for a woman to bear more than about 20 children in her lifetime, although the *Guinness Book of Records* credits a nineteenth century Muscovite with 69 live births, bearing twins, triplets or quads in each of 27 pregnancies! Astounding though that figure may be, the male record is much greater: 888 children were sired by the Sharifian emperor of Morocco, Moulay Ismail The Bloodthirsty (Daly and Wilson, *Sex, Evolution and Behavior*, 79).

5. See note 30 to Chapter 1.

6. For up-to-date summaries of the data on humans, see Symons, *The Evolution of Human Sexuality*; Symons, "The Evolution of Human Sexuality Revisited" and Symons, "The Evolutionary Approach: Can Darwin's View of Life Shed Light on Human Sexuality?" See also all the articles in Betzig, Borgerhoff-Mulder and Turke, eds., *Human Reproductive Behavior*.

7. For more detailed prediction of the criteria females will use in selecting mates, see Buss, "Sex Differences in Human Mate Selection Criteria: An Evolutionary Perspective," 339.

8. R. Thornhill and N. Thornhill, "Human Rape: An Evolutionary Analysis," 141, and "Human Rape: The Strengths of the Evolutionary Perspective," 281.

9. See Nesse, "Evolutionary Explanations of Emotions," Crawford, "Sociobiology: Of What Relevance to Psychology?," 13; and Tooby and Cosmides, "The Past Explains the Present: Emotional Adaptations and the Structure of Ancestral Environments."

"The foundation of [evolutionary psychology] is the study of the mental mechanisms of emotion/cognition/motivation/perception as adaptations—i.e., as long-term products of individual selection. The field relies on the general theory of evolution by individual selection for development of a general hypothesis of mind. Thus, the field is a subdiscipline of biology, which is the study of all life" (R. Thornhill and N. Thornhill, "The Evolution of Psychological Pain," 75).

10. See Alexander, "Ostracism and Indirect Reciprocity: The Reproductive Significance of Humor," 108." And see R. Thornhill and N. Thornhill, "The Evolution of Psychological Pain," 78; "Physical pain serves to draw an individual's attention to some aspect of anatomy that needs tending and can be fixed by the individual's attention. Mental pain seems to focus an individual's attention on the significant social events surrounding the mental pain and promotes correction of the events causing the pain and the evaluation of future courses of action."

11. See, generally, N. Thornhill and R. Thornhill, "An Evolutionary Analysis of Psychological Pain Following Rape: I. The Effect of Victim's Age and Marital Status" and N. Thornhill and R. Thornhill, "An Evolutionary Analysis of Psychological Pain Following Rape: II. The Effects of Stranger, Friend and Family-Member Offenders." See also, Scully, *Understanding Sexual Violence*, 107.

12. Texas gubernatorial candidate Clayton Williams was quoted as saying this during his 1990 campaign. *U.S. News and World Report*, November 26, 1990, p. 24. This saying has been called an "old adage" by Katz and Mazur in *Understanding the Rape Victim*, 178.

13. An example of an educational program for males in sensitive governmental positions who could benefit from knowledge of biological mechanisms in the Canadian Judicial Centre's program called gender equality begun in May 1990. It is aimed at judges "to ensure that their actions and decisions are based on a fair and informed assessment of each situation, not on preconceived notions of gender roles" (*The Toronto Star*, January 19, 1991, p. DI).

14. See the authorities cited above in this chapter, which will lead the reader to others.

15. Shields and Shields, "Forcible Rape: An Evolutionary Perspective;" R. Thornhill and N. Thornhill, "Human Rape: An Evolutionary Analysis."

16. Symons, *The Evolution of Human Sexuality*, 206.

17. Shields and Shields, "Forcible Rape: An Evolutionary Perspective," 120; R. Thornhill and N. Thornhill, "Human Rape: An Evolutionary Analysis," 138.

The Thornhills have recently suggested that rape behavior may be either the product of a rape-specific mechanism *or* a side effect "of two more general adaptations: (a) the psychological mechanism underlying men's general desire for sexual intercourse, and (b) the mechanisms underlying our species' general tendency to use force to attain any reward that was correlated with successful survival and reproduction in our evolutionary history. According to this side-effect hypothesis, there are no psychological mechanisms specifically designed for processing information about rape" (R. Thornhill and N. Thornhill, "The Evolutionary Psychology of Men's Coercive Sexuality," 367, 375).

And see R. Thornhill and N. Thornhill, "Human Rape: The Strengths of the Evolutionary Perspective," 286: "There is an evolutionary hypothesis for human rape that is an alternative to the evolutionary view of rape we have emphasized here. It is a possibility that human rape is an inevitable outcome of an evolutionary history in which males were selected to persist in their attempts to copulate and females were selected to discriminate among males and often refuse copulation. In this view, human rape is a maladaptive consequence of an adaptive general mating strategy of men. . . . We feel that the maladaptive consequence hypothesis is probably incorrect." The authors then give two detailed reasons for their belief.

To the effect that all men may be born capable of physically coercing copulation, but whether or not they do it depends on their learning experiences while maturing and whether or not they ever encounter the environmental conditions that cue it, see Smuts, "Psychological Adaptations, Development and Individual Differences," 402.

CHAPTER 6: REDUCING RAPE

1. Diana Scully asked 114 convicted rapists "how they would feel and what they would do if their own significant women were raped. The overwhelming reaction . . . 72 percent [of the] admitters and 75 percent [of the] deniers, was an expression of anger and violence. Additionally, the majority of those men said that rather than involving law enforcement, they would find a way to get personal revenge" (Scully, *Understanding Sexual Violence,* 168–69).

2. See Gray, *Primate Sociobiology,* 10.

3. " 'Common sense' supports the notion that the more severe the punishment for a given act, the less likely people are to engage in the proscribed behavior" (Thomas and Foster, "A Sociological Perspective on Public Support for Capital Punishment," 175).

4. After quoting this statement in 1948 by John Ellington: "The belief that punishment protects society from crime by deterring would-be law breakers will not stand up before our new understanding of human behavior," John Ball commented that such "protestations are but doctrinaire statements without adequate factual basis. The failing is in the inability or unwillingness to separate value-judgments from scientific knowledge" (Ball, "The Deterrence Concept in Criminology and Law," 352).

5. That it cannot: "One of the basic principles learned by every student of criminology is that *punishment does not deter*" (italics in original, Jeffrey, *Crime Prevention Through Environmental Design,* 215), and "[m]ost criminologists [have traditionally held the view] that rape is a 'spontaneous,' 'explosive,' 'expressive' offense, and thus not subject to deterrence" (Bailey, "Rape and the Death Penalty: A Neglected Area of Deterrence Research," 337). See also Knudten, *Crime in a Complex Society,* 13.

That punishment can deter, see Morris, "Punishment and Prisons," 25–26: "I don't think deterrence flows from the severity of conditions of imprisonment or even very much from the duration of imprisonment. I think it flows from the fact of imprisonment," and Golding, "Criminal Sentencing: Some Philosophical Consideration," 95: "A standard objection to [the theory of deterrence] is that punishment does not deter crime, as evidenced by the high rate of recidivism and by the fact that the offender now standing before the judge was obviously not deterred by the prospect. This objection can be rebutted, however."

The truth of the matter is probably contained in this statement from Zimring and Hawkins, *Deterrence*, 5: "It's a matter of common observation that men seek to avoid unpleasant consequences and that the threat of unpleasantness tends to be a deterrent. It is equally indisputable that not all criminal prohibitions are completely effective. But these propositions are not contradictory or mutually exclusive."

6. Bailey, "Deterrence and the Violent Sex Offender: Imprisonment vs. the Death Penalty." For a catalogue of similar studies, see Bedau, eds., *The Death Penalty in America*, 97.

7. See Gibbs and Erickson, "Capital Punishment and the Deterrence Doctrine," 304.

8. See R. Thornhill and N. Thornhill, "Human Rape: An Evolutionary Analysis," 141, 150.

9. Furthermore, it is possible that a general leveling of economic status in a population would result in women adjusting their mate selection criteria away from material resources, but continuing to discriminate on other bases. As a result, an essentially unchanged percentage of males would be denied consensual access to reproductive-age women.

10. See discussion of Beccaria and Bentham's positions in Gibbs, *Crime, Punishment and Deterrence*, 5–9; Gibbs and Erickson, "Capital Punishment and the Deterrence Doctrine," 301–4. Beccaria and Bentham also mentioned "celerity" (rapidity) of punishment as being an important part of the deterrent mix. On this point, see discussion in note 30 below.

11. Shields and Shields, "Forcible Rape: An Evolutionary Perspective," 132–33.

12. See, for example, Tooby and Cosmides, "The Past Explains the Present: Emotional Adaptations and the Structure of Ancestral Environments," 411: "[M]echanisms take the output of the monitoring algorithms . . . as input, and through integration, probabilistic weighing, and other decision criteria, identify situations as either present or absent (or present with some probability). . . . Given that a situation has been detected, the internal communications system sends a situation-specific signal to all relevant mechanisms; the signal switches them into the appropriate adaptive emotion mode." The limitations of the analogy between brains and computers are explored in Edelman, *Bright Air, Brilliant Fire*, 218–27.

13. On the existence of "free will" and its relevance to human behavior, see Beckstrom, *Evolutionary Jurisprudence*, 30–33.

14. "David Marr has argued that the first and most important step in understanding an information-processing problem is developing a 'theory of the computation' [citations]. This theory defines the nature of the problem to be solved; in so doing, it allows one to predict properties that any algorithm capable of solving the problem must have. Computational theories incorporate 'valid constraints on the way the world is structured—constraints that provide suffi-

cient information to allow the processing to succeed' [citation]" (Cosmides and Tooby, "Evolutionary Psychology and the Generation of Culture, Part II," 59).

15. Mary Maxwell cites William D. Hamilton for the proposition that "[i]nclusive fitness' is calculated by the number of one's genes that are *included* in future generations" (Emphasis in original; Maxwell, *Morality Among Nations*, 77).

16. Shields and Shields, "Forcible Rape: An Evolutionary Perspective," 118. Not only do honest courtship, deceptive courtship and rape grade into one another on a continuum, but the three tactics are often used in combination to obtain a single copulation. R. Thornhill and N. Thornhill, "The Evolutionary Psychology of Men's Coercive Sexuality," 369.

17. R. Thornhill and N. Thornhill, "Human Rape: An Evolutionary Analysis," 164.

18. "[A] bonded male who copulates throughout his mate's cycle has the same probability of producing offspring as does a male who practices 'complete promiscuity'" (Barkow, *Darwin, Sex and Status*, 329). See also R. Thornhill and N. Thornhill, "Human Rape: An Evolutionary Analysis," 165–66.

19. "Unfortunately, the risk of pregnancy from rape is difficult to assess. . . . This is in part because most rapes are not officially reported, and those that are reported are rarely followed up nine months later. In addition, rape victims often are otherwise sexually active around the time of the rape incident" (Ellis, *Theories of Rape*, 46).

Estimates of pregnancies resulting from rape have ranged from less than 1 percent to 11.6 percent (ibid., 47).

"For comparative purposes, a single act of voluntary sexual intercourse (without the use of contraception) for women in their 20s has about a 2%–4% probability of resulting in pregnancy" (ibid.).

20. R. Thornhill and N. Thornhill, "Human Rape: An Evolutionary Analysis," 148–49.

21. "When female mate choice is based importantly on resources and male striving for resources produces losers and winners, forced copulation may be a viable alternative for losers. There is increasing evidence outside humans that forced copulation often has evolved in resource-based polygynous systems [citations]" (R. Thornhill and N. Thornhill, "Human Rape: The Strength of the Evolutionary Perspective," 277).

22. "Different components of [male] sexual behavior decline at different rates. The characteristic that apparently declines most dramatically is the capacity for repeated orgasm within a short period of time, which reaches its peak in pre-adolescence and declines very rapidly after 20 years of age. The most slowly declining characteristic is the frequency of extramarital intercourse; this behavior may be influenced by increased stimulus pressure" (Hafez, *Human Reproduction*, 385).

23. "The behavior of psychotics is not expected to fit an adaptive model. For example, we expect that largely it is psychotic men who rape prereproductive and postreproductive females. . . . Present data do not allow the separation of the

behavior of psychotic and nonpsychotic rape offenders" (R. Thornhill and N. Thornhill, "Human Rape: An Evolutionary Analysis," 164).

Not all human behavior is "adaptive" in the sense of following the dictates of economic logic with gene reproduction as the goal—just as computers do not do what is planned for them all the time (virus gets put into them, parts wear out, etc.). But through the generations, there will be a prevailing tendency for most human behavioral output to be "adaptive," because repeatedly malfunctioning (nonadaptive) units will not reproduce well, if at all.

24. See note 18 to Chapter 3.

25. The two principal theories on the identity of human ancestors both trace them back to Africa during the Pleistocene. Stringer and Andrews, "Genetic and Fossil Evidence for the Origin of Modern Humans."

26. "[E]volutionary biology is best used as an heuristic [problem solving aid], providing models of the adaptive problems the psyche had to be able to solve, and providing models of the conditions within which these mechanisms had to solve these adaptive problems: Pleistocene conditions" (Tooby and Cosmides, "Evolutionary Psychology and the Generation of Culture, Part I," 30). See also ibid., 34, 37–38, and Tooby and Cosmides, "The Past Explains the Present: Emotional Adaptations and the Structure of Ancestral Environments," 406–7.

27. But see Barkow, *Darwin, Sex and Status*, 240, where he observes that natural selection is very likely to put constraints on flexible behavior programming just as the rules of a chess game put limits on what can be done while making the game itself possible.

28. Compare Turke, "Just Do It," 455, where the author counters the suggestion that printed pornography is novel for a Pleistocene-created mechanism by noting that "Pleistocene males probably imagined female images, and they certainly drew and sculpted them, so artificial images are not as novel as implied."

29. Ball, J. C. "The Deterrence Concept in Criminology and Law," 352.

30. Jack Gibbs notes that both Jeremy Bentham and Cesare Beccaria suggested that punishment must be relatively quick if it is to have a deterrent effect on the population. Gibbs expressed doubt and puzzlement as to whether and why this should be true. Gibbs, *Crime, Punishment and Deterrence*, 9. If Bentham and Beccaria were right, the explanation could have to do with the likelihood that society was not so well organized during the Pleistocene so as to delay punishments—perhaps one was either immediately punished for an offensive act or not at all. If so, maybe the resulting human mechanism does not respond to threats of possible *eventual* imprisonment or death as readily as it does to threats of possible scratching, body blows, etc., concurrent with the contemplated anti-social act.

See also Andenaes, "The General Preventive Effects of Punishment," 961: "The time element is important. Threats of punishment in the distant future are not as important in the process of motivation as are threats of immediate punishment."

31. "An obviously armed female, or one that is capable of vigorous manual self-defense, and who signals either fact is less likely to be raped at least by a lone, unarmed rapist" (Shields and Shields, "Forcible Rape: An Evolutionary Perspective," 127). Also see ibid, 133.

"[D]eterrence ... precedes the evolution of mind. Plants and animals may be said to practice deterrence. . . . The thorns of the rose deter predators" (Barash and Lifton, *The Caveman and the Bomb*, 167).

32. The *Statistical Abstract of the U.S.* (1990) shows that people who have an educational attainment of high school or less are more than twice as likely to never read a newspaper, magazine or book in a given period of time than those with at least some college education; the same is true for 16 to 20 year olds versus older age groupings up to 40 and over. Those in households with less than $15,000 annual income are almost twice as likely to be "nonreaders" than those in households with annual income of $15,000 to $24,999. And those in the latter groupings are, in turn, more than twice as likely to be "nonreaders" than those in households with an annual income of $25,000 to $39,999 and so on (p. 228).

On the other hand, people with a high school education or less viewed television essentially in the same percentages as those with a college education (in the 91 to 92 percent range); almost the same could be said for 18 to 24 year olds versus other age groupings up to 55 and for males versus females. Those in households with annual income of $10,000 to $19,999 viewed television in a higher percentage (94.2 percent) than any other annual household income groupings (p. 550).

33. R. Thornhill and N. Thornhill, "Human Rape: An Evolutionary Analysis," 150–53.

34. See U.S. Department of Justice, *Sourcebook of Criminal Justice Statistics—1988*, 640. "[M]ost rapes are not officially reported" (Ellis, *Theories of Rape*, 46).

35. See Barkow, *Darwin, Sex and Status*, 374.

36. In the United States, the question of whether a statute prescribing castration as a penalty for rape would be permissible—constitutional—has not been directly addressed by the courts. But see *Weems* v. *United States*, 217 U.S. 349, 377 (1920); *State* v. *Brown*, 284 S.C. 407, 326 S.E. 2d 410 (1985); *People* v. *Gauntlett*, 134 Mich. App. 737, 352 N.W. 2d 310, *modified*, 419 Mich. 909, 353 N.W. 2d 463 (1984).

37. On the practical difficulties in getting enacted law known and applied in a country with low literacy rates and undeveloped communications systems, see Beckstrom, "Handicaps of Legal-Social Engineering in a Developing Nation."

38. Of sixty-five major nations surveyed in 1965, only three retained the death penalty for rape where death of the victim did not occur. *Coker* v. *Georgia*, 433 U.S. 584, 596 n. 10 (U.S. Supreme Court, 1977). See also McCahill, Meyer and Fischman, *The Aftermath of Rape*, 201; Searles and Berger, "The Current Status of Rape Reform Legislation: An Examination of State Statutes," 25, 43 notes D, F. (A few U.S. states prescribe death under limited circumstances such as when the victim is under 12 years of age.)

39. Searles and Berger, "The Current Status of Rape Reform Legislation: An Examination of State Statutes," 31.

40. The plurality opinion of the U.S. Supreme Court in *Coker* v. *Georgia*, 433 U.S. 584 598 (1977), said the following:

Rape is without doubt deserving of serious punishment; but in terms of moral depravity and of the injury to the person and to the public, it does not compare to murder, which does involve the unjustified taking of human life. Although it may be accompanied by another crime, rape by definition does not include the death of or even the serious injury to another person. The murderer kills; the rapist if no more than that, does not. Life is over for the victim of the murderer; for the rape victim, life may not be nearly so happy as it was, but it is not over and normally is not beyond repair. We have the abiding conviction that the death penalty, which "is unique in its severity and irrevocability," [citation] is an excessive penalty for the rapist who, as such, does not take human life.

41. See Hamilton and Rotkin, "Interpreting the Eighth Amendment: Perceived Seriousness of Crime and Severity of Punishment," 515, for a public survey ranking crimes in terms of "seriousness."

42. Green, "Depo-Provera, Castration, and the Probation of Rape Offenders: Statutory and Constitutional Issues," 3. See also Heim and Hursch, "Castration for Sex Offenders: Treatment or Punishment? A Review and Critique of Recent European Literature."

43. See, for example, Schabarum, "Is Castration Extreme? Perhaps, But Existing Laws Are Far Too Lenient," *Los Angeles Times*, April 27, 1986, p. 5, col. 1. (Los Angeles County Board of Supervisors orders a study of castration as a punishment for men guilty of multiple sex crimes); "Castration: Would It Stop Offenders? Experts Split," *Seattle Times*, January 15, 1990. p. A1 (Washington state senate considers bill that would mandate surgical castration for first degree rape); "Former British Judge Advocates Castration as Rape Punishment," Reuters, November 9, 1987.

44. See Heim and Hursch, "Castration for Sex Offenders: Treatment or Punishment? Review and Critique of Recent European Literature," 303. Voluntary castration as a condition of reduced sentence or parole, when instituted on top of existing penalties, could have the effect of decreasing deterrence to rape because it gives offenders the option of taking what they consider (as evidenced by their act of selecting it) a lesser penalty than they would otherwise get.

45. Ortmann, "The Treatment of Sexual Offenders," 445. However, involuntary castration for sexual offenses appears never to have been involuntarily administered in Denmark (ibid.).

46. Comment, "Castration of the Male Sex Offender: A Legally Impermissible Alternative," 388. See Daly and Wilson, *Sex, Evolution and Behavior*, 99.

47. Green, "Depo-Provera, Castration, and the Probation of Rape Offenders: Statutory and Constitutional Issues," 3–4.

48. From 1929 to 1959 a total of 738 sexual offenders, the majority of whom were charged not with rape, but with indecent behavior toward minors, were castrated in Denmark. The rate of recidivism among them was between 1.4 and

2.4 percent. The rate among noncastrated offenders was 9.7 percent. Similar findings were reported in West Germany and Norway (Ortmann, "The Treatment of Sexual Offenders," 445). In recent decades, antihormonal drug treatment appears to have replaced the irreversible surgical castration procedure in Denmark (ibid., 445, 450).

49. Shields and Shields, "Forcible Rape: An Evolutionary Analysis," 132. See also Singer, *The Expanding Circle*, 172: "Fear of punishment may deter a rapist, of course, but quite apart from this deterrent effect, many forms of punishment affect the chances of certain genes—including any which predisposed toward rape—surviving in future generations."

50. See Barkow, *Darwin, Sex and Status*, 344–45, where the author speculates that natural selection may well have operated on early human populations in isolated "islands" which had different environmental characteristics. If so, "we should expect considerable genetic variability to underlie our sexual behavior." See also Ellis, *Theories of Rape*, 54, where the author theorizes that "while rape is learned behavior, for evolutionary reasons, some males are genetically (and neurogical) [*sic*] much more disposed to acquire and to be reinforced for employing raping techniques in attempting to copulate than other males."

51. For a review of information bearing on this question see Ellis, *Theories of Rape*, 86–97. "[T]win studies of criminal and related behaviors fairly consistently provide some intriguing evidence for a genetic effect, and genetic influences warrant continued, but more rigorous study" (Fishbein, "Biological Perspectives in Criminology," 45).

Biology has provided a foundation for understanding intrapopulation variation in behavior. Intrapopulational behavior alternatives are of three general types [citations]. First, individuals may have different pure alternative strategies as a result of an evolutionary stable polymorphism in the population. The alternatives stem from genetic differences between individuals exhibiting them. Second, individuals may have a single mixed strategy with two or more alternatives and spend a fixed percentage of time in one alternative and then automatically adopt another. Here, all individuals carry the genes coding for all alternatives each of which is adopted for a fixed portion of time. Last, a single conditional strategy consisting of two or more alternatives may exist within an individual, but adoption of alternatives is condition-dependent and the alternatives are associated with different reproductive returns. All individuals in the population carry genes for all alternatives comprising a conditional strategy (R. Thornhill and N. Thornhill, "Human Rape: The Strengths of the Evolutionary Perspective," 279).

52. For example, "Lawmaker Wants Parole Rapists Castrated," United Press International, March 2, 1986 (Kentucky state senator introduces bill to require castration as a condition for parole of rapists); "Lawmaker: Hang Killers, Castrate Rapists," United Press International, November 15, 1983 (South Carolina Legislature representative introduces bill to require castration of rapists); "He Should Not Come Out (of Jail) the Way He Went In," United Press International, June 17, 1982 (Judiciary Committee of the New Jersey Assembly considers proposal to permit castration of men who rape or sexually abuse children).

53. For example, "[T]here is at least some evidence that suggests that judges and juries hesitate to apply severe sanctions to the point that there may well be an inverse relationship between certainty and severity of punishment. (Thomas and Foster, "A Sociobiological Perspective on Public Support for Capital Punishment," 176). Jeremy Bentham made a similar observation at an early date. Barash and Lipton, *The Caveman and the Bomb*, 157. Johannes Andenaes has also noted that when "the penalties are not reasonably attuned to the gravity of the violation, the public is less inclined to inform the police [and] the prosecuting authorities are less disposed to prosecute" (Andenaes, "The General Preventive Effects of Punishment," 970).

54. See Shields and Shields, "Forcible Rape: An Evolutionary Perspective," 132–33.

55. For example, Andenaes, "The General Preventive Effects of Punishment"; Cohn and Ward, eds., *Improving Management in Criminal Justice*; Cole, ed., *Criminal Justice;* Feeley, *The Process Is the Punishment*.

56. See Allen, *The Decline of the Rehabilitative Ideal*.

57. Tittle, "Crime Rates and Legal Sanctions;" Logan, "General Deterrent Effect of Imprisonment;" Baily, Martin and Gray, "Crime and Deterrence: A Correlation Analysis."

" 'It has been affirmed and reaffirmed . . . by leading penal experts and criminologists in virtually every country and ever since the beginning of the nineteenth century, that if punishment could be made certain, almost all crime would be eliminated'" (Zimring and Hawkins, *Deterrence*, 161, quoting Radzinowicz).

CHAPTER 7: REDUCING STREET CRIME

1. See Knapp, Sharkey and Metts, "The Effects of an Increase in a County Sheriff's Department's Patrol Capability," 5.

2. Ibid.

3. Chaiken, Lawless and Stevenson, *The Impact of Police Activity on Crime* (New York subway system); Press, *Some Effects of an Increase in Police Manpower in the 20th Precinct of New York* (20th Precinct, New York City); Zimring and Hawkins, *Deterrence*, 348 (25th Precinct, New York City); Trojanowicz, "An Evaluation of a Neighborhood Foot Patrol Program" (Flint, Michigan); Knapp, Sharkey and Metts, "The Effects of an Increase in a County Sheriff's Department's Patrol Capability" (Lexington County, South Carolina).

4. Kelling, *The Newark Foot Patrol Experiment* (Newark, N.J.); Kelling, Pate, Dieckman and Brown, *The Kansas City Preventive Patrol Experiment* (Kansas City, Kansas). For a listing of critiques of the Kansas City experiment, see Knapp, Sharkey and Metts, "The Effects of an Increase in a County Sheriff's Department's Patrol Capacity," 6 n. 4.

5. Trojanowicz, "An Evaluation of a Neighborhood Foot Patrol Program," 415.

6. *Criminal Justice Newsletter*, Vol. 11, No. 6, p. 1, March 17, 1980 (National Council on Crime and Delinquency, Hackensack, N.J.).

7. Trojanowicz, "An Evaluation of a Neighborhood Foot Patrol Program" 417.

8. Zimring and Hawkins, *Deterrence*, 350 n. 18.

9. Ibid., 349–50 n. 18.

10. "The psychology of an organism consists of the total set of proximate mechanisms that control behavior" (Tooby and De Vore, "The Reconstruction of Hominid Behavioral Evolution Through Strategic Modeling," 197).

11. Zimring and Hawkins, *Deterrence*, 348–49.

12. " 'Having more police officers alone does not guarantee a reducing [*sic*] in crime,' said University of Maryland professor Lawrence Sherman, president of the Crime Control Institute. 'That also depends on how the officers are used. But our best guess is that concentrating police in high-crime hot spots, and keeping them there, would reduce crime in new York' " (*Newsday*, City Edition, News Section, p. 8, October 9, 1990).

13. On the theory underlying cost-benefit calculations in aggressive conflicts between members of the same species, see Dawkins, *The Selfish Gene*, 73–74; and Popp and De Vore, "Aggressive Competition and Social Dominance Theory: Synopsis," 318–22.

14. That "morality" in humans boils down to conscious and unconscious cost-benefit calculations; see Alexander, *The Biology of Moral Systems*, 118.

15. "In contrast to computers, the patterns of [human] nervous system response depend on the individual history of each system, because it is only *through interactions with the world* that appropriate response patterns are selected" (Italics in original; Edelman, *Bright Air, Brilliant Fire*, 226).

16. Wrangham, "The Significance of African Apes for Reconstructing Human Social Evolution," 52.

17. Ibid., 53.

18. See Tooby and De Vore, "The Reconstruction of Hominid Behavioral Evolution Through Strategic Modeling," 188–90. The general, overall, modeling rule suggested by Tooby and De Vore is that "animals will be selected to behave as if they were following strategies to promote their inclusive fitness" (ibid., at 189).

19. Ibid., 214, 217 (food); 218, 222, 224 (stone tools). Chimpanzees have been observed at nutcracking sites where they use tools. Tanner, "The Chimpanzee Model Revisited and the Gathering Hypothesis," 20–21; Potts, "Reconstruction of Early Hominid Socioecology: A Critique of Primate Models," 42.

20. "I conclude that African Apes and humans share a strong tendency toward closure of their social networks. This is in contrast to orangutans, which have completely open networks [citations]. Most primates, however, have closed social networks, and, from the present data it would be surprising if the [common ancestor of chimpanzees, gorillas, bonobos, and humans] did not also have them" (Wrangham, "The Significance of African Apes for Reconstructing

Human Social Evolution," 58). But see notes 22, 23 and 24 below regarding orangutans.

21. "The major explanation for agonistic behavior [intraspecies conflict] is competition for limited resources such as food, water and mates" (Fedigan, *Primate Paradigms*, 77).

22. Goodall, *The Chimpanzees of Gombe*, 335 (chimpanzee intragroup aggression over food); Bygott, "Agonistic Behavior, Dominance and Social Structure in Wild Chimpanzees of the Gombe National Park," 437 (chimpanzee intragroup aggression over bananas); Wrangham, "The Significance of African Apes for Reconstructing Human Social Evolution," 66–67 (chimpanzee deaths from intraspecies aggression); Galdikas, "Orangutan Adaptations at Tanjung Puting Reserve: Mating and Ecology," 203 (interunit orangutan aggression). See also citations in following note.

23. Goodall, *The Chimpanzees of Gombe*, 201 (chimpanzees driving competitors from ally's food or helping ally to do so); de Waal, "Chimpanzee Politics," 125–26 (chimpanzee mothers protecting infants), 127 (chimpanzee females aiding males), 128 (chimpanzee male aiding females); Smuts, "Gender, Aggression and Influence," 407 (chimpanzee female coalitions against attacks by males); Harcourt, "Alliances in Contests and Social Intelligence," 138–39 (gorillas aiding allies); Stewart and Harcourt, "Gorillas: Variations in Female Relationships," 158–59 (dominant male gorillas supporting younger members in intragroup fights), 163 (suggestion that female gorillas associate with males in order to gain protection against predation and aggression from other gorillas); Galdikas, "Orangutan Adaptations," 208 (orangutan male aiding female); Rijksen, *A Field Study of Sumatran Orangutans*, 271 (orangutan male threatening or chasing rival over female); de Waal, *Peacemaking Among Primates*, 219 (bonobo mother protecting youngster); Nishida and Hiraiwa-Hasegawa, "Chimpanzees and Bonobos: Cooperative Relationships Among Males," 174 (implication that usual bonobo male companions acted as allies of males in aggressive encounters). More generally: Kummer, "Tripartite Relations in Hamadryas Baboons," 114 (nonhuman primate mothers protecting infants); de Waal, *Peacemaking among Primates*, 11 (nonhuman primate allies stopping attacks). Compare baboons: Strum and Mitchell, "Baboon Models and Muddles," 100 (females relying on male support during aggressive interactions, giving them access to resources and reducing interference from other troop members).

"Intergroup interactions in the living [primate] species clearly create strong selection pressures for cooperative and protective relationships within groups. The present analysis shows that the simplest hypothesis is that intergroup relationships occurred in ancestral species also, so reconstructions which ignore them are unlikely to be accurate" (Wrangham, "The Significance of African Apes," 68).

24. de Waal, "Chimpanzee Politics," 124 (chimpanzee males backing off from copulation); Galdikas, "Orangutan Adaptations," 208, 211 (orangutan males backing off from copulation); Bygott, "Agonistic Behavior, Dominance and Social Structure in Wild Chimpanzees of the Gombe National Park," 416,

423 (chimpanzees backing off in male-male conflicts); de Waal, *Peacemaking among Primates*, 211 (zoo colony male gorillas put off by female coalitions).

25. See notes 19 through 24, above. Tooby and De Vore, "The Reconstruction of Hominid Behavioral Evolution Through Strategic Modeling," 189, caution that differences between species can change the function and meaning of similarities.

26. See Eibl-Eibesfeldt, "The Myth of the Aggression-Free Hunter and Gatherer Society," regarding !Kung-Bushman children robbing one another (p. 445) and older children coming to the aid of younger children by punishing attackers (p. 448).

27. See note 23, Chapter 3.

28. "The degree of situational adaptation manifested by individuals will be a matter of a) how common in the species' evolutionary history that situation has been, b) how long (in phylogenetic terms) it has been recurring, and c) how large its fitness consequences are" (Tooby and De Vore, "The Reconstruction of Hominid Behavioral Evolution," 191).

29. Zimring and Hawkins, *Deterrence*, 124.

30. Ibid., 110.

CHAPTER 8: HELP ON THE HORIZON

1. For example, researchers at the New York Hospital–Cornell Medical Center have reported in a study of 1,328 deliveries, that mothers who postponed childbirth until they were 35 or older had infants with lower death rates and fewer birth defects than younger mothers. Early registration and careful surveillance during pregnancy was cited as contributing to this disparity. See *Chicago Tribune*, October 14, 1990, Section 5, p. 9, col. 2.

2. "The younger a new mother, the greater her future reproductive prospects, or, in the jargon of evolutionary biology, her 'residual reproductive value.' So if selection has shaped maternal psychology, we would expect a woman to be less and less inclined, as her reproductive years slip away, to devalue a present offspring in terms of its compromising effects on her future" (Daly and Wilson, *Homicide*, 52).

3. See Daly and Wilson, "Evolutionary Psychology and Family Violence," 298.

4. See ibid., 295:

The expected fitness of a pubertal individual exceeds that of an infant because the infant is likelier to die before reproducing. We thus predict that *parental motivation will have evolved so that parents will appear to value offspring increasingly with offspring age. . . .* Test of this prediction is complicated by the fact that the offspring's needs also change with age. If parents feed older offspring more, for example, that is hardly evidence that they value them more. Conversely, if a mother preferentially assists her *younger* offspring, that may be because the older is better able to fend for itself.

Despite these complications, the prediction of greater parental valuation of older offspring has been tested and widely confirmed [citations] (Italics in original; see also Daly and Wilson, *Homicide*, 77).

5. A relatively reliable 1966 study put the median age for female menopause in the United States at 49.8 years. Gray, "The Menopause—Epidemiological and Demographic Considerations," 27–28. It has been reported that the oldest age for a mother at birth for which there is reliable evidence is 57 years and 129 days. There is a report of 59 years for which the evidence appears less solid. McFarlan, ed., *Guinness Book of World Records*, 13.

6. Despite the pattern of diminished sexual activity which typically accompanies old age in men, abundant spermatozoa have often been found in the testes of men in extreme old age. See Hafez and Evans, *Human Reproduction*, 195. Indeed, birth of a child when the biological father was 94 years old has been reported. Seymour, Duffy and Koerner, "A Case of Authenticated Fertility in a Man, Aged 94."

7. Daly and Wilson, *Homicide*, 52–53.

8. Ibid., 62–63.

9. Daly and Wilson, "A Sociobiological Analysis of Human Infanticide," 492. "If physical abuse of children by parents or parent substitutes is in part the result of lapses of parental solicitude and resentment of parental obligation, then child abuse, like infanticide, should prove especially frequent when parents are confronted with predictors of low offspring contribution to parental fitness" (Daly and Wilson, "Evolutionary Psychology and Family Violence," 301).

10. Barash, *Sociobiology and Behavior*, 299, 302–3. Empirical data have shown that the risk of child abuse is greater in larger families. Lenington, "Child Abuse: The Limits of Sociobiology," 22.

11. This formulation reflects Daly and Wilson's prediction in *Homicide*, 62, but they put "dispose of" in place of "abuse" because they were discussing infanticide. See also Lenington, "Child Abuse: The Limits of Sociobiology," 27: "In any situation where a female is reasonably certain she will have only one child (for example, the child is born when the woman is near menopause, or if the birth is followed by a hysterectomy) and if the child is not severely defective, the child should be at a much lower risk of abuse than expected [*sic*] from the mother."

12. See "Panel: States Lax on Child Support," *Chicago Tribune*, January 12, 1991, Section 1, p. 3, col. 2, where it was reported that the responsible state agencies were collecting "an average of 36 percent of the support owed." See also Lieberman, *Child Support in America*, 11: "Of the 4 million women who were owed child support in 1981, only 47 percent received the full amount due, and 28 percent received absolutely nothing; the aggregate amount of child support payments due in 1981 was $9.9 billion, but only $6.1 billion was actually received." See also Sorenson and MacDonald, "An Analysis of Child-Support Transfers," 41.

13. Krause, *Child Support in America*, 422–31. See, generally, Kahn and Kamerman, *Child Support*.

In testimony before the Federal Commission on Interstate Child Support in March 1991, "[p]arents, lawyers, judges and child-advocates expressed a determination to change what one person called an 'incompetent' web of bureaucracy and unenforceable state and federal laws that have run up a bill of $18 billion in uncollected child-support payments nationwide" (*Chicago Tribune*, March 3, 1991, Section 2, p. 1, col. 2).

14. The inclination of parents to distinguish between prepubertal children on the basis of which is closer to puberty (see text at note 4, above) may be considered to be an exception to this general rule.

15. Porter, "Kin Recognition, Functions and Mediating Mechanisms," 195; Kaitz, Good, Rokem and Eidelman, "Mothers' Recognition of Their Newborns by Olfactory Cues." Studies with fathers have shown them to be unable to identify newborns from olfactory cues under similar conditions. Wells, "Kin Recognition in Humans," 399.

16. Morsbach, "Maternal Recognition of Neonates' Cries in Japan," 63–69; Wells, "Kin Recognition in Humans," 398.

17. For male-female differences in ability to recognize family resemblances in nonrelatives, see Nesse, Silverman and Bortz, "Sex Differences in Ability to Recognize Family Resemblance."

18. Alexander, *Darwinism and Human Affairs*, 174–75. See also, Barash, *Sociobiology and Behavior*, 319, and Daly and Wilson, *Sex, Evolution and Behavior*, 167.

19. Alexander, *Darwinism and Human Affairs*, 169–75; Kurland, "Paternity, Mother's Brother, and Human Sociality," 157–67; Irons, "Investment and Primary Social Dyads," 184–92; Gaulin and Schlegal, "Paternal Confidence and Paternal Investment: A Cross-Cultural Test of a Sociobiological Hypothesis." See Symons, *The Evolution of Human Sexuality*, 244, where the author notes a society in which there did not appear to be a correlation between paternity uncertainty and investment in sisters' children. But see Irons, above, at 209–13, regarding the variety of other environmental factors that could intercede in the equation while being consistent with the kin selection hypothesis.

20. Barkow has recently put forth "a hypothetical and hitherto unknown psychological trait: human males must apparently monitor the sexual activities of their wives and automatically calculate their paternity confidence" (Barkow, *Darwin, Sex and Status*, 280).

21. See Irons, "Let's Make Our Perspective Broader Rather Than Narrower," 363–70; Tooby and Cosmides, "The Past Explains the Present: Emotional Adaptations and the Structure of Ancestral Environments," 386–88; Symons, "Adaptiveness and Adaptation," 431–33; and Barkow, *Darwin, Sex and Status*, 42.

22. "Erratic confidence of paternity may lead to emphasis upon phenotypic attributes of putative offspring in determining whether or not to accept them as suitable objects of paternal care (hence, perhaps, the frequent attention to the

question of whether or not a baby resembles its father)" (Alexander, *Darwinism and Human Affairs*, 159). See also Kurland, "Paternity, Mother's Brother, and Human Sociality," 172; Sherman and Holmes, "Kin Recognition: Issues and Evidence," 448.

Regarding evolved abusive behavior toward offspring, Daly and Wilson have observed: "[W]e would . . . expect that men would be increasingly likely to behave violently toward putative *own* offspring, the more phenotypically dissimilar those children to themselves; by way of comparison, phenotypic resemblance of self would not be expected to influence mothers. To the best of our knowledge this prediction has yet to be tested" (Italics in original; Daly and Wilson, "Evolutionary Psychology and Family Violence," 304–5).

23. Tooby and De Vore, "The Reconstruction of Hominid Behavioral Evolution Through Strategic Modeling," 195–96.

"Kin selection probably operated in our hominid and protohominid ancestors prior to the appearance of both cognitive understanding of human reproduction and language" (Johnson, "Kin Selection, Socialization, and Patriotism: An Integrating Theory," 132). Regarding the development and timing of the acquisition of nonverbal and verbal communications in early humans, see Fisher, *The Sex Contract*, 155–72.

24. "Interested parties take great pains to point out phenotypic similarities between newborn offspring and the putative father [citation]. This would presumably serve the dual function of drawing the father's attention to such similarities as might exist, if his own recognition abilities were deficient, and of persuading him of the existence of similarities which in some cases may not in fact be present" (Wells, "Kin Recognition in Humans," 409).

25. The consensus of sociobiologists is that genetic programming is somewhere, even if remotely, behind all behavior. Some believe that "when individual learning is expensive and the environment fairly predictable, the mechanism selected will result in heavy reliance on social [cultural] transmission of behavioral cues" (Barkow, *Darwin, Sex and Status*, 273, summarizing Boyd and Richerson's position).

26. Compare Turke's observation in "Just Do It," 455, that printed pornography may not be as novel for Pleistocene-created mechanisms as others have suggested because "Pleistocene males probably imagined female images, and they certainly drew and sculpted them."

27. See Thompson and Ford, "DNA Typing: Acceptance and Weight of the New Genetic Identification Tests;" Kelly, "Admissibility of DNA Evidence: Perfecting the 'Search for Truth' ".

28. Some evidence of the effect that DNA typing tests may have on male paternity confidence is provided by the fact that judges have indicated they have "powerful impacts" on jurors. Kelly, "Admissibility of DNA Evidence: Perfecting the 'Search for Truth' ", 606.

29. Thompson and Ford, "DNA Typing: Acceptance and Weight of the New Genetic Identification Tests," 105.

30. One expert, estimating on the basis of a given set of factors and circumstances, said that "the chances of even siblings having identical DNA fingerprints (except for identical twins who have the same genes) are estimated at 1 in 10 trillion (*Chicago Tribune*, March 6, 1988, Section 5, p. 1, col. 2).

31. "In 1978, the [new federal program aimed at augmenting the collection of child support] located 453,620 absent fathers and established paternity in 110,714 cases. By 1983, those numbers had risen to 830,758 and 209,024 respectively. . . . The number of children born out of wedlock and living out of wedlock [in the U.S.] soared from 527,000 in 1970 to . . . 2,800,000 in 1982" (Lieberman, *Child Support in America*, 9).

32. See "DNA Test Dooms Paternity Trials, Lawyers Say," *New York Times*, July 21, 1989, Section Y, p. 19, col. 3: "New tests that compare the DNA of a child with that of the alleged father are being used with such conclusiveness that legal experts predict the demise of paternity trials. . . . [A lawyer in one nationally publicized paternity case said] "[y]ou aren't going to see many more paternity trials. It's becoming trial by test instead of trial by jury."

33. Austin Hughes has recently made the fascinating suggestion that if social engineers could increase child support payments by absent fathers, a corresponding decrease in *child abuse* by stepfathers (who have taken the absent father's place in the family) could be expected to result (Hughes, *Evolution and Human Kinship*, 137).

34. See Beckstrom, "The Use of Legal Opinions to Test Sociobiological Theory: Contract Law Regarding Reciprocal Relations in a Household," 227–30.

35. In 1987 it was noted that there had been an "explosion" in kin recognition research. Fletcher and Michener, "Introductory Remarks" to *Kin Recognition in Animals*. Following that, in 1987–88 alone, at least seventy-four articles on kin recognition in animals were published (Johnson, "The Role of Kin Recognition Mechanisms in Patriotic Socialization: Further Reflections," 63). Johnson observed that "research has now confirmed the existence of kin recognition abilities for a widely diverse range of species . . . (including primates)" (ibid).

36. See, in particular, the excellent discussion by Sherman and Holmes in "Kin Recognition: Issues and Evidence."

37. See discussion of the "green beard effect" in Dawkins, *The Selfish Gene*, 89: "It is theoretically possible that a gene could arise which conferred an externally visible 'label', say a pale skin, or a green beard, or anything conspicuous, and also a tendency to be specially nice to bearers of that conspicuous label. It is possible, but not particularly likely."

38. "As originally pointed out by [W. D.] Hamilton . . ., a 'supergene' would have to influence: (a) the development of the phenotypic signature, (b) recognition of that signature, *and* (c) appropriate social responsiveness to bearers" (Italics in original; Porter, "Kin Recognition: Functions and Mediating Mechanisms," 188).

39. "However, their existence cannot be entirely discounted, since several studies have obtained results that are at least consistent with such an ability

[citing studies]" (Johnson, "Kin Selection, Socialization, and Patriotism: An Integrating Theory," 130). And see Phillipe Rushton's discussion of recognition alleles as being an important element in understanding human behavior in Rushton, "Gene-Culture Coevolution and Genetic Similarity Theory: Implications for Ideology, Ethnic Nepotism, and Geopolitics," together with Gary Johnson's agreement but reiteration of the fact that "[W]e do not yet have empirically confirmed cases of the use of recognition alleles" (Johnson, "Some Thoughts on Human Extinction, and the Impact of Patriotism on Inclusive Fitness," 149).

40. Sherman and Holmes, "Kin Recognition: Issues and Evidence," 439–40.

41. Paleontological evidence leaves little doubt that human ancestors during the Pleistocene were hunter-gatherers. Hunter-gatherers will be nomadic. Lee and DeVore, *Man the Hunter*. "[N]on-nomadic hunter-gatherers would quickly exhaust food supplies in their area and learn to move, to farm, or to starve" (Barkow, *Darwin, Sex and Status*, 172). Humans took to farming only about ten thousand years ago. Ornstein and Ehrlich, *New World New Mind*, 27.

"[W]hat was life really like in the times of *Homo erectus* people 500,000 years ago? The only thing we know for certain is that they were nomads who moved from place to place, hunting when the time was right, gathering when it wasn't" (Fisher, *The Sex Contract*, 213).

Lewis Binford has noted that the existence of "home bases" to which our Pleistocene ancestors returned after a day of foraging is not clearly established. Thus, they may have been more or less "continuously nomadic" (Binford, "Human Ancestors: Changing Views of Their Behavior," 321).

42. "[I]f association acts as a primary kin recognition mechanism for humans, and if phenotypic matching acts as a secondary mechanism, location probably serves as a tertiary mechanism" (Johnson, "The Role of Kin Recognition Mechanisms in Patriotic Socialization: Further Reflections," 65).

43. Ibid.; Sherman and Holmes,"Kin Recognition: Issues and Evidence," 440; Barkow, *Darwin, Sex and Status*, 50.

44. See Eibl-Eibesfeldt, *Human Ethology*, 615: "The basic principles of human society are fully developed in hunter-gatherer peoples. They live in individualized bands in which each person knows every other and which are composed of families consisting of three generations."

Compare Chagnon, "Terminological Kinship, Genealogical Relatedness and Village Fissioning among the Yanomamo Indians," 491–93, where the author describes a village fissioning process in the 1960s and 1970s among the primitive, hunter-gatherer Yanomamo. When a village reached a critical size (100 to 300 people) a group would depart and set up a separate village. It seemed to be invariant that "within-group average relatedness [was] higher in the newly-formed villages than it was in the larger, pre-fission village" (493).

45. Pfeiffer places the size of hunter-gatherer groups at about twenty-five persons, "a figure that holds for Bushmen, Australian aborigines, and other present-day hunter-gatherers, and, according to estimates based on the areas of excavated living sites, for their prehistoric ancestors as well" (Pfeiffer, *The*

Emergence of Society, 33). See also Lopreato, *Human Nature and Biocultural Evolution*, 319: "Humans have typically lived in societies of 25–30 individuals for most of their evolutionary history;" Eibl-Eibesfeldt, *Human Ethology*, 294 ("30 to 50"); and Barash, *Sociobiology and Behavior*, 313.

46. "Since this mechanism is vulnerable to recognition errors (i.e., familiar individuals, regardless of whether or not they are genetically related, are treated as kin), it would be expected to be implicated primarily in contexts where nonkin do not co-occur with kin during the initial phases of the familiarization process" (Porter, "Kin Recognition: Functions and Mediating Mechanisms," 187).

47. Sherman and Holmes, "Kin Recognition: Issues and Evidence," 443, 447.

48. Ibid., 442–48.

49. "We can reasonably expect that physical phenotype matching would occur even in racially homogeneous populations on the basis of physical characteristics that determine what we often call 'family resemblance'" (Johnson, "Kin Selection, Socialization, and Patriotism: An Integrating Theory," 138 n.5).

50. Van den Berghe has suggested that gross racial distinctions would not have been useful to early human ancestors because racial groups would not have come into contact with one another until the time of mass migrations, which happened relatively late in hominid history. Van den Berghe, *The Ethnic Phenomenon*, 29–33.

51. "It . . . seems likely that empirical research will eventually confirm what theory predicts: kin recognition abilities exist for all social species, and these abilities are a crucial foundation for most social behavior" (Johnson, "The Role of Kin Recognition Mechanisms in Patriotic Socialization: Further Reflections," 63).

52. See Hinde, "Patriotism: Is Kin Selection Both Necessary and Sufficient?," 60; Johnson, "The Role of Kin Recognition Mechanisms in Patriotic Socialization: Further Reflections," 66–67. See also Alexander, *The Biology of Moral Systems*, 254.

53. See, for example, the collection of authors in Reynolds, Falger and Vine, eds., *The Sociobiology of Ethnocentrism*.

54. Johnson, "Kin Selection, Socialization, and Patriotism: An Integrating Theory;" Johnson, "Some Thoughts on Human Extinction, Kin Recognition, and the Impact of Patriotism on Inclusive Fitness;" Johnson, "The Role of Kin Recognition Mechanisms in Patriotic Socialization: Further Reflections."

55. Van den Berghe, *The Ethnic Phenomenon*; van den Berghe, "Kin, Ethnicity, Class, and the State: Of Consciousness of Kind, True and False."

56. Shaw and Wong, *Genetic Seeds of Warfare*.

57. See notes 44 and 45, above.

58. See period reconstruction by anthropologist Helen Fisher in *The Sex Contract*, 194.

59. The inclusive fitness concept, first developed by William D. Hamilton, has been described as "a property of individual organisms, equal to the focal individual's own reproductive success (classical Darwinian fitness) plus the

focal individual's incremental (or decremental) influences upon the reproductive success of kin multiplied by the degree of relatedness . . . of those kin" (Daly and Wilson, *Sex, Evolution and Behavior*, 30).

60. "The sharing of food at home bases is considered by some to be a crucial expression, and perhaps the earliest one, of human social reciprocity. . . . Others having stressed that hunter-gatherers consume food while foraging and that the extent of food-sharing at the campsite [by Plio-Pleistocene hominids] is unclear. Yet in all modern hunter-gatherer bands at least some food is exchanged among adults, and children are given food" (Potts, "Home Base and Early Hominids" 340).

61.

Cooperation can evolve only when 1) there are many situations in which individuals can benefit each other at relatively low cost to themselves . . . , and 2) the probability of two individuals meeting again is sufficiently high. The probability that two individuals will meet again is increased if the individuals are long-lived and have low dispersal rates. These life-history factors also increase the *number* of situations for mutual help that two individuals are likely to encounter. The ecological and life-history factors characteristic of the human environment of evolutionary adaptedness fulfill the conditions necessary for the evolution of cooperation. Pleistocene hunter-gatherers were not only long-lived, but they lived in small, relatively stable bands. Thus, the probability was high that an individual you had helped would be around when you needed help. Moreover, in all probability these individuals, like modern hunter-gatherers, were closely related; kin selection can promote the evolution of cooperation [citations]" (Italics in original; Cosmides and Tooby, "Evolutionary Psychology and the Generation of Culture, Part II," 57. See also Wrangham, "Mutualism, Kinship and Social Evolution").

62. Napoleon Chagnon and Paul Bugos have used detailed data they collected on the interrelatedness of Yanomamo Indians, among whom they lived in Venezuela, to analyze a fortuitously filmed ax fight between residents of a village and a small group of visitors who had formerly resided in the village. Chagnon and Bugos, "Kin Selection and Conflict: An Analysis of a Yanomamo Ax Fight." Gray has succinctly summarized their findings as follows:

First, members of the two fighting teams were, on the average, more closely related among themselves than to members of the opposing team. Second, the fighters of the host team were more closely related among themselves than they were to the village at large. Third, the fighters of the visiting team were more closely related among themselves than they were to members of the host village. And, finally, the followers of a major fight leader were more closely related to that individual than to the fight leader of the opposing team (Gray, *Primate Sociobiology*, 112).

63. Johnson, "Kin Selection, Socialization, and Patriotism: An Integrating Theory," 134; Shaw and Wong, *The Genetic Seeds of Warfare*, 91–111.

64. Betzig, "Introduction" to *Human Reproductive Behavior*, 14: "Suppose that we have evolved to recognize as closest kin those with whom we live during a critical period early in life. Suppose, too, that living arrangements are now importantly different from those which characterized our evolutionary history.

In that case, we might find individuals behaving generously toward others distantly related genetically, and not very generously toward closer kin."
65.

The anthropologist, Marshall Sahlins, working quite independently of sociobiological theory, conducted cross-cultural surveys and found that human groups list their loyalties in direct order of: household, identified relatives, one's village, one's tribe, and then, finally, members of other tribes. It is interesting to note that household is more important than blood relations. So deep are the loyalties to one's family-in-the-household that they apparently become automatically extended to include non-kin if necessary. Needless to say, an adopted child is treated the same as a natural child, and even an adult non-relative seems quickly to gain all the protective sensibilities of the host family. So in humans it is the day-to-day intimacy of the parties that evokes the emotions of loyalty, rather than any kind of chemical signalling between blood kin (Maxwell, *Human Evolution*, 155).

66. Richard Alexander has noted "the paradox that extreme within-group altruism seems to correlate with and be historically related to between-group strife." Alexander, *Biology of Moral Systems*, 233.

67. See, generally, van den Berghe, *The Ethnic Phenomenon*.

68. "[A]n animal might respond selectively to others whose visual or olfactory phenotypes closely match those of familiar kin such as its mother or littermates" (Porter, "Kin Recognition: Functions and Mediating Mechanisms," 188). With humans, scientists have not yet been able to devise a way to test people on olfactory recognition by using relatives who have never previously had contact with one another. The closest they have been able to come is by using mothers and their infants delivered by Caesarean section. Relatively little infant-mother contact takes place right after delivery in such cases. Tests have shown these mothers to be significantly accurate in identifying their infant's soiled shirts from among those of other infants (ibid., 195).

69. "The behavioral phenotype of human individuals includes their verbal and other symbolic behavior" (Johnson, "Kin Selection, Socialization, and Patriotism: An Integrating Theory," 133). Studies with newborn infants and their mothers have shown them able to recognize each other's voices with limited previous out-of-womb contact (Wells, "Kin Recognition in Humans," 398).

Studies of birds have shown parents identifying their young by their chicks' "signature calls" and young birds differentiating between the calls of their parents and other adults. Porter, "Kin Recognition: Functions and Mediating Mechanisms," 183.

70. "The ontogenetic history of individual signatures has been the subject of little research. Many morphological features that function as salient cues for kin recognition are obviously genetically mediated—e.g., bird coloration patterns; shape of mouth, nose and face in humans" (Porter, "Kin Recognition: Functions and Mediating Mechanisms," 184).

71. "Markers giving information about genetic similarities and differences can also be acquired through learning. . . . Irwin . . . computed a coefficient of dialect differentiation using linguistic data from six Eskimo tribes. . . . He found

that dialect differences between tribes corresponded with genetic differences between tribes, as calculated from genealogies" (Wells, "Kin Recognition in Humans," 403–4, citing paper presented by C. J. Irwin at fifth Meeting of European Sociobiological Society, Oxford, England, 1985).

"It should be emphasized that recognition of kin by phenotype-matching does not require that the signature be genetically determined. Rather, kin signatures that are similar as a result of shared environmental influences could also be employed as phenotypes against which others are assessed. The correlation between phenotype and genotype, *not* the underlying basis for this correlation, is critical" (Italics in original; Porter, "Kin Recognition: Functions and Mediating Mechanisms," 188).

72. Shaw and Wong, in modeling the identification mechanism, list language as one of the five "recognition markers" that link larger groups to the individual's nucleus ethnic group through cognitive processes" (Shaw and Wong, *The Genetic Seeds of Warfare*, 102).

73. See ibid., 160: "If, over a period of several generations, minorities lose their own language. . . . [a]nother recognition marker—common language— would become prevalent. Accordingly, the inclusive fitness logic of patriotism would have more ground on which to prosper."

74. See authorities cited above, in particular those in notes 54, 55 and 56.

75. See "Wanting Out of Russia," *Newsweek*, November 12, 1990, pp. 40, 41 (Republic of Tartarstan enacts law making Tatar a second language as part of move for autonomy from U.S.S.R.); "Language Law Passes Key Test in Moldavia," *Chicago Tribune*, August 30, 1989, Section 1, p. 5, col. 5 (Moldavians win first step in legislature to make Moldavian the official language of the republic in move for more autonomy from U.S.S.R.).

76. See "Separatism Tearing at Canada's Fabric", *Chicago Tribune*, December 30, 1990, Section 1, p. 1, cols. 3, 4. (Canadian federalism is being torn apart by bitter linguistic bickering, civil strife among native groups, and a growing sense of alienation among the Western and Atlantic provinces).

77. Editorial, "Is Canada Headed for Splitsville?," *Chicago Tribune*, January 3, 1991, Section 1, p. 22, col. 2.

So central is language to political organization that in many societies defining the language has become tantamount to defining nationality. The view that to be French is to speak French underlay official efforts to spread French from the capital outward, something that took over 150 years to accomplish. The official goal was to create the French people, and the French nation, by giving them the French language. Clearly, though, in another equally important sense, being French is a function of geography rather than language: there have been many residents of the Hexagon who were French—legally, socially, culturally, politically—without having French as their mother tongue. The complexity of the *questione della lingua*, an ancient and ongoing debate in Italy from the time of Dante to the present, similarly rests on the notion that defining the language is a necessary precursor to defining the group and ultimately the polity as well. In Europe, language played an essential part in the establishment of political units, both in the Renaissance and in the nineteenth century, and political turmoil in Europe during and after World War I, and at present as well, is frequently expressed in terms of calls for language rights and linguistic independence (Baron, *The English-Only Question*, 6–7).

78. "[T]he Land Acts, which set aside 87 percent of property for whites, and the Group Areas Acts, which dictated where people could live on the basis of their race, were repealed on June 5 [1991]. . . . Although the laws have now been revoked, the system they maintained has become so entrenched in South African society that breaking down its separations and compartments in every day life—at schools, jobs, or distinct neighborhoods and enclaves—remains a formidable task." *New York Times* (National Ed.), June 18, 1991, p. A1, col. 3. The law that forcibly segregated every South African neighborhood by race—the Group Areas Act—proclaimed District Six in Cape Town a "whites only" area. One resident who was forced out of the district referred to it as formerly "a beautiful melting pot of nations, a potpourri of many races" (*Chicago Tribune*, February 24, 1991, Section 1, pp. 21, 24, col. 1).

79. Shaw and Wong, *The Genetic Seeds of Warfare*, 195–204.

80. Ibid., 205.

81. See summaries in Barash, *The Whisperings Within*, 170–98; Barash, *The Hare and the Tortoise*, 151–87, 332–33.

82. See Groebel and Hinde, eds., *Aggression and War*.

"Considering . . . the theory that we evolved from hunters for whom aggression was a necessary part of life—[E. O.] Wilson has said that even if this were true, it would not follow from this alone that we have a genetic tendency to aggression, for there has been time, since the beginning of agricultural societies more than five thousands years ago, for contrary selective pressures to have altered the genetic tendencies we had when we were hunters" (Singer, *The Expanding Circle*, 172).

83. Malmberg, *Human Territoriality*. See also Durham, "Resource Competition and Human Aggression, Part I: A Review of Primitive War."

84. Dollard, et al., *Frustration and Aggression*.

85. Shaw and Wong, *The Genetic Seeds of Warfare*, 204–6.

86. Daly and Wilson, *Homicide*, 22–24.

Some people are under the impression, from press reports or otherwise, that one is at a greater risk of being killed or injured by a relative than by a stranger, and this may be true in the sense that there is generally a much higher degree of mutual access between an individual and his relatives, creating increased opportunities for violence, than between an individual and strangers. A criminologist has remarked that research shows "a person is safer in Central Park at three o'clock in the morning than in his or her bedroom." Daly and Wilson answer that this is "patent nonsense, confounding frequency with rate. At three o'clock in the morning, two hundred million Americans are in their bedrooms and a handful are in Central Park. If there were only one murder in the park per century, the bedroom would still be far the safer place." Daly and Wilson go on to show statistical evidence that people are less inclined to kill "blood" relatives than others given the same provocations. Daly and Wilson, *Homicide*, 21–24.

87. "[B]y its nature, sociobiology has less to say about the behavior of large groups than about the individuals that compose them. It provides more insight into the personal life of a nation's president (whether he has a mistress, how he treats his children, what makes him angry and so on) than it will elucidate the nation's foreign policy. In the same vein, we can expect that evolutionary biology will tell us less about war as waged by societies than about fighting as carried out by individuals. Nevertheless, war has real and direct consequences for the fitness of the individuals involved" (Barash, *The Whisperings Within*, 195–96).

88. Richard Alexander has said that "universal brotherhood" is not likely "to be brought about by means currently supposed to be adequate, or . . . likely to be advanced by people who do not understand themselves through knowledge of their biologies and evolutionary history" (Alexander, *The Biology of Moral Systems*, 233).

89. "To provide a more durable foundation for peace, political and cultural ties can be promoted that create a confusion of cross-binding loyalties. Scientists, great writers, some of the more successful businessmen and Marxist-Leninists have been doing just that more or less unconsciously for generations. If the tangle is spun still more thickly, it will become discouragingly difficult for future populations to regard each other as completely discrete on the basis of congruent distinctions in race, language, nationhood, religion, ideology and economic interests" (Wilson, *On Human Nature*, 120).

90. Language instruction was one type of project carried out by the United States Peace Corp program in some seventy countries throughout the world during the second half of the twentieth century, and 5 million people have been taught English under Peace Corps auspices. See *Chicago Tribune*, March 3, 1991, Section 4, p. 2, cols. 1, 2.

CHAPTER 9: WHEN EVOLUTIONARY LEARNING WILL NOT HELP

1. "If something is genuinely unprecedented, mechanisms will not assign it to the categories that trigger the correct decisional rules, except by chance. Mechanisms that appear able to handle novelty do so only because the apparent novelty resides in one aspect of the phenomenon, while algorithms are operating on other aspects that display subtle or relational cues based on some underlying recurrent uniformity . . . linked to . . . the evolutionary past" (Tooby and Cosmides, "The Past Explains the Present: Emotional Adaptations and the Structure of Ancestral Environments," 410).

"[S]accharine displays perceptual cues that once reliably signaled nutritional value, without the ancestrally associated nutritional value; magazine erotica displays perceptual cues of opportunities for fertile copulation, without the reality" (ibid., 407). But see Turke, "Just Do It," 455, where he explains why magazine erotica may not be as novel for Pleistocene-created behavioral mechanisms as Tooby and Cosmides imply.

"Guns and condoms and automobiles and baby bottles are novelties on an evolutionary time scale; though our inventions may reveal much about our motives, there is no reason to suppose we will achieve fitness in our use of them" (Daly and Wilson, "Evolutionary Psychology and Family Violence," 305).

2. See Barkow, *Darwin, Sex and Status*, 296: "Human beings have apparently been selected to seek copulations with partners with particular characteristics, rather than to seek pregnancy *per se*."

3. "Some environmental features which, strictly speaking, are unnatural may resemble natural features so closely that the difference is irrelevant. On the other hand, the introduction of even one unnatural feature—say vasectomy—might have a profound effect on ultimate explanations of human behavior in a given population" (Symons, *The Evolution of Human Sexuality*, 34–35).

Compare Singer, *The Expanding Circle*, 131: "The growth of modern contraceptive techniques is a splendid example of the use of reason to overcome the normal consequences of our evolved behavior. It shows that reason can master our genes." It has been speculated that the use of contraceptives could eventually be eliminated by natural selection, as it is nonadaptive. However, a selective use of contraception may result in optimal reproduction in particular environments. In any event, Singer is surely right that the use of contraception by an individual short-circuits, on an ad hoc basis, whatever ancient genetic programming is pushing humans to copulate and thereby reproduce. For further observations on the same topic, see van den Berghe, *Human Family Systems*, 182–83.

4. See Irons, "Let's Make Our Perspective Broader Rather Than Narrower," 370: "I see no reason to assume that we know in advance at what point in the path from genes to phenotype to reproduction novel environments will disrupt adaptations."

5. For an example of a social engineering attempt to increase the birthrate, see "*Chicago Tribune*, December 1, 1983, Section 1, p. 43. (A proposed law to pay $3,000 to a parent who quits his or her job to care for a third child has sparked debate on whether France, or any nation, can increase the reproductive habits of its citizens.)

For an example of a social engineering attempt to decrease the birthrate, see *Chicago Tribune*, November 19, 1979, Section 1, p. 12, col. 3. (China has begun levying a 'baby tax' against couples producing a third child as its latest weapon in a war on population growth.)

6. "Certain human behaviors are ... more rigidly constrained by genotype, while others are more flexible. Behavior that appears almost universal to all cultures, that has not varied during recorded history, and that is clearly of biological advantage, would seem likely to have [relatively direct] genetic underpinnings" (Barash, *The Hare and the Tortoise*, 52).

7. Glynn Isaac and other archaeologists have suggested that food sharing was a major "conditioner" for many human characteristics believed to have

already been present in hominids during the Plio-Pleistocene. Binford, "Human Ancestors: Changing Views of Their Behavior," 292. Binford (ibid., 321) has questioned whether adequate evidence exists for food sharing during this period.

8. For example, Tiger and Fox, *The Imperial Animal* (generally); Konner, *The Tangled Wing*, 405–6; Beckstrom, *Sociobiology and the Law*, 117–23; Alexander, *The Biology of Moral Systems* (generally); Beckstrom, *Evolutionary Jurisprudence*, 45–53; Frank, *Passions Within Reason*, 237–46, 248.

9. Barkow, *Darwin, Sex and Status*, 373–91 (Barkow has provided a tentative outline of how that ultimate enemy of humankind, unhappiness, might be better addressed by social engineers if they understood evolutionary biology); Hughes, *Evolution and Human Kinship*, 126–38.

10. Hughes, *Evolution and Human Kinship*, 137–38.

11. See Chagnon and Bugos, "Kin Selection and Conflict: An Analysis of a Yanomamo Ax Fight."

12. "The relevance of paternity assessment to violence has received [relatively little] attention. Certainly the revelation of nonpaternity can be a stimulus to violence against both wife and child, and manifest nonpaternity was offered as a justification for routine infanticide in 20 societies in Daly & Wilson's (1984) ethnographic review" (Daly and Wilson, "Evolutionary Psychology and Family Violence," 304).

13. Regarding incest, William Irons has made the correlated point that if males think they can distinguish which of their wives' children are really theirs, they should be more inclined to pursue incestuous relations with those they believe are "step-daughters," other things being equal. Irons, "Incest: Why all the Fuss?," 9.

14. In any random testing of a large population sample, some percentage of putative biological fathers would prove not to be. But any adverse effects on solicitude for the children involved might well be considerably overbalanced, from an overall societal perspective, by an increase in solicitude for the children of men whose biological paternity was confirmed to them.

15. Lionel Tiger and Robin Fox have recently commented that if, in an earlier writing, they had "understood then the motivations involved in acquiring wealth and those involved in redistributing it . . . we might have announced that the small family firm in a totally competitive economy was the ideal situation for maximizing our human potential" (Tiger and Fox, *The Imperial Animal*, xxiii; see also, Frank, *Passions Within Reason*, 238–40).

16. Researchers who spent a year or more executing well-planned studies of interactions between people in Venezuelan and Trinidadian villages have suggested that considerably more time and effort would be required to identify such interactions as incidents in ongoing aid-giving exchanges. Hames, "Relatedness and Interaction among the Ye'kwana: A Preliminary Analysis," 244; and Flinn, "Resources, Mating, and Kinship: The Behavioral Ecology of a Trinidadian Village," 236–37.

17. "Blood and Money," *Newsweek*, Vol. 114, Special Issue, Winter 1898/Spring 1990, p. 82.

18. Ibid.

19. The noted American statistician, W. Edwards Deming, who is said to be the person most responsible for the late twentieth-century success of Japanese business in global competition, reportedly "believes American business can't compete because it hasn't learned to cooperate—within the ranks of managers and employees of a particular company; between companies and their suppliers and customers; and among companies within industries." *Chicago Tribune*, December 23, 1990, Section 7, p. 3, col. 3.

EPILOGUE: WHAT ARE WE LIKELY TO DO TOMORROW?

1. "Man conquered the physical world not by insisting that he could *outdo* nature but by finding out exactly what it was he could *do with* nature. Our modest suggestion is that the same demanding principle should be applied to our attempts at social transformations" (Italics in original; Tiger and Fox, *The Imperial Animal*, 237).

Bibliography

Aafjes, J., and J. Vreeburg. "The Hypothalamo-Pituitary-Testis Axis at Puberty." In *Handbook of Human Growth and Developmental Biology*, Vol. II: Part A, eds. E. Meisami and P. Timiras. 1989. CRC Press, Boca Raton, Fla.

Alexander, R. "The Evolution of Social Behavior." 5 *Ann. Rev. Ecology and Systematics* 325 (1974).

_____ . "Evolution and Culture." In *Evolutionary Biology and Human Social Behavior*, eds. N. Chagnon and W. Irons. 1979. Duxbury Press, North Scituate, Mass.

_____ . *Darwinism and Human Affairs*. 1979. University of Washington Press, Seattle.

_____ . "Ostracism and Indirect Reciprocity: The Reproductive Significance of Humor." 7 *Ethology and Sociobiology* 105 (1986).

_____ . *The Biology of Moral Systems*. 1987. Aldine de Gruyter, New York.

Allen, F. *The Decline of the Rehabilitative Ideal*. 1981. Yale University Press, New Haven.

Altmann, S. "Altruistic Behavior: The Fallacy of Kin Deployment." 27 *Animal Behavior* 958 (1979).

Andenaes J. "The General Preventive Effects of Punishment." 114 *U. of Penn. L. Rev. 949 (1966)*.

Bailey, W. "Rape and the Death Penalty: A Neglected Area of Deterrence Research." In *Capital Punishment in the United States*, eds. H. Bedau and C. Pierce. 1976. AMS Press, New York.

_____ . "Deterrence and the Violent Sex Offender: Imprisonment vs. the Death Penalty." 6 *J. Behavioral Econ.* 107 (1977).

Bailey, W. J. Martin and L. Gray. "Crime and Deterrence: A Correlation Analysis." 11 *J. Res. in Crim. and Delinquency* 124 (1974).

Ball, J. "Memes as Replicators." 5 *Ethology and Sociobiology* 145 (1984).

Ball, J. C. "The Deterrence Concept in Criminology and Law." 46 *J. of Crim. L., Criminology and Police Science* 347 (1955).

Banks, S., and M. Kahn. *The Sibling Bond.* 1982. Basic Books, New York.

Barash, D. *The Whisperings Within.* 1979. Harper & Row, New York.

―――― . *Sociobiology and Behavior.* 2d ed. 1982. Elsevier, New York.

―――― . *The Hare and the Tortoise: Culture, Biology and Human Nature.* 1986. Viking, New York.

Barash, D., and J. Lipton. *The Caveman and the Bomb.* 1985. McGraw-Hill, New York.

Barkow, J. *Darwin, Sex and Status.* 1989. University of Toronto Press, Toronto.

―――― , ed. *Evolved Constraints of Cultural Evolution.* Vol. 10, Nos. 1–3, *Ethology and Sociobiology* (1989).

Baron, D. *The English-Only Question.* 1990. Yale University Press, New Haven.

Beckstrom, J. "Handicaps of Legal-Social Engineering in a Developing Nation." 22 *Am. J. of Comp. L.* 697 (1974).

―――― . *Sociobiology and the Law: The Biology of Altruism in the Courtroom of the Future.* 1985. University of Illinois Press, Urbana, Ill.

―――― . "The Use of Legal Opinions to Test Sociobiological Theory: Contract Law Regarding Reciprocal Relationships in a Household." 8 *Ethology and Sociobiology* 221 (1987).

―――― . *Evolutionary Jurisprudence: Prospects and Limitations on the Use of Modern Darwinism Throughout the Legal Process.* 1989. University of Illinois Press, Urbana, Ill.

Bedau, H., ed. *The Death Penalty in America.* 1982. Oxford University Press, New York.

Bertram, B. "Social Factors Influencing Reproduction in Wild Lions." 177 *J. of Zoology* 463 (1975).

―――― . "Problems with Altruism." In *Current Problems in Sociobiology*, eds. Kings College Sociobiology Group, Cambridge. 1982. Cambridge University Press, Cambridge.

Betzig, L., M. Borgerhoff-Mulder and P. Turke, eds. *Human Reproductive Behavior: A Darwinian Perspective.* 1988. Cambridge University Press, Cambridge.

―――― . "Mating and Parenting in Darwinian Perspective." In *Human Reproductive Behavior: A Darwinian Perspective*, eds. L. Betzig, M. Borgerhoff-Mulder and P. Turke. 1988. Cambridge University Press, Cambridge.

―――― . "Introduction" to *Human Reproductive Behavior: A Darwinian Perspective*, eds. L. Betzig, M. Borgerhoff-Mulder and P. Turke. 1988. Cambridge University Press, Cambridge.

Binford, L. "Human Ancestors: Changing Views of Their Behavior." 4 *J. of Anthropological Archaeology* 292 (1985).

Bishof, N. "Comparative Ethology of Incest Avoidance." In *Sociobiology and Human Development*, ed. R. Fox. 1975. Malaby, London.

Bishop, W., and J. Clark, eds. *Background to Evolution in Africa*. 1967. University of Chicago Press, Chicago.

Bixler, R. "Incest Avoidance as a Function of Environment and Heredity." 22 *Current Anthropology* 639 (1981).

Boggess, J. "Infant Killing and Male Reproductive Strategies in Langurs." In *Infanticide: Comparative and Evolutionary Perspectives*, eds. G. Hausfater and S. Hrdy. 1984. Aldine Publishing Co., New York.

Bowler, P. *Evolution: The History of an Idea*. Rev. ed. 1989. University of California Press, Berkeley.

Boyd, R., and P. Richerson. *Culture and the Evolutionary Process*. 1985. University of Chicago Press, Chicago.

Brandon, R., and R. Burian, eds. *Genes, Organisms, Populations: Controversies over the Units of Selection*. 1984. MIT Press, Cambridge, Mass.

Breuer, G. *Sociobiology and the Human Dimension*. 1982. Cambridge University Press, Cambridge.

Burgess, R., and J. Garbarino. "Doing What Comes Naturally? An Evolutionary Perspective on Child Abuse." In *The Dark Side of Families: Current Family Violence Research*, eds. D. Finkelnor, R. Gelles, G. Hotaling and M. Strauss. 1983. Sage Publications, Beverly Hills.

Buss, D. "Sex Differences in Human Mate Selection Criteria: An Evolutionary Perspective." In *Sociobiology and Psychology*, eds. C. Crawford, M. Smith and D. Krebs. 1987. L. Erlbaum Associates, Hilldale, N.J.

Bygott, J. "Agonistic Behavior, Dominance, and Social Structure in Wild Chimpanzees of the Gombe National Park." In *The Great Apes*, eds. D. Hamburg and E. McCown. 1979. Benjamin/Cummings Publishing Co., Menlo Park, Calif.

Campbell, D. "The Two Distinct Routes beyond Kin Selection to Ultrasociality: Implications for the Humanities and Social Sciences." In *The Nature of Prosocial Development: Theories and Strategies*, ed. D. Bridgeman. 1983. Academic Press, New York.

Cavalli-Sforza, L., and M. Feldman. *Cultural Transmission and Evolution*. 1981. Princeton University Press, Princeton, N.J.

Chagnon, N., and P. Bugos, Jr. "Kin Selection and Conflict: An Analysis of a Yanomamo Ax Fight." In *Evolutionary Biology and Human Social Behavior*, eds. N. Chagnon and W. Irons. 1979. Duxbury Press, North Scituate, Mass.

———. "Terminological Kinship, Genealogical Relatedness and Village Fissioning among the Yanomamo Indians." In *Natural Selection and Social Behavior*, eds. R. Alexander and D. Tinkle. 1981. Chiron Press, New York.

Chaiken, M., M. Lawless and K. Stevenson. *The Impact of Police Activity on Crime: Robberies on the New York City Subway System* (New York City Rand Institute, January 1974).

Cohn, A., and B. Ward, eds. *Improving Management in Criminal Justice*. 1980. Sage Publications, Beverly Hills.

Cole, G. *Criminal Justice: Law and Politics*. 2d ed. 1976. Duxbury Press, North Scituate, Mass.

Comment. "Castration of the Male Sex Offender: A Legally Impermissible Alternative." 30 *Loy. L. Rev.* 377 (1984).

Contemporary Studies Project. "A Comparison of Iowans' Dispositive Preferences with Selected Provision of the Iowa and Uniform Probate Codes." 63 *Iowa L. Rev.* 1041 (1978).

Cosmides, L., and J. Tooby. "Evolutionary Psychology and the Generation of Culture, Part II." 10 *Ethology and Sociobiology* 51 (1989).

Crawford, C. "Sociobiology: Of What Relevance to Psychology?" In *Sociobiology and Psychology*, eds. C. Crawford, M. Smith and D. Krebs. 1987. L. Erlbaum Associates, Hillsdale, N.J.

Daly, M., and M. Wilson. "Abuse and Neglect of Children in Evolutionary Perspective." In *Natural Selection and Social Behavior*, eds. R. Alexander and D. Tinkle. 1981. Chiron Press, New York.

_____ . *Sex, Evolution and Behavior*. 1978. Wadsworth Publishing Co., Boston.

_____ . *Sex, Evolution and Behavior*. 2d ed. 1983. PWS Publishers, Boston.

_____ . "A Sociobiological Analysis of Human Infanticide." In *Infanticide: Comparative and Evolutionary Perspectives*, eds. G. Hausfater and S. Hrdy. 1984. Aldine Publishing Co., New York.

_____ . "Child Abuse and Other Risks of Not Living with Both Parents." 6 *Ethology and Sociobiology* 197 (1985).

_____ . "Evolutionary Psychology and Family Violence." In *Sociobiology and Psychology*, eds. C. Crawford, M. Smith and D. Krebs. 1987. L. Erlbaum Associates, Hillsdale, N.J.

_____ . *Homicide*. 1988. Aldine de Gruyter, New York.

Dawkins, R. *The Selfish Gene*. 1976. Oxford University Press, Oxford.

_____ . *The Extended Phenotype*. 1982. Oxford University Press, Oxford.

_____ . *The Blind Watchmaker*. 1986. W. W. Norton Co., New York.

Degler, C. *In Search of Human Nature: The Decline and Revival of Darwinism in American Social Thought*. 1991. Oxford University Press, Oxford.

de Waal, F. "Chimpanzee Politics." In *Machiavellian Intelligence: Social Expertise and the Evolution of Intellect in Monkeys, Apes and Humans*, eds. R. Byrne and A. Whiten. 1988. Oxford University Press, Oxford.

_____ . *Peacemaking Among Primates*. 1989. Harvard University Press, Cambridge, Mass.

Dictionary of Biology (Warner Books). 1985. Time Warner Publishing, New York.

Dollard, J., N. Miller, L. Doob, O. Mowrer, R. Sears, C. Ford, C. Hovland and R. Sollenberger. *Frustration and Aggression*. 1939. Yale University Press, New Haven.

Dorne, C. *Crimes Against Children*. 1989. Harrow to Heston, New York.

Dunbar, R. "Darwinizing Man: A Commentary." In *Human Reproductive Behavior: A Darwinian Perspective*, eds. L. Betzig, M. Borgerhoff-Mulder and P. Turke. 1988. Cambridge University Press, Cambridge.

Durham, W. "Resource Competition and Human Aggression, Part I: A Review of Primitive War." 51 *Quarterly Rev. of Biology* 385 (1976).

————. "Toward a Coevolutionary Theory of Human Biology and Culture." In *Evolutionary Biology and Human Social Behavior*, eds. N. Chagnon and W. Irons. 1979. Duxbury Press, North Scituate, Mass.

Dwyer, J. *Human Reproduction: The Female System and the Neonate.* 1976. F. A. Davis Co., Philadelphia.

Edel, A. "Attempts to Derive Definitive Moral Patterns from Biology." 1955. Reprinted in *The Sociobiological Debate*, ed. A. Caplan. 1978. Harper & Row, Hagerstown, Md.

Edelman, G. *Bright Air, Brilliant Fire: On the Matter of the Mind.* 1992. Basic Books, New York.

Eibl-Eibesfeldt, I. "The Myth of the Aggression-Free Hunter and Gatherer Society." In *Primate Aggression, Territoriality and Xenophobia*, ed. R. Holloway. 1974. Academic Press, New York.

————. *Human Ethology.* 1989. Aldine de Gruyter, New York.

Einstein, A. *Out of My Later Years.* 1950. Philosophical Library, New York.

Ellis, L. *Theories of Rape.* 1989. Hemisphere Publishing, New York.

Encyclopedia Britannica, Vol. 14, Populations, Human. 1980.

Fedigan, L. *Primate Paradigms.* 1982. Eden Press, Montreal.

Feeley, M. *The Process Is the Punishment: Handling Cases in a Lower Criminal Court.* 1979. Russell Sage Foundation, New York.

Fellows, M., R. Simon and W. Rau. "Public Attitudes about Property Distribution at Death and Intestate Succession Laws in the United States." 1978 *Am. B. Found, Research J.* 319.

Fellows, M., R. Simon, T. Snapp and W. Snapp. "An Empirical Study of the Illinois Statutory Estate Plan." 1976 Ill. L.F. 717.

Fishbein, D. "Biological Perspectives in Criminology." 28 *Criminology* 27 (1990).

Fisher, H. *The Sex Contract: The Evolution of Human Behavior.* 1982. W. Morrow, New York.

Fisher, R. *The Genetical Theory of Natural Selection.* 1958. Dover Publications, New York.

Fletcher, D., and C. Michener. "Introductory Remarks" to *Kin Recognition in Animals.* 1987. Wiley, New York.

Flew, A. "From Is to Ought." 1967. Reprinted in *The Sociobiological Debate*, ed. A. Caplan. 1978. Harper & Row, New York.

Flinn, M. "Resources, Mating, and Kinship: The Behavioral Ecology of a Trinidadian Village." 1983. Ph.D. diss., Northwestern University.

Foley, R. "How Useful Is the Culture Concept in Early Hominid Studies?" In *The Origins of Behavior*, ed. R. Foley. 1991. Cambridge University Press, New York.

Fox, R., ed. *The Red Lamp of Incest.* 1980. E. P. Dutton, New York.

Frank, R. *Passions Within Reason.* 1989. W. W. Norton, New York.

Freedman, D. *Human Sociobiology.* 1979. Free Press, New York.

Galdikis, B. "Orangutan Adaptation at Tanjung Puting Reserve: Mating and Ecology." In *The Great Apes,* eds. D. Hamburg and E. McCown. 1979. Benjamin/Cummings Publishing Co., Menlo Park, Calif.

Gaulin, S., and A. Schlegal. "Paternal Confidence and Paternal Investment: A Cross-Cultural Test of a Sociobiological Hypothesis." 1 *Ethology and Sociobiology 301 (1980).*

Geovannoni, J. "Definitional Issues in Child Maltreatment." In *Child Maltreatment: Theory and Research on the Causes and Consequences of Child Abuse and Neglect,* eds. D. Cicchetti and V. Carlson. 1989. Cambridge University Press, New York.

Gibbs, J. *Crime, Punishment and Deterrence.* 1975. Elsevier, New York.

Gibbs, J., and M. Erickson. "Capital Punishment and the Deterrence Doctrine." In *Capital Punishment in the United States,* eds. H. Bedau and C. Pierce. 1976. AMS Press, New York.

Golding, M. "Criminal Sentencing: Some Philosophical Considerations." In *Justice and Punishment,* eds. J. Cederblom and W. Blizek. 1977. Ballinger Publishing Co., Cambridge.

Goldsmith, T. *The Biological Roots of Human Nature.* 1991. Oxford University Press, New York.

Gondos, B. "Oogenesis and Spermatogenesis." In *Handbook of Human Growth and Development Biology,* Vol. II: Part A, eds. E. Meisami and P. Timiras. 1989. CRC Press, Boca Raton, Fla.

Goodall, J. *The Chimpanzees of Gombe.* 1986. Belknap Press, Cambridge, Mass.

Gould, S. *Ever Since Darwin: Reflections in Natural History.* 1979. W. W. Norton, New York.

Grafen, A. "A Geometric View of Relatedness." In *Oxford Surveys in Evolutionary Biology,* vol. 2, eds. R. Dawkins and M. Ridley. 1985. Oxford University Press, New York.

Gray, J. *Primate Sociobiology.* 1985. HRAF Press, New Haven.

Gray, R. "The Menopause-Epidemiological and Demographic Considerations." In *The Menopause,* ed. R. Beard. 1976. University Park Press, Baltimore.

Green, W. "Depo-Provera, Castration, and the Probation of Rape Offenders: Statutory and Constitutional Issues." 12 *U. Dayton L. Rev.* 1 (1986).

Groebel, J., and R. Hinde, eds. *Aggression and War: Their Biological and Social Bases.* 1989. Cambridge University Press, New York.

Hafez, E. *Human Reproduction.* 2d ed. 1980. Harper & Row, Hagerstown, Md.

Hafez, E., and T. Evans. *Human Reproduction.* 1973. Harper & Row, Hagerstown, Md.

Hames, R. "Relatedness and Interaction among the Ye'kwana: A Preliminary Analysis." In *Evolutionary Biology and Human Social Behavior,* eds. N. Chagnon and W. Irons. 1979. Duxbury Press, North Scituate, Mass.

Hamilton, V., and L. Rotkin. "Interpreting the Eighth Amendment: Perceived Seriousness of Crime and Severity of Punishment." In *Capital Punishment in the United States*, eds. H. Bedau and C. Pierce. 1976. AMS Press, New York.

Hamilton, W. "The Evolution of Altruistic Behavior." 97 *Am. Naturalist* 354 (1963).

———. "The Genetical Evolution of Social Behavior, I & II." 7 *J. Theoret, Biol.* 1 (1964).

Harcourt, A. "The Social Relations and Group Structure of Wild Mountain Gorillas." In *The Great Apes*, eds. D. Hamburg and E. McCown. 1979. Benjamin/Cummings Publishing Co., Menlo Park, Calif.

———. "Alliances in Contests and Social Intelligence." In *Machiavellian Intelligence: Social Expertise and the Evolution of Intellect in Monkeys, Apes and Humans*, eds. R. Byrne and A. Whiten. 1988. Clarendon Press, Oxford.

Hausfater, G., and S. Hrdy, eds. *Infanticide: Comparative and Evolutionary Perspectives*. 1984. Aldine Publishing Co., New York.

Heim, N., and C. Hursch, "Castration for Sex Offenders: Treatment or Punishment? A Review and Critique of Recent European Literature." 8 *Archives Sexual Behavior* 281 (1979).

Hinde, R. *Individuals, Relationships and Culture*. 1987. Cambridge University Press, Cambridge.

———. "Patriotism: Is Kin Selection Both Necessary and Sufficient?" 8 *Pol. and Life Sci.* 58 (1989).

Hofstadter, R. *Social Darwinism in American Thought*. Rev. ed. 1959. G. Braziller, New York.

Hrdy, S. *The Woman That Never Evolved*. 1981. Harvard University Press, Cambridge, Mass.

———. "Assumptions and Evidence Regarding the Sexual Selection Hypothesis: A Reply to Boggess." In *Infanticide: Comparative and Evolutionary Perspectives*, eds. G. Hausfater and S. Hrdy. 1984. Aldine Publishing Co., New York.

Hughes, A. *Evolution and Human Kinship*. 1988. Oxford University Press, New York.

Irons, W. "Investment and Primary Social Dyads." In *Evolutionary Biology and Human Social Behavior*, eds. N. Chagnon and W. Irons. 1979. Duxbury Press, North Scituate, Mass.

———. "Kinship." In *Evolutionary Biology and Human Social Behavior*, eds. N. Chagnon and W. Irons. 1979. Duxbury Press, North Scituate, Mass.

———. "Natural Selection, Adaptation and Human Social Behavior," 29–31. In *Evolutionary Biology and Human Social Behavior*, eds. N. Chagnon and W. Irons. 1979. Duxbury Press, North Scituate, Mass.

———. "Incest: Why All the Fuss?" Paper presented at the Evolution and Human Behavior Meeting, University of Michigan, April 1986.

_____ . "Incest: Can It Be a Viable Strategy for Humans?" Paper presented at the 1989 meeting of the Animal Behavior Society at Northern Kentucky University.

_____ . "Let's Make Our Perspective Broader Rather Than Narrower." 11 *Ethology and Sociobiology* 361 (1990).

Jay, M., and S. Doganis, *Battered: The Abused Children.* 1987. St. Martin's Press, New York.

Jeffrey, C. *Crime Prevention Through Environmental Design.* 1971. Sage Publications, Beverly Hills.

Johnson, G. "Kin Selection, Socialization, and Patriotism: An Integrating Theory." 4 *Pol. and Life Sci.* 127 (1986).

_____ . "Some Thoughts on Human Extinction, Kin Recognition, and the Impact of Patriotism on Inclusive Fitness." 4 *Pol. and Life Sci.* 149 (1986).

_____ . "The Role of Kin Recognition Mechanisms in Patriotic Socialization: Further Reflections." 8 *Pol. and Life Sci.* 62 (1989).

Kahn, A., and S. Kamerman. *Child Support: From Debt Collection to Social Policy.* 1988. Sage Publications, Newbury Park, Calif.

Kaitz, M., A. Good, A. Rokem and A. Eidelman. "Mothers' Recognition of Their Newborns by Olfactory Cues." 20 *Developmental Psychobiology* 587 (1987).

Katz, S., and M. Mazur. *Understanding the Rape Victim, A Synthesis of Research Findings.* 1979. Wiley, New York.

Kelly, C. "Admissibility of DNA Evidence: Perfecting the 'Search for Truth'. " 25 *Wake Forest L. Rev.* 591 (1990).

Kelling, G., T. Pate, D. Dieckman and C. Brown. *The Kansas City Preventive Patrol Experiment: A Technical Report.* 1974. (Washington D.C.: Police Foundation).

_____ . *The Newark Foot Patrol Experiment.* 1981 (Washington D.C.: Police Foundation).

Kitcher, P. *Vaulting Ambition: Sociobiology and the Quest for Human Nature.* 1985. MIT Press, Cambridge, Mass.

Knapp, F., J. Sharkey and J. Metts. "The Effects of an Increase in a County Sheriff's Department's Patrol Capability." 8 *J. of Police Sci. and Admin.* 5 (1980).

Knudten, R. *Crime in a Complex Society.* 1970. Irwin-Dorsey, Homewood, Ill.

Konner, M. *The Tangled Wing: Biological Constraints on the Human Spirit.* 1982. Holt, Rinehart & Winston, New York.

Krause, H. *Child Support in America: The Legal Perspective.* 1981. Michie Co., Charlottesville, Va.

Kummer, H. "Tripartite Relations in Hamadryas Baboons." In *Machiavellian Intelligence: Social Expertise and the Evolution of Intellect in Monkeys, Apes and Humans,* eds. R. Byrne and A. Whiten. 1988. Clarendon Press, Oxford.

Kurland, P. "Paternity, Mother's Brother, and Human Sociality." In *Evolutionary Biology and Human Social Behavior*, eds. N. Chagnon and W. Irons. 1979. Duxbury Press, North Scituate, Mass.

Lansky, V. *Welcoming Your Second Baby*. 1984. Bantam, New York.

Leavitt, G. "Sociobiological Explanations of Incest Avoidance: A Critical Review of Evidential Claims." 92 *Am. Anthropologist* 971 (1990).

Lee, R., and I. DeVore. *Man the Hunter*. 1968. Aldine Publishing Co., Chicago.

Lenington, S. "Child Abuse: The Limits of Sociobiology." 2 *Ethology and Sociobiology* 17 (1981).

Lewontin, R., S. Rose and L. Kamin. *Not in Our Genes*. 1984. Pantheon Books, New York.

Lieberman, J. *Child Support in American: Practical Advice for Negotiating—and Collecting—a Fair Settlement*. 1986. Yale University Press, New Haven.

Lightcap, J., J. Kurland and R. Burgess. "Child Abuse: A Test of Some Predictions from Sociobiological Theory." 3 *Ethology and Sociobiology* 61 (1982).

Logan, C. "General Deterrent Effect of Imprisonment." 51 *Soc. Forces* 64 (1972).

Lopreato, J. *Human Nature and Biocultural Evolution*. 1984. Allen & Unwin, Boston.

Lumsden, C., and E. Wilson. *Promethean Fire: Reflections on the Origin of the Mind*. 1983. Harvard University Press, Cambridge, Mass.

Lynche, M. "Ill Health and Child Abuse." 2 *The Lancet* 317 (1975).

Malmberg, T. *Human Territoriality*. 1980. Mouton, New York.

Masters, R. "Is Sociobiology Reactionary? The Political Implications of Inclusive-Fitness Theory." 57 *Quarterly Rev. of Biology* 275 (1982).

———. "Evolutionary Biology and Political Theory." 84 *Am. Pol. Sci. Rev.* 195 (1990).

Maxwell, M. *Human Evolution: A Philosophical Anthropology*. 1984. Columbia University Press, New York.

———. *Morality among Nations: An Evolutionary View*. 1990. State University of New York Press, Albany.

———, ed. *The Sociobiological Imagination*. 1991. State University of New York Press, Albany.

Mayr, E. *Toward a New Philosophy of Biology*. 1988. Belknap Press, Cambridge, Mass.

McCabe, J. "FBD Marriage: Further Support for the Westermarck Hypothesis of the Incest Taboo?" 85 *Am. Anthropologist* 50 (1983).

McCahill, T., L. Meyer and A. Fischman. *The Aftermath of Rape*. 1979. Lexington Books, Lexington, Mass.

McFarlan, E., ed. *Guinness Book of World Records*. 1989. Guinness Superlatives, New York.

Morris, N. "Punishment and Prisons." In *Justice and Punishment*, eds. J. Cederblom and W. Blizek. 1977. Ballingter Publishing Co., Cambridge, Mass.

Morris, R. *Evolution and Human Nature*. 1984. Seaview/Putnam, New York.

Morsbach, G. "Maternal Recognition of Neonates' Cries in Japan." 23 *Psychologia: An International Journal of Psychology in the Orient* 23 (1980).

Murphy, J. *Evolution, Morality, and the Meaning of Life*. 1982. Rowman & Littlefield, Totowa, N.J.

Nesse, R. "Evolutionary Explanation of Emotions." 1 *Human Nature* 261 (1990).

————, A. Silverman and A. Bortz. "Sex Differences in Ability to Recognize Family Resemblance." 11 *Ethology and Sociobiology* 11 (1990).

Newton, P. "Infanticide in an Undisturbed Forest Population of Hanuman Langurs." 34 *Animal Behavior* 785 (1986).

Nishida, T., and M. Hiraiwa-Hasegawa. "Chimpanzees and Bonobos: Cooperative Relationships among Males." In *Primate Societies*, eds. B. Smuts, D. Cheney, R. Seyfarth, R. Wrangham and T. Struhsaker. 1986. University of Chicago Press, Chicago.

Nomura, F. "The Battered Child 'Syndrome,' A Review." 25 *Hawaii Med. J.* 387 (1966).

Olds, D., and C. Henderson, Jr. "The Prevention of Maltreatment." In *Child Maltreatment: Theory and Research on the Causes and Consequences of Child Abuse and Neglect*, eds. D. Cicchetti and V. Carlson. 1989. Cambridge University Press, Cambridge.

Ornstein, R., and P. Ehrlich. *New World New Mind*. 1989. Doubleday, New York.

Ortman, J. "The Treatment of Sexual Offenders." 3 *Int'l. J. L. Psychiatry* 443 (1980).

Parker, S. "Precultural Basis of the Incest Taboo: Toward a Biosocial Theory." 78 *Am. Anthropologist* 285 (1976).

Pastner, C. "The Westermarck Hypothesis and First Cousin Marriage: The Cultural Modification of Negative Sexual Imprinting." 42 *J. of Anthropological Res.* 573 (1986).

Pereira, M. "Abortion Following the Immigration of an Adult Male Baboon." 4 *Am. J. of Primatology* 93 (1983).

Pfeiffer, J. *The Emergence of Society*. 1977. McGraw & Hill, New York.

Plomin, R., J. DeFries and G. McClearn. *Behavioral Genetics: A Primer*. 1990. W. H. Freeman, New York.

Popp, J., and I. De Vore. "Aggressive Competition and Social Dominance Theory: Synopsis." In *The Great Apes*, eds. D. Hamburg and E. McCown. 1979. Benjamin/Cummings Publishing Co., Menlo Park, Calif.

Porter, R. "Kin Recognition: Functions and Mediating Mechanisms." In *Sociobiology and Psychology*, eds. C. Crawford, M. Smith and D. Krebs. 1987. L. Erlbaum Associates, Hillsdale, N.J.

Potts, R. "Home Base and Early Hominids." 72 *Am. Sci.* 338 (1984).

————— . "Reconstruction of Early Hominid Socioecology: A Critique of Primate Models." In *The Evolution of Human Behavior: Primate Models*, ed. W. Kinzey, 1987. State University of New York Press, Albany.

Press, J. *Some Effects of an Increase in Police Manpower in the 20th Precinct of New York (New York Rand Institute, October 1971)*.

Reynolds, V., S. Falger and I. Vine, eds. *The Sociobiology of Ethnocentrism.* 1986. University of Georgia Press, Athens.

Rijksen, H. *A Field Study of Sumatran Orangutans.* 1978. Veerman & Zonen, Wageningen.

Ruse, M. *Taking Darwin Seriously: A Naturalistic Approach to Philosophy.* 1986. Blackwell, New York.

Rushton, J. "Gene-Culture Coevolution and Genetic Similarity Theory: Implications for Ideology, Ethnic Nepotism, and Geopolitics." 4 *Pol. and Life Sci.* 144 (1986).

Russell, R., and P. Wells. "Estimating Paternity Confidence." 8 *Ethology and Sociobiology,* 215 (1987).

Scully, D. *Understanding Sexual Violence: A Study of Convicted Rapists.* 1990. Unwin Hyman, Boston.

Searles, P., and R. Berger. "The Current Status of Rape Reform Legislation: An Examination of State Statutes." 10 *Woman's Rights L. Rep.* 25 (1987).

Seemanova, E. "A Study of Children of Incestuous Matings." 21 *Human Heredity* 108 (1971).

Seymour, F., C. Duffy and A. Koerner. "A Case of Authenticated Fertility in a Man, Aged 94." 105 *J.A.M.A.* 1423 (1935).

Shaw, R., and Y. Wong. *The Genetic Seeds of Warfare.* 1989. Unwin Hyman, Boston.

Shepher, J. "Mate Selection among Second-Generation Kibbutz Adolescents and Adults: Incest Avoidance and Negative Imprinting." 1 *Archives Sexual Behavior* 293 (1971).

————— . *Incest: A Biosocial View.* 1983. Academic Press, New York.

Sherman, P., and W. Holmes. "Kin Recognition: Issues and Evidence." In *Experimental Behavioral Ecology and Sociobiology. In Memoriam Karl von Frisch 1886–1982*, eds. B. Holldobler and M. Lindauer. 1985. Sinauer, Sunderland, Mass.

Shields, W., and L. Shields. "Forcible Rape: An Evolutionary Perspective." 4 *Ethology and Sociobiology* 115 (1983)

Singer, P. *The Expanding Circle.* 1981. Farrar, Straus & Giroux, New York.

Smuts, B. "Gender, Aggression and Influence." In *Primate Societies*, eds. B. Smuts, D. Cheney, R. Seyfarth, R. Wrangham and T. Struhsaker. 1986. University of Chicago Press, Chicago.

————— . "Psychological Adaptations, Development and Individual Differences." 15 *Behavioral and Brain Sci. 401 (1992)*.

Sociobiology Study Group of Science for the People, "Sociobiology—Another Biological Determinism." 1976. Reprinted in *The Sociobiological Debate*, ed. A. Caplan. 1978. Harper & Row, New York.

Sorenson, A., and M. MacDonald. "An Analysis of Child-Support Transfers." In *The Parental Child-Support Obligation*, ed. J. Cassetty. 1983. Lexington Books, Lexington, Mass.

Spiro, M. *Children of the Kibbutz*. 1958. Schocken Books, New York.

Statistical Abstract of the U.S. (1990).

Steele, B., and C. Pollock. "A Psychiatric Study of Parents Who Abuse Infants and Small Children." In *The Battered Child*, eds. R. Helfer and C. Kempe. 1974. University of Chicago Press, Chicago.

Stephens, W. *The Family in Cross-Cultural Perspective*. 1963. Holt, Rinehart & Winston, New York.

Sterelny, K., and P. Kitcher. "The Return of the Gene." 85 *J. of Philosophy* 339 (1988).

Stewart, K., and A. Harcourt. "Gorillas: Variation in Female Relationships." In *Primate Societies*, eds. B. Smuts, D. Cheney, R. Seyfarth, R. Wrangham and T. Struhsaker. 1986. University of Chicago Press, Chicago.

Strahlendorf, P. "Evolutionary Jurisprudence: Darwinian Theory in Juridical Science." Draft S.J.D. thesis, University of Toronto, 1991.

Stringer, C., and P. Andrews. "Genetic and Fossil Evidence for the Origin of Modern Humans." 239 *Science* 1263 (1988).

Strum, S., and W. Mitchell. "Baboon Models and Muddles." In *The Evolution of Human Behavior: Primate Models*, ed. W. Kinzey. 1987. State University of New York Press, Albany.

Sussman, M., J. Cates and D. Smith. *The Family and Inheritance*. 1970. Russell Sage Foundation, New York.

Symons, D. *The Evolution of Human Sexuality*. 1979. Oxford University Press, New York.

———. "The Evolution of Human Sexuality Revisited." 3 *Behavioral and Brain Sci.* 203 (1980).

———. "The Evolutionary Approach: Can Darwin's View of Life Shed Light on Human Sexuality?" In *Theories of Human Sexuality*, eds. J. Greer and W. O'Donohue. 1987. Plenum Press, New York.

———. "Adaptiveness and Adaptation." 11 *Ethology and Sociobiology* 427 (1990).

Tanner, N. "The Chimpanzee Model Revisited and the Gathering Hypothesis." In *The Evolution of Human Behavior: Primate Models*, ed. W. Kinzey. 1987. State University of New York Press, Albany.

Taylor, C., and M. McGuire, eds. *Reciprocal Altruism: 15 Years Later*. Vol. 9, Nos. 2–4, *Ethology and Sociobiology* (1988).

Thomas, C., and S. Foster. "A Sociobiological Perspective on Public Support for Capital Punishment." In *Capital Punishment in the United States*, eds. H. Bedau and C. Pierce. 1976. AMS Press, New York.

Thomas, M. "Child Abuse and Neglect Part I: Historical Overview, Legal Matrix, and Social Perspective." 50 *N.C. Law Rev.* 293 (1972).

Thompson, W., and S. Ford. "DNA Typing: Acceptance and Weight of the New Genetic Identification Tests." 75 *Vir. L. Rev.* 45 (1989).

Thornhill, N. "The Evolutionary Significance of Incest Rules." 11 *Ethology and Sociobiology* 113 (1990).

Thornhill, N., and R. Thornhill. "Evolutionary Theory and Rules of Mating and Marriage Pertaining to Relatives." In *Sociobiology and Psychology*, eds. C. Crawford, M. Smith and D. Krebs. 1987. L. Erlbaum Associates, Hillsdale, N.J.

_____ . "An Evolutionary Analysis of Psychological Pain Following Rape: I. The Effect of Victim's Age and Marital Status." 11 *Ethology and Sociobiology* 155 (1990).

_____ . "An Evolutionary Analysis of Psychological Pain Following Rape: II. The Effect of Stranger, Friend and Family-Member Offenders." 11 *Ethology and Sociobiology* 177 (1990).

Thornhill, R., and N. Thornhill. "Human Rape: An Evolutionary Analysis." 4 *Ethology and Sociobiology* 137 (1983).

_____ . "Human Rape: The Strengths of the Evolutionary Perspective." In *Sociobiology and Psychology*, eds. C. Crawford, M. Smith and D. Krebs. 1987. L. Erlbaum Associates, Hillsdale, N.J.

_____ . "The Evolution of Psychological Pain." In *Sociobiology and the Social Sciences*, eds. R. Bell and N. Bell. 1989. Texas Tech University Press, Lubbock, Tex.

_____ . "The Evolutionary Psychology of Men's Coercive Sexuality." 15 *Behavioral and Brain Sci.* 363 (1992).

Tiger, L., and R. Fox. *The Imperial Animal.* Owl Book edition, 1989. McClelland & Stewart, Toronto.

Tittle, R. "Crime Rates and Legal Sanctions." 16 *Soc. Probs.* 409 (1969).

Tooby, J., and L. Cosmides. "Evolutionary Psychology and the Generation of Culture, Part I." 10 *Ethology and Sociobiology* 29 (1989).

_____ . "The Past Explains the Present: Emotional Adaptations and the Structure of Ancestral Environments." 11 *Ethology and Sociobiology* 375 (1990).

Tooby, J., and I. De Vore. "The Reconstruction of Hominid Behavioral Evolution through Strategic Modeling." In *The Evolution of Human Behavior: Primate Models*, ed. W. Kinzey. 1987. State University of New York Press, Albany.

Trepper, S., and M. Barrett. *Systematic Treatment of Incest, A Therapeutic Handbook.* 1989. Brunner/Mazel, New York.

Trivers, R. "The Evolution of Reciprocal Altruism." 1971. Reprinted in *Readings in Sociobiology*, eds. T. Clutton-Brock and P. Harvey. 1978. W. H. Freeman, San Francisco.

_____ . *Social Evolution.* 1985. Benjamin/Cummings Publishing Co., Menlo Park, Calif.

Trojanowicz, R. "An Evaluation of a Neighborhood Foot Patrol Program." 11 *J. of Police Sci. and Admin.* 410 (1983).

Turke, P. "Just Do It." 11 *Ethology and Sociobiology* 445 (1990).

Uniform Probate Code (U.S.) (6th ed. Official 1982 Text).

Uniform Probate Code (U.S.) (10th ed. Official 1991 Text).

U.S. Department of Justice. *Sourcebook of Criminal Justice Statistics.* 1988. National Criminal Justice Info. and Stats. Washington, D.C.

van den Berghe, P. *Human Family Systems: An Evolutionary View.* 1979. Elsevier, New York.

_____ . *The Ethnic Phenomenon.* 1981. Elsevier, New York.

_____ . "Human Inbreeding Avoidance: Culture in Nature." 6 *Behavioral and Brain Sci.* 91 (1983).

_____ . "Kin, Ethnicity, Class, and the State: Of Consciousness of Kind, True and False." 4 *Pol. and Life Sci.* 142 (1986).

_____ . "Incest Taboos and Avoidance: Some African Applications." In *Sociobiology and Psychology*, eds. C. Crawford, M. Smith and D. Krebs. 1987. L. Eralbaum Associates, Hillsdale, N.J.

Weiss, J. *Your Second Child.* 1981. Summit Books, New York.

Wells, P. "Kin Recognition in Humans." In *Kin Recognition in Animals*, eds. D. Fletcher and C. Michener. 1987. Wiley, New York.

West Eberhard, M. "The Evolution of Social Behavior by Kin Selection." 50 *Q. Rev. Biology* 1 (1975).

Westermarck, E. *3 The History of Human Marriage.* 1925. Macmillan and Co., London.

Williams, G. *Adaptation and Natural Selection.* 1963. Princeton University Press, Princeton, N.J.

_____ , ed. *Group Selection.* 1971. Aldine/Atherton, Chicago.

Wilson, E. *On Human Nature.* 1978. Harvard University Press, Cambridge, Mass.

_____ . "The Relation of Science to Theology." 15 *Zygon* 425 (1981).

_____ . *Biophilia.* 1984. Harvard University Press, Cambridge, Mass.

Wolf, A., and C. Huang. *Marriage and Adoption in China, 1845–1945.* 1980. Stanford University Press, Stanford, Calif.

Wrangham, R. "Mutualism, Kinship and Social Evolution." In *Current Problems in Sociobiology*, ed. King's College Sociobiology Group, Cambridge. 1982. Cambridge University Press, Cambridge.

_____ . "The Significance of African Apes for Reconstructing Human Social Evolution." In *The Evolution of Human Behavior: Primate Models*, ed. W. Kinzey. 1987. State University of New York Press, Albany.

Young, R. "Meaning of 'Issue' and 'Descendants'." 13 *American College of Probate Counsel Probate Notes* 225 (1988).

Zimring, F., and G. Hawkins. *Deterrence.* 1973. University of Chicago Press, Chicago.

Index

Why a Gay Person
Can't Be Made
Un-Gay

WHY A GAY PERSON CAN'T BE MADE UN-GAY

❖

The Truth About Reparative Therapies

Martin Kantor, MD

 PRAEGER

AN IMPRINT OF ABC-CLIO, LLC
Santa Barbara, California • Denver, Colorado • Oxford, England

Library of Congress Cataloging-in-Publication Data

Kantor, Martin.
  Why a gay person can't be made un-gay : the truth about reparative therapies / Martin Kantor, MD.
    pages cm
  Includes bibliographical references and index.
  ISBN 978-1-4408-3074-7 (alk. paper) — ISBN 978-1-4408-3075-4 (ebook)
  1. Homosexuality—Treatment—Evaluation.   2. Sexual reorientation programs—Evaluation.   3. Gays—Mental health services—Evaluation.   I. Title.
  RC558.K36   2015
  616.85'83—dc23   2014024280

ISBN: 978-1-4408-3074-7
EISBN: 978-1-4408-3075-4

19 18 17 16 15   1 2 3 4 5

This book is also available on the World Wide Web as an eBook.
Visit www.abc-clio.com for details.

Praeger
An Imprint of ABC-CLIO, LLC

ABC-CLIO, LLC
130 Cremona Drive, P.O. Box 1911
Santa Barbara, California 93116-1911

This book is printed on acid-free paper ∞

Manufactured in the United States of America

For Michael

"This is our great obligation: To enable the human animal to accept nature within himself, to stop running away from it and to enjoy what now he dreads so much."[1]

Contents

Introduction

Even today, gays still put themselves in harm's way by seeking psychotherapy to turn straight. They apply to semiprofessional and professional practitioners who call themselves variously reparative therapists, SOCE ("sexual orientation change efforts") therapists, ex-gay therapists, or conversion therapists, all who have one thing in common: viewing gays as broken, they set out to fix them. However, instead of making them straight, they waste their time, squander their money, and cause them to lose out on some of the best years of their lives. For these therapists are the rough equivalent of modern-day alchemists. Promising what they can't deliver, they offer to turn gay dross into heterosexual gold. But as Besen quoting Troy Perry, founder of the Metropolitan Community Church, notes: "Not one person who [goes] through one of these programs [doesn't] end up coming back home."[1] Because sooner or later gays discover that being gay is inborn and immutable; that being gay is nonpathological and normative; that being gay is moral not sinful and requires no absolution; and that being gay is not criminal and needs no punishment.

Laymen and professionals alike believe that reparative therapy (the term I prefer for this group of therapeutic interventions), having been finally discredited, is a thing of the past. All concerned respond incredulously when I tell them that, "Yes, even today they are still attempting to cure homosexuality psychotherapeutically." Just recently I asked a local shopkeeper in the gay-oriented town where I live if he knew of anyone

who had undergone ex-gay therapy and might agree to be interviewed. The other patrons in the shop overheard me, and with one voice naively responded, "Good thing they don't do that anymore!"

Even those who recognize that reparative therapy is, no relic of a bygone era and still practiced today, know little about its theory and practice. In fact, more than a few continue to recommend it, both in the psychiatric literature and in the lay press, where they advocate it, if not for all gays, then for those who are unhappy about being homosexual. Even though the high-profile recent defection of Exodus, a major arm of the ex-gay movement, was an abjectly apologetic one full of self-recrimination, it has still, according to The Baptist Press,[2] failed to convince everyone that the ex-gay industry is not fully reputable.

And the methods and practices of therapists currently doing ex-gay therapy are very much the same today as they were yesterday. Worse, Exodus's exit seems to have merely allowed other extant ex-gay "trees" to grow stronger in the enhanced light that newly streams down in the recently denuded ex-gay forest. As a result, those reparative therapists who remain behind still practicing their "craft" have become more powerful, more organized, more alert to the nature and consequences of their errors, and better able to hide the mistakes they make. For example, few reparative therapy organizations these days fall into the trap of retaining a spokesperson who, after publicly claiming he is cured, goes on to put himself in the compromising position of getting caught cruising the local gay bars and haunting the giveaway gay trysting places. So, even after the recent closing of Exodus, coupled with sincere apologies for the lives ruined, there are still many, or even more, therapists doing what Exodus people did—except that they have now become ever more sophisticated and savvy in explaining themselves, in warding off criticism, and in protecting themselves from the forces that "seek to oppress" them and suppress, or outright ban, what they do. Sophisticated spin such as "we no longer speak of cure, we speak of palliation" has taken over. By better convincing the disbelievers, they threaten to entice even more innocent gays and lesbians into buying into their propaganda. As a result, more than ever before, they jeopardize the well-being of even more already perfectly normal individuals simply because they just happen to be homosexual.

As with perhaps no other form of therapy, emotional reasoning drives reparative therapy's "scientific" theory and practice. All concerned allow hope, irrationally springing eternal in their breasts, to sway their judgment, as they buy into exactly what they wish to believe: that reparative therapy is a viable form of treatment for those who want to go straight.

So they continue to seduce gays into at best participating in nothing, and at worst exposing themselves to damage by a form of psychotherapy that is at the very least ineffective, and at its very worst seriously harmful.

Although in this book I set forth a firmly antireparative therapy agenda, I nevertheless recognize that, with the numerous exceptions outlined throughout the text, reparative therapy, like almost any other form of therapy, is not entirely bad, or always completely harmful. Rather, even though routinely poorly executed and predictably done to little or no avail, it can nevertheless do some good for some people. For the goal of making gays ungay and lesbians unlesbian is in and of itself at least theoretically not an entirely irrational one. Only it is mostly unachievable, and this is in part because it uses methods that are so untraditional that they make a mockery of the basic tenets of traditional psychiatric intervention. For reparative therapy, unlike most other traditional therapies, is generally unsupportive; and it is disrespectful of the patients' true wishes, for rather than its practitioners acting the part of the proverbial therapeutic blank screen, reparative therapists push a distinct ideological agenda on those most vulnerable to its influence. Misusing mainstream psychiatric methods, reparative therapists distort them to become a form of wild psychoanalysis as they apply psychoanalytic principles excessively, and when inappropriate. They also use behavioral interventions that are too trivial to work at all, and methods that work only in vitro (in the lab) but are ineffective in vivo (once the subject leaves the lab and enters the real world). To me, the most damaging aspect of reparative therapy is that it doesn't just bastardize some of psychotherapy's basic principles. Rather it retains them but uses them where, and in such a way, that they don't apply. Thus reparative therapy not only makes a mockery of psychiatry's oaths to respect patients' wishes, to help patients self-realize to "be me and to be free," to enhance mental health rather than to damage it in the therapeutic pursuit of curing some imaginary mental illness, and to retain the Holy Grail of therapeutic neutrality, but it also uses good techniques (such as developing insight) in a bad cause. Sometimes its goals are admirable; sometimes its objectives are achievable; and sometimes its methods are rational. But these things never seem to go together. For, so to speak, it's as if having discovered the value of peeling a banana, reparative therapists now use the exact same technique but in a way that is completely inapplicable to the task at hand: in order to peel a grape.

This is the only *psychiatric* book I know of on the subject of reparative therapy written from the *negative* perspective. Beson's thorough exposé of reparative theory is written not by a psychiatrist but by a reporter, and so while his excellent book effectively condemns the practice, and makes for

easy reading and humorous, when not frightening, exposition, it doesn't deal with the serious structural problems associated with the actual techniques of reparative therapy, problems that make it such a flawed scientific enterprise. The few other books that exist on the subject, in particular those by Joseph Nicolosi and Gerard van den Aardweg, are written by (Christian) men who are unabashedly pro-reparative therapy. Nicolosi is not only a practicing reparative therapist, he is also a founder and past president of National Association of Research and Therapy of Homosexuality (NARTH) (described in the text). He seems to be a generally honest, hardworking, and sincere doctor who nevertheless in my opinion uses a rational analytically oriented approach for irrational reasons, in all the wrong places, and in all the wrong ways, with a one-size-fits-all mentality coupled by timing that sometimes seems to be more off than in sync with his patients' interventional needs and treatment readiness. And worse, though a therapist treating homosexuals, he is, like the fireman who turns out to be the arsonist, when it comes to homosexuals and homosexuality, a man with a truly dark homophobic side, consisting of a homohating attitude and philosophy that contaminates his supposedly sanguine approach to helping gays. This side of him emerges in homonegative rants that I go on to quote in this text, thus making much of what he does less of a cure for, than a brutalization of, gays.

And van den Aardweg, the author of *The Battle for Normality*, one of the nastiest books about gays ever written, presents an approach heavily reflective of that unfortunate convergence (not infrequently found in reparative therapists) of irrational dogma, blind faith in themselves, full acceptance of the non- or pseudoscientific hypotheses of the uncritical believer, the missteps of the anarchistic (and sometimes poorly trained) psychotherapist who is at the same time scrupulously far to the right in his or her religious beliefs, and the blatant homophobic orientation of the un-healer whose proclaimed theoretical posture is not so much *ex-* as it is *antigay*—all mixed in with some surprisingly brilliant insights that impel one to read on, but ultimately only at one's peril. His book, which I recommend reading (but for all the wrong reasons, e.g., to be exposed to the many myths reparative therapists subscribe to and prejudices reparative therapists advance about gays), not only propounds wild (often psychodynamic-type) theories but at the same time bashes those the theories are meant to help, and in a most unsatisfying, insulting, and ultimately ridiculous manner.

In too many cases reparative therapists' beliefs and practices consist of self-serving ideology where postulated theory is but a handmaiden of desired outcome. In particular, when reparative therapists aver that homosexuality is not an inborn sexual preference but is rather an acquired

behavior, they seem to be doing so for a specific reason: so that they can tout how they can cure it much as they can cure a phobia or an obsession—through psychotherapy, which just happens to be the bread and butter of what they do (for a living). There are, it must be admitted, some exceptions to the rule that "homosexuality cannot be cured." Bisexuals can at least theoretically make a choice, and that choice can be influenced by psychotherapy. But otherwise the goal of treating ego-dystonic (unwelcome) homosexuality psychotherapeutically cannot be reparation. It must rather be, if anything, stabilization—through helping gay patients joyfully accept being gay—in the recognition that they don't have a sickness that needs to be cured, are not enduring a lifestyle that needs to be changed, and don't need to become something that they are not.

After discussing reparative therapy, and why mostly psychiatric practitioners ought not to even try it, I describe a preferred alternate form of therapy for gays: affirmative eclectic psychotherapy, which in my opinion is the treatment of choice for gays who feel the need to enter treatment to make non-ex-gay changes. Here the therapeutic goal shifts from gays' need to change from gay to straight to gays' need to understand and repair not being gay but their irrational desire to go straight, which too often is itself a sign of emotional problems that can in their turn make gay life harder than it is and has to be. For emotional problems do not discriminate based on sexual orientation. And so to this end reparative therapists ought to explore how, though being homosexual can alter the superficial presentation of an emotional disorder, gays and lesbians fundamentally suffer from the same emotional disorders as anyone else, and it is at these that a reparative therapist might be advised to look. Thus I recommend reparative therapists do an about-face and explore the general psychology/psychopathology of the gay men and lesbians who apply to them for treatment: their guilt, their regrets, their vulnerabilities to inner and social criticism, and their negative, self-destructive attitudes and actions that turn what they are into a conviction that they must become something else. Put another way, for gays and lesbians what often needs to be repaired is not their being gay and broken, but their feeling broken because they are gay, and so their desire to fix something that is already intact, a desire that is less authentic than it is symbolic—of a general problematic (neurotic) mindset. Those who believe that sexual orientation needs to be changed should consider the possibility that, instead, emotional dysphoria needs to be lifted. What gays might be encouraged to do is to discuss any and all aspects of their lives both related and unrelated to their being gay. And if it is determined that they need therapy, it should be the same mainstream therapy, with an affirmative emphasis, that is

suitable for anyone else, straight or gay, potentially wanting or needing psychiatric treatment.

The internal inconsistencies of reparative theory and practice as well as its varieties make any attempt to fully describe the approach a difficult to impossible task. For example, reparative therapists who overall insist that homosexuality is acquired nevertheless assert that one of its causes, temperament, is inborn (and so of biological origin). Thus many reparative therapists believe that a sensitive temperament (which is of genetic origin) explains the pre-homosexual child's proneness to respond to negative parental input. Yet overall their "science" is otherwise focused almost entirely on advancing developmental theories that (it would seem) entirely preclude the possibility that gays are born that way, in any conceivable manner. At odds with one another are the different schools of thought that go into making up the panoply of extant reparative therapy practices, so that no one-size-fits-all definition of what those who are part of the reparative therapy industry do can ever exist. Calling themselves "reparative therapists" are therapists from all schools of thought, ranging among the psychoanalytic, interpersonal, cognitive-behavioral, religious, and biological (pharmacotherapeutic) disciplines. Additionally, working under the reparative therapy rubric are not only therapists of different theoretical psychotherapeutic persuasions but also therapists of various personal profiles ranging from the sincere to the opportunistic, from the straight to the newly ex-gay (who are at times recloseted gays changing others in an attempt to change themselves); from the insightful to the unknowledgeable; from the orthodox to the flexible; from the humanitarian to the cruel (for they advocate great suffering now for some theoretical and generally elusive future good); from the permissive and respectful to the coercive and devaluing; and from the wishful to the realistic. As their methods differ, so do their goals, ranging from complete cure ("pray the gay *away*") to the altering of superficial behavior by encouraging gays to be celibate without eviscerating all (or any) of their deep sexual fantasies. Therefore, any description of reparative therapy has to be specific to the individual therapist, not generalized to the practice itself.

Reparative therapy is often religiously oriented. Patients tend to seek cure because they view being gay/lesbian as being both at odds with their personal religious beliefs and as being authentically anti-God. Religious counselors may urge gays to use their religion to get better. Religious therapists, flaunting the basic rule of mainstream psychotherapy, departing from the path of therapeutic neutrality, may encourage gays not to self-realize and embark on a quest for self-fulfillment, but to go straight—and not by exploring their mental health issues but by uncovering their

sins and paying for their crimes—to cleanse their souls of the evil that supposedly lives now, and that will surely live after them.

While this text is written both for and about gay men wanting, or being told, to go straight and lesbians wanting, or being told, exactly the same thing, in reality reparative therapy is prevailingly employed by white therapists in order to treat white, often religious, Christian gay men. This is the case for at least several reasons. First, a woman's sexuality seems to be less the subject of approbation and therefore the need for change less urgent than a man's. Second, because women are devalued when compared to men; determining what they do wrong in bed and finding ways to make it right become of less interest/importance than doing the same thing for men. And third, a religious cloak in many ways hangs like the sword of Damocles over the entire reparative industry, and religion, often used as a tool of reparation, is, it would sometimes seem, in an all-too-common practice, a man thing.

In an attempt to be fair and balanced, I do present some of the upsides of reparative therapy. Unfortunately most of these fall into the category of unintended (positive) consequences. No gay individual should plan in advance to enter reparative therapy to turn it into a helpful experience. Help, however, might come about by chance, that is, serendipitously, for some emotionally strong gays stuck in reparative therapy have been able to turn it to their advantage. But affirmative therapeutic approaches constitute better, more direct, ways to get needed help without unnecessarily expending time and effort hoping that by accident one will have stumbled across some sort of magic. One should never attempt to make reparative therapy, a red light of a therapeutic procedure, into a shining beacon—a green light meant to lead the therapeutic way.

Part I

Description

1

An Historical Overview

In this chapter I briefly review some highlights of the history of reparative therapy. At least the idea of reparative therapy is not new. Thus Jones quoted a letter that Freud wrote in 1935 to a mother of a gay son, setting forth his assessment of what today we might call "psychoanalytic Christian reparative therapy." The letter said, when it comes to changing homosexuality to heterosexuality "in the majority of cases it is [not] possible."[1]

Ross writing in the *New Yorker* in an article entitled "Love on the March" says that many "leading psychiatrists, abandoning Freud's relatively nonjudgmental position [soon began to] describe . . . homosexuals as 'sexual psychopaths.' There were experiments in electric and pharmacological shock treatment"[2] in an attempt to decondition the homosexual responses. Some practitioners gave cocaine and strychnine injections (the latter amounting to the spiritual equivalent of chemical waterboarding) while their patients viewed pictures of men engaged in homosexual acts. (These practitioners overlooked that a reaction in response to a pornographic picture is a superficial indicator of same-sex, or any form of sexual, attraction, since much that is "sexual attraction" is not strictly physical but is rather also mental, that is, aroused by what other individuals say and do that makes them attractive, and so constitutes a response to one or more qualities other than appearance.) Others recommended "hormone injection, [chemical and physical] castration, and lobotomy."[3] Some clinicians were even tempted to try transplanting sex organs into humans simply because some experimenters had tried it in neutered rats. According

to Besen some therapists (like Nunberg) virtually advocated male circumcision in the belief that fewer Jews than non-Jews were gay.[4] I have heard that for lesbians some recommended liberation of adherent clitorises. Some believing that homosexuality was due to nervous exhaustion recommended curative exercise, for example, bicycle riding. Others recommended attending a brothel. (A patient of mine simultaneously in therapy with me and another therapist, a reparative therapist, tried a therapeutic home visit from a prostitute—but any hope he may have had for a cure faded because he was chagrined that he had no cash and so could only offer to pay by check, which, as he feared, would be readily traceable back to him.)

I have heard that still other reparative therapists considered preventive intervention upon the growing fetus (they believed that changing the fetus's intrauterine hormonal bath might change its sexual orientation). Some suggested, and still suggest, overdosing on homosexuality (flooding). Many recommended (and still recommend) prayer, with for them "praying the gay away"—less a psychological technique than an actual appeal to God Himself hoping He might intervene—not an attempt to harness the power of suggestion but an attempt to obtain the actual intervention of a kindly, concerned, helpful Almighty. Some developed their methods after reading books on psychotherapy and/or receiving formal training in its art and adapting its procedures to treating and curing the "syndrome of homosexuality" which they saw as a form of insanity. Others developed their methods more implicitly: out of their religious beliefs and tenets leading to the use of religiously-oriented theory and procedure that in some ways resembled the laying on of hands and in other ways reminded of a miracle cure—with the therapist himself or herself a Savior, perhaps assisted by the therapeutic intervention of a Higher Power. Some used to recommend (and still recommend) abstinence (celibacy), which they viewed not (as it actually is) as helping the patient develop a new problem, but as coming up with a new and desirable solution to an old one (the "problem of homosexuality").

Today, Freudian mainstream concepts are still being regularly and steadfastly applied, or misapplied, in an attempt to *analyze* the gay away—with homosexuality equated with such mental illnesses as borderline personality disorder, and to be treated, like these illnesses, with a mostly unmodified form of psychoanalytically-oriented psychotherapy/psychoanalysis. It is as if there is still some enduring truth in David Reuben's 1969 assertion as claimed in his popular book *Everything You Always Wanted to Know About Sex (But Were Afraid to Ask)*, "If a homosexual who wants to renounce homosexuality finds a psychiatrist who knows how to cure

homosexuality, he has every chance of becoming a happy, well-adjusted heterosexual."[5] Some psychoanalytically oriented therapists feel that all homosexuals are borderline psychotic or outright psychotic so that, as Isay tells us, "homosexuals have been said to suffer from a large variety of ego defects . . . including 'primitive features of the ego' similar to those found in schizophrenia . . . and sociopathy."[6] Some agree with Socarides who, according to Isay, believed that "half of the patients who engage in homosexual practices have a 'concomitant schizophrenia, paranoia, are latent or pseudoneurotic schizophrenics or are in the throes of a manic-depressive reaction. The other half, when neurotic, may be of the obsessional or, occasionally, of the phobic type. . . . Most of the patients he labeled as schizophrenic would probably be classified in his later formulation (1978) as belonging to the class of Preoedipal Type #2, 'Suffering from a transitional condition lying somewhere between the neuroses and psychoses.'"[7] Such therapists then recommended "the appropriate treatment" for gays' supposed "severe underlying disorders," often and mostly unmodified but sometimes an adapted form of psychoanalytically oriented psychotherapy/psychoanalysis.

Bieber and colleagues believed that he had determined the psychological underpinnings of homosexuality and so could cure it via a causally oriented psychoanalytic method. However, according to Besen, Bieber interviewed not the gay patients themselves but those treating them, all of whom had learned from each other and were passing on what they had heard,[8] resulting perhaps in a kind of skinny-based group psychosis where all concerned shared the belief that being gay was clearly a developmental problem, in the main one due to specific types of faulty parenting, and that being gay could presumably be reversed by a process central to which was discovering exactly why the problem had arisen in the first place. For example, by wresting memories of early trauma out of repression and revealing the trauma to the patient then determining (and taking measures to reverse) its exact effects, psychoanalysts could presumably deal with/reverse the impact of the trauma and so its sequelae precisely responsible for a boy becoming gay/a girl becoming a lesbian. (How the mysterious leap between developing intellectual insight and changing sexual orientation actually occurred was never, and still has not, been fully explained.) Santayana said "those who cannot remember the past are condemned to repeat it."[9] But few traditional analytically oriented therapists say with any degree of conviction that "those who do remember and understand the past may still be doomed to repeat it if the past has been significantly traumatic." That's because past traumas generally take permanent hold, and so an irreversible toll, on the present, and therefore cannot easily be

"unseen," no matter what one does or how hard one tries to put the trauma out of one's mind.

As Isay put it in 1993, "Because psychoanalysts generally regard homosexuality as a pathological and a psychologically uneconomical solution to early conflict, they tend to believe that, whenever possible, it is in the best interest of the patient to change his sexual orientation or sexual behavior [by bringing] warring intrapsychic structures . . . into greater harmony. [Gay men in harmony become] less inclined to act out 'unacceptable impulses' as the[se] become increasingly tolerated by a strengthened ego and successfully sublimated. [Furthermore, the] literature is replete with recommendations in support of modifications of analytic technique that are deemed appropriate to the treatment of homosexual patients [such as] suggesting to a homosexual patient that he seek out women or discussing with a patient how to engage in heterosexual sex,"[10] as well as behavioral (aversive) and inspirational (religious-oriented) treatments.

(Recommendations on how to modify analytic technique in order to be able to cure homosexuality are found scattered throughout the psychoanalytic literature in the form of "parameters" involving breaking away from the strict development of insight to the recommending of action-oriented treatments such as a man's having trial sex with a woman. Such recommendations typically focusing on sex overlook the broader picture including issues related to desired lifestyle, especially the comfort gays take in same-sex *platonic* relationships/shared activities.)

Ultimately, as Pinsker notes in a personal communication in September 2013 referring to psychotherapy in earlier times, "back in the days when we thought psychotherapy could deal with everything, I had patients who wanted to be free of their homosexual desires." Then he adds, "We failed to meet that goal."

Nicolosi still believes in, and is currently a leading practitioner of, psychoanalytically-oriented reparative psychotherapy. In his books, among the few full texts written on the subject (many papers, often anecdotal or otherwise less than scientific, exist about reparative therapy or its equivalent), Nicolosi in essence describes at great length a method for treating homosexuality based on the assumption that being homosexual having been acquired developmentally and along lines that can be explained psychodynamically can be analyzed like any other acquired psychological symptom such as a phobia, or an obsession, and thus "de-acquired" predominantly through developing insight into its causation. In what seems to be his attempt to bridge the gap between mind and body, Nicolosi throughout his texts focuses on raw feelings occurring in the here and now,[12] perhaps because, as he sees it, emotions have both a cerebral and a

somatic component, and so, being a two-in-one compound, can presumably function as gateway phenomena opening up the door to bridging the gap and fording the mysterious leap between gay mind (what makes one think queer) and gay body (what makes one feel and act queer). (Nicolosi helped found and was at one time a leader in NARTH—the National Association of Research and Therapy of Homosexuality—an organization devoted to curing homosexuals of their supposed affliction. A NARTH Web site[13] details this organization's philosophy and defends itself against its critics.)

I myself, trying to go straight, partly out of a (misplaced) sense of profound guilt about being gay, partly because at the time a distinguished professor of Psychiatry at a large urban hospital where I applied for a position was reluctant to have me as a gay man on the staff, and partly because I wanted to be an analyst but the Boston Psychoanalytic Institute I aspired to join wouldn't accept gays (because they believed that we were too psychologically damaged to be able to treat others with psychological problems), underwent an extensive five-year psychoanalysis that in spite of the high level and sincerity of my motivation and freely avowed conscious need and desire to convert didn't touch my sexuality one bit. What I seemed not to have appreciated was this: according to the Boston Psychoanalytic Institute, ex-gay wouldn't have been good enough for them anyway; for once gay, always broken. So no matter what I went on to become I remained damaged goods, unable to ever myself do psychoanalysis because, as my analyst (thinking of herself as sparing me criticism for being incompetent) suggested, "I couldn't handle the stress." (As she in essence so unhelpfully opined, "You should spare your sensibilities and emotions pain and look elsewhere for your career.") As a result of this analysis, instead of being a practicing homosexual I became a practicing sexually abstinent relationally deprived alcoholic. Unfortunately, the sequelae of this dark period of my life reverberated for years after my psychoanalysis was over: until I went back into therapy to be reworked/reprogrammed by a therapist who treated me, this time helpfully, for what I call my postanalytic failed ex-gay-syndrome. This consisted of loneliness—both residual and consequential (to some extent I had passed my prime time for meeting a lifetime partner); constant acting out—due to being angry and resentful of the misguided and endless treatment I had received for no purpose and to no avail; and depression—due to my having been first seduced (with analysis) then abandoned professionally with, for now, nowhere to go, having to take what I considered to be a second-rate job because I had been excommunicated by the "first rate big boys" of whose regime I so desperately wanted to be a part.

Today the tide is turning, but not fast enough. As recently as 2006 "Alan Chambers of Exodus International . . . and other pro-family groups protested at the American Psychological Association over its stance that homosexuality cannot be changed." Then, as Roach in the *Baptist Press* on June 20, 2013, wrote in an article titled, "Exodus Int'l closes after Chambers' apology" "Chambers [apologized] to people who have been hurt by the organization."[14] To which R. Albert Mohler Jr., president of Southern Baptist Theological Seminary, continuing to defend reparative therapy, replied, "Sadly, it appears that this rethinking has resulted in something like a surrender to the cultural currents of the day."[15]

2

·❖·

Problems with Research Studies

In this chapter I discuss, and raise questions about, the positive research outcomes reported for reparative therapy. These positive outcomes are questionable for the reasons described in the following sections.

THE RESEARCH OVERLOOKS THE SUBJECTS' PRIOR SEXUAL HISTORY

Some research studies enhance positive results (to allow for even more and better, however insubstantial, claims of cure) by stacking the deck with individuals most likely to change: individuals who at some earlier time in their lives having had heterosexual impulses and a prior history of heterosexual behavior are more likely to "progress" from homosexual to heterosexual than those who in the first place have never had prior heterosexual experiences. Many bisexuals can shift from homosexuality to heterosexuality in their sexual fantasies, their sexual attractions, and their sexual behaviors (and back again). Bisexuals previously in a heterosexual relationship/marriage with an occasional homosexual dalliance are among the most likely to show "progress" from "gay" to "straight." Also likely to report this change are "opportunistic homosexuals" such as incarcerated, basically straight, men; as well as younger men and women, for this group generally includes many individuals who, being polymorphous, that is, in the formative and so changeable stages of their sexuality, are prone to change over from homosexuality to heterosexuality as if spontaneously,

making it look like a given therapeutic intervention really worked to produce that change, even though the change from homosexual to heterosexual was simply an enhancement of one's extant heterosexuality.

THE STUDIES ACCEPT ONLY AFFILIATED PATIENTS

Ideally subjects should be chosen at random. Limiting participants to white, European, middle-class, religious, Christian men referred by organizations that favor, or actually do, reparative therapy, and those gotten through advertisements on right-wing/evangelical radio stations, is as distortive as would be limiting participation to those enlisted from other special cohorts such as patients in an AIDS clinic, or patients recruited by handing questionnaires out in gay sports bars, making such study results inapplicable to the general gay population. Incarcerated patients spiritually resemble those who have been forced into treatment by their parents. The inclusion of these incarcerated individuals, as well as those referred by the courts for sex crimes, almost always distorts the results of therapy because these individuals, less interested in being cured than in being exonerated, will predictably skew their stories in a direction that keeps them out of jail or, if they are currently in jail, satisfies the demands of a probation panel deciding whether or not to release them. Such subjects are likely to be the most "cooperative" of all, telling their therapists what their therapists want to hear so that they, as subjects, can be considered cured, be let out of jail, and be discharged from therapy—and in my experience to go on to be as gay as ever, perhaps at most this time more discretely than before.

THE STUDY IS DONE BY RESEARCHERS
WHO ARE AFFILIATED

Affiliated interviewers with ties to pro-ex-gay organizations consciously or unconsciously fudge outcome studies by how they ask questions, by omitting questions that might lead to data that contradict a primary premise they wish to advance and prove, and by omitting data that subvert a study outcome that they hold in favor.

THE STUDIES ARE OVERLY SIMPLISTIC

I believe that lab measurements of sexual arousal made by using a penile circumference gauge at most apply in, but not outside of, the lab, particularly when one is looking for a reliable measurement of one's sexual

orientation/arousal patterns in everyday life. Sexual arousal and sexual response, being based on multiple cues that light up/enhance, or suppress, sexual fantasies are not accurately measurable in vitro (experimentally in the lab). In particular, experiments that measure sexual arousal in vitro by measuring arousal by pornographic pictures are missing the point: that much sexual arousal in the real world takes place not to an inanimate picture of an object (to its physicality) but to a real person involved in a real-life situation, which of necessity would have to include emotional cues. For example, for one patient a man eating alone became sexually attractive because he, the patient, resonated with the other man's loneliness and/or neediness; and for another man, a person with a slight limp became attractive because it aroused the first man's mothering instincts (and, for those analytically inclined, his oedipal castration fantasies). Also, pornographic pictures alone don't properly measure whole fantasies in part because they fail to elicit the many senses besides sight involved in sexual arousal, particularly smell and touch. A *lack* of response to pornography proves little or nothing at all. Men and women with a religious bent might not approve of pornography on moral grounds, making it a major turn-off for them. Therefore, they will likely show a neutral or negative response to pornography not because they have changed their sexual orientation but because they have a ratcheted-up morality that at least in this particular venue squelches their sexual responsiveness.

THE STUDY OVERLOOKS THE ROLE PLAYED BY PREEXISTING/COEXISTING EMOTIONAL DISORDER

Preexisting/coexisting emotional disorder determines a patient's responsiveness (or lack thereof) to reparative therapy. Emotional problems of individuals in the study group will predictably determine motivation (or lack thereof) to enter the study, the content of the participant's communications in therapy, the participant's responses to therapy, the participant's willingness to complete the study, and the participant's eagerness to afterward offer (positive or negative) testimonials as to the study's results. The *passive-dependent* patient is more likely to be "cooperative" (and claim positive results) than the patient who is *passive-aggressive*. Passive dependent gays who desire to please, and to meet their therapists more than halfway, tend to call even temporary celibacy "a cure." They might even marry heterosexually to meet their therapists' expectations (although many of those who do so continue to have relationships with same-sex partners throughout their marriages). Passive dependents, being more submissive than rebellious, and desirous of being liked, give their

interviewers what they perceive the interviewers want—a determination that they can easily make by being alert to the therapist's cues, including, or especially, nonverbal signs. In contrast, passive-aggressives being rebellious calculate ways to defeat the interviewer's obvious purposes and/or undermine the authority of the interviewer/abrade the interviewer's sensibility. As cold call interviews such as those made to assess outcome tend to elicit false positive results from passive dependents who want to please the interviewer, perhaps because they are flattered about being asked to participate in a survey, such calls tend to elicit false negatives from passive-aggressives who wish to displease the interviewer, likely because they are annoyed about having been bothered. In the realm of wanting to please eliciting false positives, passive patients often answer questions (even when there is a promise of strict anonymity) in a way that they anticipate will put them in a favorable light with questioners with whom they routinely become personally involved, for even in these impersonal situations patients want to be liked, admired, and remain shamefree. In the realm of wanting to displease eliciting false negatives, many patients are not only annoyed about being bothered but also have preexisting negative feelings about the content of the study, or about the questioner's manner or approach. In my follow-up (anecdotal) studies of cold call responses, I found that a very large group of cold call subjects reacted to being called at random by "goofing" on their interviewers, either by not telling them what they seemed to want to hear, or by giving them outrageous replies to provoke them to become perplexed or get mad and hang up.

Patients with a *borderline* predisposition/personality disorder, being likely to be "as-if" personalities—most want to identify with impressive figures (like their reparative therapists), feel close to them, then morph accordingly just to become "one with their idol." Thus one gay borderline man was acting out his "as-if" personality by virtually deluding himself into thinking he actually was straight. He had made "being straight" his preferred identity since his therapist was straight, and as well clearly wanted his patients to be straight like him.

Depressed interviewees, disturbed about who they are and hating themselves for what they stand for, can be more highly motivated to report positive change than patients who are normothymic (not depressed) simply because those in the former group so long for a cure that, grasping at straws, they speak of minor as major improvements. Depressives also tend to be excessively moral, often erotophobic, individuals who welcome sexual retreat into celibacy, which they in turn confound with sexual orientation change. Those who are *bipolar* and are presently in the hypomanic state

either deny relapses or rationalize them as understandable and unexceptional (e.g., "I'm not really gay any more, I only revert to being gay when things get tense at home," or "I am not being gay once again, for though I did, and still do, a few gay things a few times, I was drunk each time, so what's the big deal, let's just forget about it"). Also patients in the hypomanic state, almost always reluctant to admit defeat, are especially likely to report positive outcomes that reflect little more than their being euphoric about minor accomplishments that do not indicate any significant accomplishment along the lines of true sexual orientation change.

The *psychopathic* gay individual might claim "I am straight now" only to be better able to secretly go behind the therapist's back to continue to be nonstop gay, as before. Psychopaths don't simply consciously lie. Rather, focused on outcome instead of process, ends rather than means, such as the prospect of immediate gain rather than on the virtues of being honest, they confound "what is" with "what they wish it to be," a distortion of self-reporting which for these individuals is the rule rather than the exception. As an example, I know of one candidate in training at a Psychoanalytic Institute who unconsciously falsely avowed throughout his entire psychoanalysis that he had become ex-gay just so that he could get admitted to the Institute and begin and finish the training he so eagerly sought. (At the time no gay man was provisionally accepted to be trained as an analyst, and this patient's analyst was a training analyst who could make or break him, i.e., he could or could not recommend him for graduation from the Psychoanalytic Institute depending on a number of factors, one of which was his sexual orientation.)

Psychopaths' need to achieve a specific goal/objective for personal (reputational) or professional reasons leads them to quickly learn the purpose of any study they are involved in and, if they have a personal stake in its outcome, to readily and willingly distort their reporting to serve their personal needs, and especially to tell the world that they turned straight to hide that they are having a sexual relationship with outside partners, or even with the therapist.

Gays who dissimulate may, however, not be psychopathic. They may do so out of guilt and shame about being gay. Such individuals are not psychopathic in the sense that they are fooling themselves for purposes of personal gain. Rather they are fooling themselves to satisfy/appease others they depend on (such as parents who are openly pushing them to become ex-gay) or their therapists. Also, patients who are uninsightful may not understand the true meaning of some or all of what researchers are getting at, their reporting corrupted by their not comprehending many of the subtleties both of what goes on in therapy and of what constitutes actual

improvement. And most gays like anyone else asked about sexuality tend to reflexively downplay and underreport their continuing attractions and minimize the frequency of their same-sex contacts due to what amounts to their negative bias directed to all forms of sexuality—a bias that is not uncommon even in gays who are out fully themselves.

Masochists dissimulate in order to self-punish via overcompliance. Some knowingly continue in a nonworking therapeutic relationship with a homonegative therapist for months or even years without any real progress being made. These patients often think, and say, that they are getting a good result when in fact theirs is a bad outcome, as they deliberately wish to participate in, and even thoroughly enjoy participating in, nothing—self-punitively involving themselves in what, though they call it therapy, is in fact more like substitute living.

PROBLEMS EXIST IN THE FABRIC OF THE STUDIES

Even studies based on interviewing patients directly, in my opinion the best way to do a study, are so labor intensive that it becomes difficult to correctly evaluate large enough numbers of subjects to constitute a reliable sample. Also, most, or possibly all, studies have difficulty obtaining satisfactory (e.g., large enough, or otherwise suitable) control groups.

THE STUDY CONFUSES SUPPRESSION WITH ACTUAL CHANGE

Many patients are able to suppress their homosexuality, at least in the short run. Some report this suppression, though it's a short-term development, as being indicative of long-term progress. Such patients though they speak of diminished sexual interest in the same sex are just as homosexually inclined as before, only now they have become less compelled to have sex of any sort. Individuals who are guilty about their erotic feelings routinely view present suppression of sexuality as a welcome outcome, although most observers would see relinquishing so important an aspect of one's self/identity as requiring such severe emotional and physical sacrifice that it constitutes not a cure but the development of a new disease.

THE STUDY CONFUSES SYMPTOM SUBSTITUTION WITH REAL PROGRESS

Patients often view symptom substitution as a positive outcome, especially when the substitution involves replacing life with therapy. I lived

such a substitute life for five years when I was in psychoanalysis. During this time I was mostly celibate, having managed to force myself to feel more asexual than sexual. My analysis had become my life—central to its nature and direction—and I worked hard, several nights a week at a second job taken to pay for my treatment, in part to please my analyst even though that meant displeasing myself. As far as I was concerned at the time, my therapy was working, although in fact it was not helping me at all, and was likely making me worse.

THE SCOPE OF OBSERVATION IS LIMITED

Very few research studies go into more than a few aspects of an individual's sexual life, omitting, for example, masturbatory fantasies; yet it is these that are the strictest measure of sexual orientation, and hence the best indicator of progress, or (mostly) the lack thereof.

HOMOPHOBIC BIAS EXISTS AND TAKES OVER

Too often reparative advocates have overt or secret homophobic affiliations (consciously or unconsciously determined) and skew studies along lines that satisfy their homophobic agendas. Perhaps Regnerus's study of children of gay marriages[1] reflects a homophobic mindset in its averring that such children are likely to be damaged in some significant way. For Regnerus proves his point by using subjects who are not children brought up in the context of a gay marriage but children brought up in marriages where one or both parties have had outside relationships with same-sex partners. He thus manages to select out a study population characterized by those who have been unfaithful to their partners—likely indicating that relational trouble of a general sort exists within the marriage. Perhaps in his eagerness to prove that gays shouldn't raise children he overlooks the obvious: that children of parents with relational problems often do less well than children of parents who get along at least relatively well with, and are faithful to, each other.

DIFFERING (MORE OR LESS STRINGENT) OUTCOME CRITERIA ARE EMPLOYED

Studies' goals and objectives are inherently flexible. Many reparative therapists recognize that the fantasy life of the committed homosexual doesn't change much if at all through treatment. Therefore, many researchers accept as their goal a less than definitive change in a patient's

fantasy life, such as becoming less homosexually active, or less sexually (homosexually) compulsive, or acting heterosexually although with continuing homosexual urges and behaviors. (Mostly those reparative therapists who are willing to thus compromise are more ethical, more honest, more in touch with reality, and ultimately more helpful to their patients than those who are dishonest from the start about what they can and cannot do for their patients.)

EFFECTIVENESS IS EXAGGERATED

Reparative therapists' studies exaggerate their success rates by speaking of the likes of "partial success" as being satisfactory enough. Some excuse present failure by pushing success into the future (e.g., "it takes at least five years for this to work").

BLAME IS DISPLACED

Not a few reparative therapists account for their obvious failures by blaming not their reparative tools themselves but how one uses them, saying, for example, "reparative therapy only works for those trained to use it," or "reparative therapy only works when patients with an oedipal fixation are selected for treatment." Many spin by blaming their victims. If reparative therapy is ineffective it is only so because the patient doesn't have "the right kind of homosexuality."

ANCILLARY HARM GETS OVERLOOKED

So often one gay man's cure is another person's relapse. Reparative therapists often define cure in a way that fails to take other people in the patient's life into account/overlooks the harm done to them. I have been deeply hurt by friends who leaned heavily on me in the beginning of their cure only to eventually drop me when halfway into therapy their therapists told them their cure involved giving up all contact with homosexuals. Wives of supposedly ex-gays who married are often in a state of limbo, constantly wondering where their husbands are, and if they are going to run off with a boyfriend. I'd much rather be a child of two out gays or lesbians than a child in a family where the "cured" husband or wife is cheating, and the straight partner knows it, feels resentful, and (likely) takes his or her disappointment/anger out on the children.

In conclusion, it is true that some people who have undergone reparative therapy have benefitted in some way. But one must be specific about

how; both as to who these cured people are, and as to how exactly they did benefit—how much, and in what areas, and whether temporarily or permanently. Too often "cure" means forcing oneself into a heterosexual charade, including a straight marriage, with homosexual liaisons on the side. Almost no one goes from gay to straight fantasies, and the few who do are likely originally bisexual. Behavior can change but only at great personal sacrifice usually made to please, appease, or even displease some authority figure.

On a more positive note, SOCE while generally ineffective and potentially harmful, is still therapy, and can, like almost any form of therapy, in favorable cases induce some degree of improvement/change. As the APA says, "enduring change to an individual's sexual orientation is uncommon: most continued to experience same-sex attractions (SSA) following SOCE and did not report significant change to other sex attractions . . . though some showed lessened physiological arousal to all sexual stimuli."[2] But some, a very few, become "skilled in ignoring or tolerating their same-sex attractions. Some individuals report that they went on to lead outwardly heterosexual lives developing a sexual relationship with another sex partner, and adopting a heterosexual identity."[3]

Celibacy, partial or full, is probably the only realistic change that reparative therapists can regularly expect to bring about in their patients. Celibacy can be a choice. But it's not a healthy choice, and it can't generally be sustained for long, as resentment and feeling deprived eventually undermine the attempt to give up on sex entirely. The definition of celibacy varies from observer to observer and no one definition is uniformly applicable. I think I was celibate for about five years, but even here as I look back I see that certain escapades that occurred at the time do not conform to what people usually think of as being celibate. I think that rather I was acting dissociatively, that is, in two alternating personalities, so that one part of me didn't know, or acknowledge, or even care, what the other part of me was doing. One part of me was undergoing the cure as the other part of me was in San Francisco doing business as usual (e.g., seeing my old lover for old and good times' sake), reverting to my true former self, but not reporting it, not only not to my analyst, but not even to myself.

As noted throughout, homosexuality cannot be cured because it's not inherently a disease, because being inborn it's innate, and because it tends on some level to be ego-syntonic—that is, the homosexual basically likes to be homosexual, and therefore doesn't have, or denies having, problems associated with homosexuality. It is very difficult for a therapist to take away something the patient down deep really wants to keep. Reparative

therapy doesn't work when gays enjoy being gay and the gay/lesbian life-style, that is, they enjoy not only the sex but also the social accoutrements of being gay. Pushing/forcing gays into a foreign lifestyle basically causes them to become depressed. Like Ruth, according to Keats, they are "sad of heart . . . sick for home, stand in tears amid the alien corn."[4]

As mentioned throughout, when homosexuality is ego-dystonic (un-wanted), the ego-dystonicity is often reflective not of a fundamental wish to change but of a fear of being who one is, with not the homosexuality but the guilt about it the most troublesome thing of all, and is therefore that which needs to be therapeutized away.

One can say in defense of reparative therapists that most outcome stud-ies in all forms of psychotherapy depend to some extent on collecting data as to improvement (or the reverse) from unreliable sources such as tele-phone interviews and/or patient reports as given not to a therapist but to a statistician performing the study. This predictably introduces elements of distortion into psychiatric studies. And studies other than ones involv-ing reparative therapy also come up with creative ways to report positive results. For example, they report positive testimonials only, overlook what patients with negative experiences have said, or cherry-pick a few positive phrases out of all the many negative comments made, perhaps quoting what patients say without qualifiers, thus being able to present outcome as positive simply by removing crucial negative modifiers.

3

━━━━━━━━━━━━━━━ ❖ ━━━━━━━━━━━━━━━

Outcome

In this chapter I describe issues relevant to determining the outcome of reparative therapy. (Outcome issues are also discussed in the previous chapter on problems with research studies.) It is difficult to measure the outcome of reparative therapy in part due to the unreliability of/absence of fixed parameters for measuring (homo)sexual orientation/(homo)sexual arousal. For example, the penile plethysmograph (a device placed on the penis to measure the circumference of the penis as reflective of arousal) measures arousal only crudely in part because, while it measures tumescence, it fails to uncover why exactly the tumescence has occurred. When it comes to arousal, *subjective* attitudes reign, including those involving the patient's overall attitude toward sexuality in general and about homosexuality in particular. Thus, religiously inclined gays decrying pornography for moral reasons will not be fully aroused by pornographic pictures—making it seem as if they are "cured" when in fact they are just offended. The nature of the transference to the researcher also affects arousal, as interviewees who are flattered about having been selected for the attention become more aroused than those who are annoyed at having been bothered by the intrusion.

Moreover, outcome researchers are rarely of one mind about the definition of sexual orientation and what elements of sexuality go into making it up. Should sexual orientation be defined by fantasies or by behavior? A particularly gray area involves the relationship between sexual orientation and sexual identity with the latter a construct that overlaps with, but

is not identical to, the former, and relates not only to *having* fantasies but also to the degree of *tolerance* and *acceptance* of the fantasies that one has.

Adding to the unreliability surrounding outcome studies is that so often research subjects are initially referred by ex-gay advocates. The subjects are often individuals who, having already contacted NARTH, are primed by what goes on in that initial consultation.

Many outcome studies are flawed due to being anecdotal. Some are secretly based not on the patient population the study (vaguely) identifies but on observations from the researchers' own life, such as those made on the researchers' children. Often the so-called case histories of patients of ex-gay therapists are in fact tales told by friends and relatives, or stories from the therapists' own life written up as if they are episodes in the lives of actual patients. Secret affiliations leading to biases also prevail. Since orthodox analysts were, and perhaps many still are, routinely both theoretically and actually/personally homophobic, for years no homosexual could be admitted to orthodox psychoanalytic institutes on the grounds that gay candidates, being sick themselves, would be in no condition to treat patients who were equally troubled and in much the same way. Therefore, it is entirely likely that analytic studies of outcome will represent a level of bias since, with psychoanalysis only done—at least until the later years—by straights (or closeted gays), no openly gay individual could, at least officially, contribute much of fundamental value to the psychoanalytic literature.

The individual pathology of those doing the study can color their evaluation of outcome. Classic distortions leading to biased outcome often result when researchers allow their psychopathological wish fulfillment to surpass their scientific judgment.

Outcome studies can be flawed due to the design of the study. Thus according to Besen, Bieber's study supposedly proving the relationship between unique mother/father-versus-child interpersonal disturbances leading to a homosexual outcome was based on a sample that not only didn't include direct interviews of the patients themselves but also "included 28 schizophrenics."[1]

Many reparative therapists determine outcome without first fully describing/determining the "premorbid" sexual orientation of those who seek to go straight, that is, without exactly defining what these individuals' sexuality was like before consultation/treatment, and especially to what extent they were heterosexual albeit in a rudimentary form. Many of those who apply for therapy, though they seemed to be fully homosexual, were in fact just experimenting/going through a phase, that is, "developing," and were prematurely at odds with themselves as to who they would

permanently turn out to be. Such individuals are more likely to respond to reparative therapy by changing their sexual orientation than are individuals whose homosexuality is more ingrained than it is transient.

GOALS/OBJECTIVES/PROGNOSIS

Few reparative therapists have as their goal a patient's complete sexual reorientation/reconstruction. Even the most optimistic of ex-gay therapists rarely claim to be able to change homosexuals fully into heterosexuals. Instead, they generally hedge about what constitutes possible/desirous outcome, promising improvement but saying "It takes a long time," knowing that by then you have probably quit. Or they say, "We at least do better than oncologists treating cancer," selecting self-protectively what they believe to be a devalued form of therapy to make that comparison. When progress is nonexistent they nevertheless call it slow and aver, "Things are going well, but there is still work to be done." They emphasize how favorable alterations have occurred if not in content then in intensity—a hard-to-measure impressionistic indicator of outcome reliant upon the claim that through therapy patients have come to experience fewer homosexual feelings and decreased frequency of sexual activity. Most, recognizing that sexual fantasies will remain hardly modified or unchanged no matter what they do, overstress the value of resolutely *partial* and often *superficial* alterations to sexual orientation, such as "inculcating a new attitude about one's old self, short of creating a new self." Some, realizing that it can be easier to give up an activity entirely than to alter its performance somewhat, settle for complete suppression of sexual activity in the form of celibacy, perhaps (hopefully at least) accompanied by the morphing of erotic same-sex feelings (sinful) into feelings of same-sex closeness (acceptable).

Because there are core elements of SOCE that overlap with the core elements of other, more traditional forms of therapy, many patients will, not surprisingly, experience some general benefits from reparative treatment even though their sexual orientation doesn't basically change. And when these individuals give testimonials (e.g., when they are asked if they got better) they refer mainly if not exclusively to these ancillary benefits, for example, those related to the reduction of guilt about their sexuality. (Testimonials are discussed at greater length below.) Some gays feel that they get more from the therapist's desire to help than they lose by being told that they need that help because they are broken and have to change. One generally positive effect of this otherwise notoriously ineffective therapy is the fantasy that "just being in treatment means that I have tried

everything," allowing patients to be gay with less guilt because they have at least given, and continue to give, going straight a try. Gays often improve when reparative therapists include, as many do, an element of life-coaching as part of their treatment protocol. Reparative therapists who tell their patients what they need to know in order to survive, not only because they are gay but also because they are human, often offer their patients palpable benefit.

Another salutary outcome of reparative therapy might involve helping patients who are in conflict with their faith reconcile their sexual orientation with their religious beliefs. Sometimes reparative therapy can help patients review their relationship with God and the church to hopefully become less confused and conflicted about extant tensions between their religion and their sexuality. Patients can now better reconcile what they are with what they ought to be, thus bringing religion and sexuality into synch if only by the use of such "unhealthy" defense mechanisms as dissociation, selective repression, and convenient partial denial, and such healthy mechanisms as conscious reframing, allowing patients to see their religion versus their sexuality in a different, more sanguine, light. Disapproval from others on religious grounds, a severe threat to one's self-comfort, is, unfortunately, part of living as a homosexual. In a relatively good outcome, patients are helped to not let their self-esteem sag just because homonegative (at times conservative religious) people seem to want to reset it to a lower level.

For lonely gays even a therapist who says and does the wrong thing but is nonetheless inherently interested in/fond of them has benefits, as they disregard his or her therapeutic mistakes in favor of being thankful for, and so responding positively to, the proffered help. Gays who are disorganized in their thoughts and their lives may experience the regularity of therapy sessions as calming and the effects of therapeutic caveats, however ill-considered, as organizing, since both add a sense of structure to their lives, with the structure at the minimum having the salutary effect of setting limits on undesirable behaviors. Gays who are masochistic can thrive on the punishment they get out of the harsh attempts made to repair them. Depressed gays who feel hopeless often see even false hope as better than none. Gays who are obsessional can shift from obsessions that are malignant, such as a constant unwarranted worry about STDs (sexually transmitted diseases), to at least seemingly more benign obsessions such as those involving religious scrupulosity (e.g., "have I or have I not said my prayers exactly right?"). If gays are hypomanic, the calm of the therapist and the steady immutably repetitious drone of the therapeutic input often has a normalizing effect on elevated mood.

Many therapists doing ex-gay therapy have under the guise of repairing homosexuality helped gays become less *sexually compulsive*. They have helped them reduce the level of their sexual activity by keeping an initial sexual response from kindling; that is, they have taught them to quash their spontaneous sexual responses by in effect taking an emotional on-the-spot "cold shower" (e.g., through supplication and/or mortification). It does often happen that men and women who have undergone reparative therapy improve somewhat in terms of their overall emotionality enough for unwanted pressured sexual obsessions to fade away, although not entirely, leading patients to considerably restrict their homosexual contacts—thinking as gay as, but without being as actively sexual as, before. Treating compulsive sexuality is discussed further in chapter 19.

I personally interviewed some gays who have learned how to have heterosexual sex by thinking of a same-sex partner while performing with a member of the opposite sex, or by selecting an opposite-sex partner with a crossover gender presentation. Such outcomes, neither firmly satisfactory nor completely unsatisfactory, are right for some but wrong for others depending on the individuals involved: their personalities and the extent to which they willingly compromise themselves for an (imagined or real) greater (often social) good—all things involving very personal choices. Therapists should not make such choices for their patients. They should rather leave these choices entirely up to their patients to make.

Patients in group therapy likely experience the standard benefits of being in a helpful group: those arising out of having peers to relate to, reduction of loneliness; consensual validation that can help them cope with/ diminish anxiety and guilt; learning to develop more and better interpersonal relationships; enhancement of self-esteem leading to a better relationship with oneself and a fresh development of new and better ways of relating to/dealing with others, including one's family; learning new methods for dealing with/coping with external and internal homophobia; and reduction of depressive-spectrum symptoms particularly of grief over what they can never have or accomplish—as they come to grips, and to terms, with what can and what can't, and what will and what will never, be.

In summary, I feel that it is unlikely that any positive statistics relative to cure rate hold true, even if the process is given the three to five years of trying to get it to work that many reparative therapists recommend. I believe that such statistics, like all statistics about cure rates, while they seem precise, in fact don't do justice to the complexities of collecting such data. I believe that outcome studies reveal that ex-gay therapy, while it can lead to a change of sexual *behavior*, cannot lead to a change of sexual

orientation. In a rare instance of my agreeing with van den Aardweg, I applaud his saying that antigay treatment doesn't work because those who undergo it mostly "slip back to a self-tormenting double-life (including the usual lying)."[2]

TESTIMONIALS

Positive testimonials sometimes tout the benefits even of this most misguided form of therapy. Some reparative therapists who are theoretically challenged can nevertheless still be personally inspiring—wrong about psychodynamics, but right in their overall concern for their patients' well-being and sufficiently motherly/fatherly in their respect and love for their patients to have a positive impact on them. Patients of a sanguine temperament often see the good while ignoring the bad in reparative therapy—hearing the helpful things that their therapists say while ignoring the therapeutic errors that their therapists make. Patients who are lonely can feel less so just because they have someone to talk to. Masochistic patients can thrive on the abusiveness that is embedded in the concept of being repaired—accepting punishment as deserved and turning it into an inspiration to do better, while welcoming the therapists' bullying as a kind of tithing that allows them to reject other bullying in their lives because they already have fulfilled their quota. Bisexuals often choose to tip to straight as a conscious response to therapeutic urging. It is not unheard of for patients to change in some superficial respects to satisfy parents who through a therapist are pressing their child to become heterosexual.

While few patients thankfully and tearfully speak of a cure, not a few speak of improved happiness and fulfillment due to a diminution of homosexual urges. Thus a man wrote an anonymous letter to LifeSiteNews saying that "he . . . would not want anyone pressured or forced to go into therapy, but that without it, he would not be as happy and as fulfilled." He ended with this plea: "I ask you, please, let's make sexual orientation change efforts better and more responsible; but please don't eliminate it, please don't criminalize it, and please don't say that it is a violation of human rights, because for me, it made me the person I am today."[3]

Positive testimonials about successfully having become ex-gay often reflect gays' positive transference, a thing that leads gays to respond irrationally to their reality. If in their positivity they don't fully recommend their treatment to others they at least gloss over how their treatment has failed to accomplish its main goal. They do so simply because they like or love their therapists. Such patients in giving "falsely positive" testimonials don't consciously lie, but they do confound feeling good about their

therapist with having good results from their therapy. If they have an overall positive relationship with their therapist they have a "transference improvement" where they feel and do better because, motivated to see improvement where there is little or no progress, they take what good exists in their bad therapy out of context, and, maximizing that, simultaneously minimize what went wrong with their treatment.

Testimonials about the positive outcome of reparative therapy often represent a confounding of feeling positively about therapy's ancillary effects with feeling positively about therapy's reparative effects. Sometimes patients' positive reports represent little more than the wish-fulfilling fantasies of activists wanting to make a case for reparative therapy due to their being highly invested, for personal and financial reasons, in reparative therapy's working. For some, being ex-gay (and criticizing others who condemn the ex-gay movement) has become their calling in life/their job/ their paid profession, that is, they have become official spokespersons for the movement. Many actually prepare and rehearse what they have to say so that they can, dissimulating for desired effect, more persuasively overstate the positive results of their reparative therapy. A few, as *narcissists*, are so in love with themselves that they tout any minor movement as consisting of a major accomplishment. Some are gushing *histrionics*, overawed with their minor achievements. Histrionic patients, being excessively suggestible, tend to view in fact unhelpful therapists as "infallible healers." Others are *passives* who gladly say what others wish them to say, and give others what they want, in the hopes that they will be rewarded for their compliance. Still others because they are *depressed* will say and do almost anything to elicit positive feedback from their therapists or from a group to which they aspire to, or actually do, belong. As depressives they are especially apt to respond to pressure from advocacy groups attempting to bully them into compliance.

Gays who have experienced past abuse might give positive testimonials because they fear that if they didn't say and do what powerful others wanted them to they would experience more of the same abuse. In many cases, theirs is a history of past traumatization that has been sufficiently serious to have left them with a *PTSD-like syndrome* for which (much as war survivors walk out of a movie when war scenes are being portrayed) the best and only remedy is doing everything they can to avoid being retraumatized once again along similar lines.

OCD patients experience lack of progress as a deserved/welcome atoning for past sins, which allows them to forgive themselves for the bad they believe they have done. *Borderline* patients often alternatively angrily denounce everything their therapists stand for, and merge with their

therapists to become one with a therapist who, coming to represent their ideal, has, at least for the moment, become the logical, perfect object to emulate. *Psychopathic* patients feloniously hoodwink others into believing for reasons of their own involving personal satisfaction and gain. *Delusionally* mentally ill patients revel in their belief that by advocating reparative therapy they do God's will. In some respects they are like an acquaintance of mine who was tipped into a psychosis by both therapy and drugs, including hallucinogens (LSD), that led him to believe that he was one of the seven straight disciples and well on his way to becoming the leader of the pack, touting his cure so that as many others as possible would join his Movement. *Bisexuals* do "change," but not because their therapists cured them, but because their therapists presided over an essentially spontaneous shift of emphasis (rather than a true sexual orientation change) onto straight from formerly gay. (Many of the individuals described in Regnerus's study[4] were bisexual in the sense of being prevailingly heterosexual with occasional homosexual contacts.)

I have had a number of bisexual gay friends who "through therapy" proceeded to get married and have children. But the reasons for their conversion always remained as perplexing as the depth and nature of their new sexual orientation remained suspect. Many on (informal) follow-up were spotted in compromising positions in cruising places such as men's rooms on the highway or in the dunes of gay beaches. Studies based on testimonials that include many individuals like these, persons who would in any event progress to some heterosexuality on their own, can claim excessively favorable results. Though some might assert that these individuals have been cured, in my opinion reparative therapeutic interventions have not created the turnaround. Rather the therapist has simply presided over a bisexual patient's natural (and often temporary) progression from gay to straight.

SUGGESTIONS

My hope then is that reparative therapists will soften their goals in the direction of changing not their patients but themselves. Reparative therapists need to recognize that their calling is not to control or to intimidate patients into going straight, but to develop the doctor-patient relationship to include meaningful consensus as to what direction their patients might want, and need, to take. A big part of the reparative therapist's job is to tell his or her patients what they need to know about reparative therapy. Therapists should say "I don't know if I can make you straight, but I can work with you toward the self-understanding that can help you

become more comfortable with yourself as we work with who you are, instead of trying to change you into someone else." They might note, "There are, of course, exceptional cases where change does occur," but what they should also mention is that such change is often due to innate spontaneous progression from gay to straight (one can also change spontaneously from straight to gay) involving the shifting gay-straight imbalance in someone who is innately bisexual. In such cases, the therapist's job is not to prevail but to preside over a natural development, and to do so without taking undeserved credit for a positive outcome.

All therapists should admit that some patients, having obtained collateral benefits that are the result of the positive, unintended consequences of reparative therapy, do "improve" with reparative therapy, but only in ways their therapists do not generally anticipate, recognize, or even acknowledge. Some patients who are excessively guilty about being gay though they stay gay at least feel better about not being able to change, perhaps because now they can boast about how they tried to do something about themselves (only it wasn't humanly possible to accomplish their sincere, reparative goal). Gays who are sufficiently masochistic to endure/enjoy the punishment inherent in tithing can benefit from giving up a tenth (or more) of their lives so that they can allow themselves to enjoy what remains (nine-tenths), once and for all convinced "I did give all I could to change, but this is who and what I am, and changing it is, as I discovered after long hard work, if not unnecessary, then at least impossible."

Perhaps the best reparative therapists can do is to save gays from an STD, or from a life of constantly cruising, compulsively looking for sex, thus from either disastrous serial gay polygamy or equally destructive serial gay monogamy. This outcome does not represent some unfortunate but necessary compromise compared to full cure. Rather this outcome represents a valid, and good enough, result in and of itself.

Reparative therapists can help considerably by correcting any of their patients' misunderstandings about gay life that lead their patients to having unfulfilling lifestyles. For example, the misinformation that "you can never be happily married as a gay man or lesbian" facilitates serial monogamy as well as compulsive sexuality as gays think, "why bother trying to be monogamous, it won't work" or "even if I try, and try again, I will never finally succeed." I almost failed to connect to my lifetime partner and live happily ever afterward because my therapist told me "gay marriage doesn't work"—a myth about gay life that too often permeates professional reparative practices and spills over to affect an individual patient's decision making thus to negatively affect his or her love life.

Gays themselves must think through their wish to change at least enough to fully consider what they, and their life, will be like should change actually come about. They must ask themselves, "Will I be comfortable in my new skin, or will my ego-syntonic homosexuality evolve to become ego-dystonic heterosexuality?"

Reparative therapists not only have to deal with gays resisting becoming straight, they also have to help gays who are overzealous in regard to changing become more realistic about actually doing so. They need to understand the motivation of individuals who seem just too eager to become straight but are motivated to do so for all the wrong reasons. Too many gays want to go straight not because they are unhappy with being gay but because they are at odds with their being at all sexually active, or because they have picked the wrong people to get involved with and have been hurt, or because they have brought their personality problems to being gay—and so have problems that are not an integral part of their being homosexual but an integral part of their being troubled people, that is, those beset by difficult interpersonal relationships, with the difficulty only remotely related, if related at all, to their sexual orientation. These individuals need to be helped not to stop being broken but to stop having broken relationships and to start to live healthy, fulfilling lives, still gay, but now far happier than formerly.

Reparative therapists might deal with ultraconservative religious gays not by urging them to go straight as God declares they ought, but by helping them tolerate the inevitable problems associated with significant degrees of religious nonconformity on their parts. The goal is for these individuals to synthesize their religious, spiritual, and physical needs into a rational, workable whole—a philosophy where each of their (often warring) sides gets some say, but no one side seizes all the power and takes over completely. Many scrupulously religious gays, just to function at all, have to settle for a degree of hypocrisy as a necessary part of living a good life. They should seek not change but an increase in self-acceptance as to who they are, however much hypocrisy that that full self-acceptance may involve. In psychoanalytic terms, their goal ought to be not the modification of their id (their sexual desire) but the softening of their superego (their excessive reactive guilt), allowing their ego (their "self") to forge a better compromise within as they authenticate and integrate such disparate forces as those of immutable instinct clashing both with irrational guilt and the simultaneous machinations of external and internalized disapprobation. This is a rational goal for all gays, religious or not; for while psychotherapy can't induce sexual orientation change—for sexuality is a force of nature—it can modify the superego, for the superego is a force of

one's psychology, accompanied by an often too highly developed sado-masochistic relationship with one's own sexuality, so that individuals can bring their identity into line with their desire as they enhance their self-acceptance by exploring the relationship between what they think about themselves and what others think of them, evaluating and selectively integrating the feedback they get from within and from without—always factoring in how expert (and godly) others actually are, as distinct from how learned they claim to be, though they know nothing. Thus, according to Throckmorton and Yarhouse the therapist's job involves resolving various conflicts that sexual attraction to the same sex elicit such as "irreconcilable differences between religious beliefs and sexual feelings and practices,"[5] as well as other conflicts such as those between sexual attraction and gender identity and between self-acceptance and social and moral approval.

Reparative therapists should consider helping the parents of their patients as well as the patients themselves. They might be especially involved in helping parents who made the original referral to them. They might attempt to convince such parents to stop trying to make their children straight and to start trying to help their children be happier and more successful—to reclaim themselves as the originals they in actuality are, and no longer the forgeries some parents, and their reparative therapists, would have them become.

4

※

Ethical, Moral, and Practical Considerations/on Banning Reparative Therapy

This chapter hypothesizes that the question "should reparative therapy be banned?" cannot be answered in its original form, and suggests that instead the question be reformulated by being broken down into its component parts—separate questions that all concerned—gays, reparative therapists, and involved laymen, including activist politicians—can better understand and respond to both rationally and meaningfully.

These component sub-questions consist of the sub-questions described in the following sections, whose possible answers are either stated or implied.

Who is to make decisions about the dilemmas I describe in this chapter? How knowledgeable and well trained do individuals have to be in order to make and formalize them? Should a professional organization such as the American Psychiatric Association make these decisions because this is a doctor issue, or should the politicians make these decisions because those most affected are citizens whom they have a sworn duty to protect?

How can we resolve the conflict between therapists' individual freedom to treat others as therapists see fit, and the imposition of outside, often governmental, control of the therapeutic process, which is put into place, at least ostensibly, for the individual and general good? How can we protect

patients from malpractice while protecting their therapists' rights to freely practice their profession as they believe it should be practiced, without overzealous law enforcement outsiders interfering? If there are therapists who seek to resolve their patients' conflicts about sexual orientation by reorienting their patients sexually through conversion therapy, should we prevent these therapists from trying to do so? Should we, or should we not, make laws regulating what consenting adults talk about in private?

When dealing with medical issues shouldn't science trump political correctness, so that reparative therapy ought be evaluated solely on scientific grounds then offered, or banned, much as any therapy—even lobotomy for same-sex orientation—ought to be either offered or banned on its scientific merits, independent of the popular view/its popular appeal?

Is determining what is practice and what is malpractice in this field especially difficult to impossible to do because there are no clear end points, with no one school of thought the immutable gold standard, and thus the only one that ought to prevail?

Should all reparative therapy be outlawed? Should everyone subscribe to the official hard line (of the American Psychological Association) that "for a psychotherapist to persist with [the reparative] approach to treatment given the strong data regarding lack of successful outcome of the so-called conversion therapies constitutes ethically problematic action"?[1] Or is the issue more nuanced than that? For, after all, there are many different reparative therapists and reparative therapies practiced by either more or less rigidly orthodox reparative practitioners—that is, some practitioners who are more, and some practitioners who are less, mainstream (some of whom do, and some of whom don't, actually practice what they preach). There are different reparative therapies with different goals and different methods practiced differently by variously competent, sensitive, caring, well- or poorly motivated practitioners, making some reparative therapies more valid and hence more acceptable than others, and so theoretically at least less subject to being biased or less subject to the ravages of incompetency, and therefore less in need of being officially banned. For some reparative therapists, the goal may involve attempting to change sexual orientation outright: an unlikely outcome. For other reparative therapists the goal may be to leave sexual orientation intact but to assist the patient to become less sexually active/celibate, via managing, altering, diminishing, or lysing strong homoerotic impulses without converting them to heterosexual ones. For some, the goal is to enable a patient to marry heterosexually (this form of therapy according to Throckmorton and Yarhouse is referred to as "sexual identity management")[2] through harnessing a patient's conscious will, enabling the patient to adapt without

actually having changed. For others, those more traditionally inclined, the idea is first to offer help in a general way by using a therapeutic method that is overall traditional and only then veer into reparative therapy not as a matter of routine but only in those individual cases where therapists sense they have a willing, malleable patient who, in the midst of receiving mainstream treatment, requests that the therapist give reparative methods a try.

Are there black-and-white extremes of effectiveness? Is reparative therapy *either* harmful *or* helpful, a helping hand *or* a form of child abuse, a loving endeavor *or* a hate crime, with nothing in between, such as "somewhat harmful yet somewhat helpful"? If done judiciously and performed with sensitivity, can reparative therapy ever have more advantages than disadvantages, some positive features as well as many negative ones, as it actually works to some extent, and in some ways, but mostly leaves sexual orientation untouched? So many gay men feel less *broken* than they feel *excluded* (a typical reaction on the part of this group, as other, minority groups to a lack of acceptance by mainstream society), so that even if reparative theory is misguided overall or simply not right for a given individual, just the presence of an accepting, even welcoming, therapist is enough to help the patient feel better, even without what the therapist says and does having a lot to do with the improvement.

How can we protect patients, and especially younger patients, from predatory practitioners who might do them harm not only by wasting their time and money but also by exposing them to an ineffective and possibly (perhaps likely) dangerous procedure?

What if we ban reparative therapy outright for those patients who, for reasons of their own, "swear by reparative therapy" and beg not to have it outlawed because they believe it will help others like it helped them? Shouldn't we allow these patients, after giving them the information they need to make informed decisions, to make their own choices, perhaps in consultation with experienced traditional therapists and/or pastoral care providers? As Throckmorton and Yarhouse say, quoting Haldeman, "psychology's role is to inform the profession and the public, not to legislate against individuals' rights to self-determination," and "we must respect the choices of all who seek to live life in accordance with their own identities," and not add on pressures that keep patients from determining where they want to go with their therapy and their lives, a decision that should be based on their own perhaps unique values, desires, and ideology as they develop their own goals and objectives—so that they can apply for "treatment as they see fit without undue interference from the [nay-saying] practitioners"[3] Reparative therapists like Nicolosi ask, "if a person isn't satisfied being gay, why shouldn't he have help to reduce his

unwanted homosexuality and develop his heterosexual potential?"[4] In his view "*outlaw[ing]* treatment for unwanted same-sex attraction (SSA) is in striking violation of contemporary liberalism's own professed commitment to diversity."[5] Is it also "anti-research, anti-scholarship, and antithetical to the quest for truth"?[6] As Norman Siegel, who "led the New York Civil Liberties Union from 1985 to 2000" says in what amounts to a testimony about the viability of informed consent for instituting bans on reparative therapy: "While the government has every right to educate people on the harmful potential of huge drinks . . . [when it comes to banning them] let people make their own decision."[7] But others, disagreeing, claim that the government's job is to protect the people by discouraging them from doing what is bad for them, even if that means limiting individual freedom for the individual and the collective good.

Are some forms of reparative therapy not so much psychotherapy as they are simply religious practice by another name? Being faith based, is reparative therapy protected by our Constitution's assurance of religious freedom, or not protected due to aspects of the doctrine of separation of church and state?

Are reparative therapists actually doing harm by purveying not only a useless remedy but additionally one that allows emotional disorder to fester and spread while the ineffective therapy is going on? Is reparative therapy comparable to Wilhelm Reich's peddling likely worthless orgone boxes to naive impressionable people, not only wasting their time and money, but also discouraging them from getting effective other treatment in a timely fashion? Does reparative theory in practice and effect parallel a treatment given to a nurse I worked with who as a cancer patient was receiving an infusion of Laetril (a worthless cancer remedy made from apricot pits) and accepting this remedy from a renowned University physician, in the conviction that it would cure her, while her uncomprehending misguided physician allowed a cancer that likely could once have been fully excised surgically to spread and become inoperable?

Should a therapy be permitted if only a small number of people benefit even though many or even most experience no benefit, or undergo actual harm?

Can therapists reasonably be accused/convicted of malpractice if after obtaining informed consent they nevertheless prescribe a certain treatment after having made it clear to their patients that it probably wouldn't work? Does obtaining informed consent award carte blanche to therapists who can simply by dint of being honest thereby go ahead and do what they like and think is right, however misguided they happen to be, and others think, or know, that they are?

These questions cannot generally speaking be answered definitively. All one can do is delineate the (few) positive and (mostly) negative features of reparative therapy. Extremes of government intervention are both protective and coercive while extremes of nonintervention are both (helpfully) noncoercive and (dangerously) overpermissive. Should each encounter be evaluated separately and on its own merits, according to who specifically is doing and recommending a practice, and who specifically is receiving and accepting its ministrations? The question is not only "what therapy?" but also "for whom?" In particular, the individual patient's pre-therapy sexual makeup has to be, but too rarely is, taken into account. It's one thing to offer core homosexuals reparative therapy and another to offer bisexuals the same thing. For some bisexual patients *can* be helped to make changes in what seems to be their primary sexual orientation, though even in cases of bisexuality, the therapy should, though not banned, generally nevertheless be discouraged.

I believe that all therapists, and particularly reparative therapists, should seriously consider and reconsider the meaning and nature of informed consent as a core part of their therapy. Therapists should begin sessions with patients who say they wish to change by exploring precisely why their patients aren't satisfied with who and what they already are. That is, reparative therapists need to consider the differential diagnosis of "self-dissatisfaction." A preliminary evaluation should attempt to discover whether the wish to change is preferential or symptomatic, that is, is it "what I want" or "what I feel I must have" or "something that somebody else is forcing upon me"? Too often therapists assume that an antigay identity is primary and that the desire to forfeit one's gay identity is the true, ego-syntonic position. But for some it may not be that way at all, with the need to forfeit one's *overall* identity being primary, and part of the problem, as well as taken out on (displaced onto) one's *partial* identity, one's sexual identity/orientation. Which self are we to encourage and which self are we to protect the patient from? The one that says "I am who I am" or the one that irrationally thinks, due to the presence of an identity problem, "I may be gay but I want to, and can, go straight with just a little judicious help"?

Most adults when they first come for therapy are under too much duress over the prospects of internal upheaval, religious damnation, and social exile to choose their therapy and their therapists wisely, no matter how much informed consent is (supposedly) operative. Gays bullied and threatened from the outside by the possibility of becoming outcasts are the ones most likely to seek therapy, really a therapist's protection (against "persecution" for being gay). But being as desperate as they are, they are

extremely vulnerable to misguided ministrations they don't fully understand, however much the questionable workings may be legally sanctioned. (Informed consent is discussed further in chapter 11.)

I believe that an important focus of any anticipated "ban" on reparative therapy should be not so much on the therapy itself but on its theoretical underpinnings: the (often patently false) theories that constitute the core basis of reparative practice. Then what happens next will take care of itself. If we definitively discredit the idea that being gay is a disease, then fewer further actions will need to be taken to ban its cure. If we discredit theories about how developmental, interpersonal, and intrapersonal events cause being gay—with being gay necessarily a manifestation of fixation-regression, anxiety that is the result of wish-fear conflicts, problematic early and current interactions with parents, and the like, there will clearly not be nearly the same thrust for doing reparative therapy to change homosexuals to heterosexuals as there once was, meaning that the need for outside/governmental intervention will collapse of its own lack of support. We all have parents, no parent is perfect, and all parents to some extent parent pathologically. The pictures of the ideal mother who is not too close and not too distant, the ideal father who is manly but not so grotesquely so as to become a satire on himself, and the like are stereotypes that either don't exist in real life or do exist but have nothing to do with homosexuality's failure to develop. For while all gays, as well as everyone else, have a somewhat marred developmental history, only a few have a marring developmental disorder. While not a few gays have problematic parents, few have really serious parental problems. While many gays to some extent feel ashamed of who and what they are because they were brought up to resonate with shame, few have a shame-based disorder based on this particular aspect of deficient upbringing. In short, those who are homosexual are not homosexual because of anything, good or bad, ideal or less than ideal, in their psychological background. No important scientific connection has ever been established between the formative years and a later gay outcome, if only because the outcome generally occurs before the completion (or even the start) of the formative years. No firm intellectual bridge has ever been built between everything that goes wrong or right in gays' early life and men and women becoming, and being, gay. If the developmental view were taken less seriously, then reparative therapy wouldn't be so widespread, and so we shouldn't need to ban it so vociferously.

Also, a ban on reparative therapy might inadvertently apply, with disastrous results, to other forms of therapy that have serious parallels with reparative therapy. For example, perhaps any form of therapy where there

is a big power imbalance (between a therapist who, though in some ways misguided, knows what is right for the patient and a patient who, as all patients supposedly are, is too broken to reliably decide on his fate) should be banned, whether the technique is reparative, or psychoanalytic, or behavioral. But that might effectively mean banning not only reparative, but all forms of talk psychotherapy.

I believe that politicians should look elsewhere for a sphere of influence for themselves. For political interventions should be founded entirely upon medical considerations, and these are not part of politicians' proper armamentarium. To the extent that reparative therapy is technically and spiritually speaking *medically* incorrect it should be officially discouraged, like any other form of bad medical treatment. But that should be decreed by medical people. The prospect of discouraging a medical treatment politically is unappealing. Politicians commonly don't know what being gay really involves, or what is and what is not properly in a psychotherapist's bailiwick. Or when they do know it to some extent, they are too often merely speaking from the so-called knowledge they have gained from their own personal experiences/therapy, which are far too limited to have universal meaning and significance for everyone else.

I believe that limiting bad treatment only to those over 18 is a face-saving conceit. It makes those who are against reparative therapy feel as if they are doing something helpful by protecting youngsters. While it is true that younger people are more vulnerable than those who are older, limiting malpractice to those under 18, while it would give the illusion of action, in fact would mostly constitute inaction or bad action, since it is possible that any form of therapy that harms teenagers would also harm adults (and conversely any form of therapy that helps adults might well help teenagers). Adults are as defenseless as are children against self-proclaimed experts who are sufficiently persuasive to impress patients (and their parents) into believing everything they hear, and even what is wrong and potentially dangerous. Transference makes this true, with so many adults so childishly dependent on, in the transference to their therapists that they lose some of their powers of mature reason, to the point that many adults, like the youngest of children, have to be protected from themselves. Reparative therapists who apply pressure to frightened adults vulnerable to being pushed into going along with the reparative ministrations of "experts" can effectively infantilize those who are otherwise mature. Persuasive authority can make everyone feel childlike. And adults made by authority to feel like children are in need of the same protection as any other "youngster." For everyone in transference is in some ways a youngster, if not chronologically then at least intellectually and emotionally.

5

❖

Harm

INTRODUCTION

This chapter sets forth some of the ways reparative therapy can, and does, actually harm patients. These days, most reparative therapists are aware of the challenges to reparative therapy in the media, by such professional organizations as the American Psychological Association and by patients who have unsuccessfully undergone ungay therapy. Reparative therapists know that they have been, and still are being, accused of doing a great deal of harm to their patients. So they come up with clever ways to spin/hide the harm they do. They defend themselves against accusations of hurting their patients through the use of sophistry whose goal is to convince others, as well as themselves, that what they do is not only effective but also comes with minimal negative side effects/complications. They protest, "Changing is the patient's idea, not ours." They say, "We obtain informed consent, so we are giving our patients what they want." They claim, "We don't think that anyone was hurt by our efforts." They aver, "Sexual orientation change therapy is only talk, and talk can never really harm anyone." But as described below, my personal experience as a patient in psychoanalytic reparative therapy, although it was not called that at the time, underscores my belief that misguided therapy, however much it is all talk, being all wrong, can still do significant damage as it did to me, causing me to waste years of my youth denying my physicality for many so-called greater goods, mostly promised but never delivered—including

being able to become an analyst; pleasing my parents; kowtowing to homophobic bosses like the professor of psychiatry who didn't want me on the staff because "I was . . .," and meeting my own need to fully yield up to my personal grossly demanding and punitive conscience that, like a misguided advocate for all the wrong causes, put principles over people—myself in particular, as I zealously prioritized the transcendent value of renouncing deserved and palpable secular pleasures for the chimera of questionable future rewards. Reparative therapy can wreak its harmful effects, as described in the following section.

PATHOLOGIZING HOMOSEXUALITY

Two trends existing in psychiatry today clash with one another: depathologizing, or demedicalizing, homosexuality that *reduces* guilt (and external criticism) about who and what one is, and pathologizing, or medicalizing, homosexuality that *increases* guilt about who and what one is (in preparation for making someone over into something else). The (reparative) process of *pathologizing* homosexuality includes viewing normative homosexual behaviors (ranging from same-sex sexual attraction itself to specific forms of same-sex sexual activity including, or especially, anal sex) as abnormal, while downplaying, or entirely overlooking, both the actual pleasures and meaningful interpersonal involvements that characterize aspects of homosexual love and sex, doing so in order to be able to view all homosexual behaviors as deviant, or, worse, as disgusting, ungodly, sinful, and/or criminal. Thus reparative therapists dividing homosexuals into tops and bottoms pathologize those who indulge in anal sex as tops as sadistic, and those who indulge in anal sex as bottoms as masochistic, and, expanding that view, go on to condemn homosexuality as a symptom of an underlying illness (in this case a sadomasochistic personality disorder), in addition to being an actual, active, illness in and of itself (the sickness we call "homosexuality"). Reparative therapists who do harm by building this (rickety) bridge between certain forms of sexuality and sexual and personal psychopathology (without any real evidence that such a connection exists) in effect wrongly describe and explain a healthy biological function in psychopathological terms. They then, after postulating this false equivalency, pursue trivial, misguided, therapeutic interventions aimed at curing "the illness" of which they speak. For example, they mistreat homosexuality as if it were a phobia or a compulsion, something they do without any scientific evidence that their presumed equivalency actually constitutes a true identity.

BRINGING ABOUT NEGATIVE SIDE EFFECTS/ COMPLICATIONS

Any form of psychotherapy, like any medical intervention, can have both beneficial and harmful effects. Adverse consequences and undesirable complications of therapy of any form, such as the creation of excessive dependency, or the induction of problematic anxiety and/or depression, occur commonly. Even good therapy well done can on occasion be harmful, for the best therapists make bad mistakes from time to time. But harmfulness is practically assured when therapy starts with the unsupportive premise that all gays have problems and need help because they are broken, then routinely embraces and acts upon the unacceptable belief that, as a consequence of being broken, all gays need to be fixed.

DELAYING THE COMING OUT PROCESS

Reparative therapists shackle gays by discouraging them from coming out and instead encouraging them to stay in, or to go back into, the closet. They want gays to become "un-gay" by hiding who they are—by acting straight, or by becoming celibate, which amounts, at the very least, to encouraging them to become inauthentic, and at the worst, leads to fostering their becoming a nonperson.

CATCHING PATIENTS UP IN THE POLITICS OF REPARATIVE THERAPY

With predictable consequences, reparative therapists sometimes seem to represent less a school of therapy than an activist political party. They ask their patients to be, like them, committed industrialists in the "ex-gay" business. For example, they drag their patients into the limelight so that they may give testimonials as to the effectiveness of the cure they have just undergone. They often also ask their patients to accept the hypocrisy of the therapists' politics. Thus reparative therapists who condemn the American Psychiatric Association for depathologizing homosexuality for political (nonmedical) reasons themselves pathologize homosexuality for equally political, nonmedical reasons and expect their patients to do the same.

ACTING BASED ON FLAWED STATISTICS

Reparative therapists tell their patients that the reason they set out to cure them of their homosexuality is that statistically gays have a markedly

higher incidence of emotional problems than do heterosexuals and that the problems that they have are more severe than those of their straight counterparts. But reparative therapists are especially poorly positioned to judge the prevalence of emotional and behavioral problems in the gay population overall, because they only see, and so only become familiar with, those individuals who have sufficient difficulties to lead them to request medical attention.

LOWERING SELF-ESTEEM/INDUCING DEPRESSION

Reparative therapists often lower their patients' self-esteem by demeaning them—as when they call their gay patients defective—of less value than, and having a second-class status as compared to, straights. This renders their gay patients guilt ridden and intensifies their already negative broken self-image—the very same broken self-image that prompts gays to come for treatment in the first place. Patients get depressed and become despondent because they are told that they are "inferior people, no different from children who are adopted, illegitimate, love children, children from some lesser race or ethnic group, and/or pathetic people due to some significant disability that disfigures and cripples them, if not physically then emotionally." In a vicious cycle, depression causes gays to do what they are being accused of having done: to act out sexually, becoming promiscuous to feel better about themselves, only to then, despising themselves even more for being gay, seek even more sexual contact to once again try to improve their self-image so that they can come to despise themselves less. Some feel sufficiently deaffirmed to become suicidal, and of these some even make a suicidal attempt, which is too often successful, as they destroy themselves as a way to act on the feeling that they are "despicable people"—only to have their therapists make things even worse for them by, instead of admitting that their proffered therapy is what is defeating them and stopping treatment, advising them that therapy is their last and only hope for feeling better, so that they need to undergo even more of the same.

BULLYING PATIENTS

Focusing on problems (which admittedly is something that all therapists do) can by itself be a form of bullying via, as Ben Brantly says in another context, reviving "the particular dread that runs like an icy rivulet through even the happiest childhoods . . . the nagging awareness of the monster under the bed, the bully on the bus, the first day of school and the teacher who lurks there to make your life a humiliating hell."[1]

The deaffirmative premise "You are broken" is particularly brutalizing for gays who, as have many, been subjected to bullying in the past. Hearing from a therapist that one is flawed in one's core because one is suffering from an illness called "homosexuality" is especially problematic for gays who have already long been exposed to the very same message every day from friends, family, and strangers. Gays told that they should not accept who they are/condone what they do with their bodies/excuse what they do with their lives will likely be less helped by this so-called "constructive criticism" than they will be harmed by being destructively picked on all over again.

COERCING PATIENTS (ESPECIALLY HARMFUL TO THE YOUNGER PATIENT)

Some inpatient facilities, often at the request of parents and/or the courts, shanghai younger patients into SOCE. They might even force outpatient adolescents into accepting nonresidential (outpatient) treatment. Coercion is supposedly acceptable for adolescents because adolescents by virtue of their age and inexperience are unable to know what is good for them and so are unable to readily participate in their own therapy. Indeed, many reparative therapists believe that not only adolescent gays but all gays are so incapable of knowing their own minds that others *must* decide for them what they need and want, tell them how to get it, and order them how, where, and when to comply.

Coercion generally has a paradoxical effect by causing gays to become more resistant, as they stubbornly, deliberately do the opposite of what others expect of them and order them to do. Thus as many reparative therapists who set out to force a cure of adolescents' homosexuality "before it becomes ingrained" too soon discover, reparative treatment can have the reverse effect—as the adolescent entrenches just to be stubborn and get revenge. (In my opinion, a better approach might be to observe gay adolescents for a prolonged period of time before even considering intervening—the goal being to see what happens, possibly to determine if they do or do not remain prevailingly gay/become prevailingly straight on their own.)

Of course, mainstream psychiatry itself is not entirely free of the use of coercive methods, such as some of those that involve behavioral remediation. But, generally speaking, mainstream psychiatry reserves coercion for patients who are very ill and cannot actively participate in their own cure, and/or are a danger to themselves or others, and/or need physical restraint. The upshot is that mainstream psychiatry generally reserves coercive

treatments for such serious emotional disorders as schizophrenia or severe psychopathy. But coercive techniques are anathema to treating homosexuality first, because homosexuality is not an emotional disorder and second, because one's sexual orientation is, and should remain, one's personal choice, not subject to being told what orientation one ought to embrace.

Reparative therapists justify their coercive methods by citing something other than the homosexuality as the reason why their patients should be treated involuntarily. Some exaggerate "associated" symptoms to concoct a reason to force intervention/incarceration—in particular the so-called symptoms of conduct disorder and adjustment reaction of adolescence. As the APA notes, "private psychiatric hospitals . . . use alternative diagnoses—such as GID, conduct disorders, oppositional defiant disorders, or behavior identified as self-destructive to justify forced SOCE" thereby undercutting "the adolescents' rights to consent to outpatient and inpatient mental health treatment."[2] So often doctors tell their patients that the reason they as patients fail to cooperate is because they are so delusional that they are just imagining therapeutic intrusion into their lives and upon their personal integrity. Doctors may do this to back up parents who want their children locked up and forced to take the cure. The only thing that saves many normal children from even more damaging, forced physical "cures" is the faith people still seem to have that *verbal* psychotherapy can cure almost everything.

Parents tempted to force a child or an adolescent to undergo a treatment that is unlikely to work need to explore their own motivations for buying into the possibility that sexual reorientation is possible and that forced therapy can accomplish that end. These motivations would include the following: acting based on lack of information; shame that a child is embarrassing them by being gay; fears of social isolation as punishment for being parents of gay children; and the outcome of parents' personality problems including severe passivity, major control issues, and paranoia about what others might be saying about them—all together leading them to conclude that the only way to "do right by their children" is to force them into therapy. Thus motivated they readily buy into the propaganda of advocacy programs as they place their sons or daughters in inpatient programs with benign-sounding, charming, seemingly loving names that these organizations use to cleverly spin malignant intent and practices, as they cunningly pass themselves off as officially licensed treatment centers, although they are in fact cults attempting to indoctrinate not only the child but also his or her parents.

No parent should allow a child/adolescent to be psychiatrically hospitalized just for being gay. And every parent should see to it that any facility that hospitalizes children/adolescents for any reason must allow the hospitalized children to retain outside contacts/obtain outside support—for example, by maintaining established relationships with peers and, when necessary, obtaining consultation with outside neutral, affirmative practitioners. Essentially only violent/acutely highly suicidal patients should be medicated against their will and/or put into seclusion. Before parents bite and send a child off to an organization whose avowed purpose is to treat homosexuality on an inpatient basis, they should learn everything they can about the organization, especially whether or not its therapists are sound and whether the organization has in place a positive attitude that respects what their patients want for themselves. Do they spin coercion as valid behavioral therapy consisting of a remedial, "I will let you see your friends if you agree not to touch your, or someone else's, genitals?" The young, being emotionally immature and socially, if not personally, dependent, need others and especially their parents, to use their *good* judgment *protectively*. That means parents must reality-test for their children by sifting what they hear about reparative therapy, and the organizations that do it, through their own observing ego. Because children are especially vulnerable to being deprived of their civil rights and having their physical and mental health damaged, Mom and Dad need to protect them from being exploited—which too often leads to the equivalent of their being emotionally raped.

Residential therapy predictably does harm when it facilitates sexual contact with other "aspiring ex-gay patients" and at times even with the patients' own therapists. The resultant boundary violations further catalyze and intensify the guilt and self-questioning that is in the first place a somewhat predictable complication of any form of treatment.

USING TECHNIQUES THAT DEVIATE TOO GREATLY FROM MAINSTREAM PSYCHIATRY

Reparative therapists harm their patients by failing to treat them in a way that respects mainstream psychiatric principles. It sometimes seems as if the cumulative knowledge of hundreds of years of psychiatry has taught some reparative therapists nothing, as if basic traditional tenets, such as the importance of therapeutic moral neutrality, can be simply discarded to good effect and without negative consequences.

Too few reparative therapists seem to feel the need for making a diagnosis and determining the nature of, in preparation for working with,

idiosyncratic, pathological transference/transference resistances. Too few identify and correct for their own personal countertransference distortions. Too few incorporate mainstream therapy's hard-won principle of only imparting insight when the patient is ready, that is, of only making interpretations when the therapist can impart insight safely without unduly creating more anxiety than he or she relieves—by stirring up more emotionally laden material than patients can handle at any given point in their therapy, and at any given time in their lives.

TREATING ALL GAYS AS THE SAME

Reparative therapists apply the same methods to a wide swath of gay patients without considering patients' differing personalities, needs, and (especially) vulnerabilities such as the individual's ability (or inability) to handle the release of repressed material. Few avoid digging too deeply with gay borderlines/schizophrenics. Instead many too readily open up their wounds causing such brittle patients either to fall silent and stop cooperating or to decompensate emotionally due to induced intense anxiety. Reparative therapists assuming that all gay patients are the same, that they all have the same needs, and therefore that they will all respond equally well to the same regimens/interventions, treat one patient like another when they treat OCD gay patients as if they are not much different from borderline gay patients, in effect advocating one-size-fits-all diagnosis-independent techniques for everybody gay, without recognizing that some gay patients are more, and some less, amenable to a given therapeutic approach than others. Homosexuals are not all alike. They are not a homogenous group all properly assignable to the same therapeutic pool, having all grown up in exactly the same way, with similar parental problems resolved in a depressingly identical, highly pathological, manner. Moreover, they are not all equally motivated, intelligent, and insightful, meaning that insight-oriented therapy is right for some but wrong for others, such as those who might instead benefit more from therapy that is supportive than therapy that is by nature revelatory.

BEING INADEQUATELY TRAINED/UNTRAINED AS THERAPISTS

SOCE is often practiced by licensed mental health practitioners (LMHPs) who are inadequately trained and therefore unable to deal with the complexities involved in treating anyone, let alone those who are seriously

unhappy with who and what they are. Too many reparative therapists if trained at all are trained only in the area of their primary interest and concern (reparation), and so are completely unprepared to work with the whole patient. To avoid having their patients decompensate fully, well-trained LMHPs know to pull back when they seem to be making the patient anxious. In contrast, poorly trained reparative therapists tend to routinely view surfacing high degrees of anxiety as a sign of progress—as if a patient's decompensating is proof that their therapy is working. So they continue as before, or even ratchet up their efforts, likely causing their patients added pain as they double down therapeutically at a time when they should instead be pulling back judiciously. Untrained therapists not alert to the timing of their interventions overlook the dimension of a patient's readiness to hear what they have to say. So they either speak prematurely without holding off until the time is right to say something specific, or are consistently brutally honest—saying things whether or not their patients are prepared at any given moment to hear them, and doing things that their patients are not prepared to accept and embrace. They forge ahead convinced that opening up all the psychological sluice gates is the royal road to reversing "pathological homosexuality," although doing that is more likely than not to be a direct pathway to intensifying emotional problems. In particular, untrained reparative therapists not uncommonly dwell on a patient's sexuality without being mindful of how "just talk" about sex can threaten patients who are erotophobics, that is, those who are uncomfortable with their own sexual nature due to being embarrassed by many, most, or all of their sexual fantasies.

Untrained reparative therapists too rarely consider how transference distorts a patient's view of who the therapist is and creates a misunderstanding of what the therapist says and does. Transference leads patients to come up with imaginary interpretations of much of what goes on in therapy, as patients simultaneously don't respond to that which is actually occurring and respond to something that is not happening at all. Transference leads many patients to specifically introduce distortions into therapist-patient communications thus turning a mutual endeavor into an adversarial situation. I know of patients who have had engulfing mothers respond to an actually non-engulfing therapist by feeling uncomfortably pressured because they viewed an actually permissive therapist as overwhelming and controlling. I have known patients with rejecting fathers who have come to feel rejected by an actually accepting, respectful therapist whom they see, irrationally as it turns out, as Dad dissing and as otherwise saying mean things to and about them.

Conversely, sometimes therapists excuse the bad therapy they actually do do by citing nonexistent negative transference distortions on their patients' parts. I have known many patients who have actually suffered at the hands of rejecting therapists complain that their therapists rejected them only to have the rejecting therapists tell them that they only *feel* rejected because they are misinterpreting their therapists' intent transferentially. Reparative therapists who become overly close to their patients in order to be supportive of them too often blame their patients for feeling inappropriately engulfed—*as if* they are being raped—when in fact they *are* actually, at least on some level, being emotionally or even physically violated.

When therapists act like amateurs/incompetents in these and other ways, the healthier patients quit and the sickest patients (who are unable to defend themselves) stay. With amateur/incompetent therapists, patients dropping out of therapy is not a sign of incorrigibility, but is rather a healthy option, for it is indicative of the patient's getting better and a sign that the patient is becoming stronger.

USING TRIVIAL/BIZARRE TECHNIQUES

Simplistic behavioral modification approaches, almost always rooted in simplistic concrete thinking, lead to seemingly remedial, but in actuality ineffective, therapeutic actions whose effects are as negligible as their theoretical underpinnings are shaky. For example, therapists who don't allow gay patients to use men's rooms anywhere because just entering one might relight (homo)sexual fantasies tend to forget that, as any person gay (or straight) knows, (homo)sexual fantasies, more than representing a simple conditioned response, also occur in response not only to place but also to person, and can just as easily be aroused on a public street as in a public men's room. (In my opinion, making a patient agoraphobic is no more an adequate therapy for being gay than making a patient celibate is an adequate substitute for a patient's having gay sex.)

Aversion therapy, a form of behavioral therapy that might consist of administering electric shocks that leave burn marks, or having the patient swallow a bitter pill/smell rotten meat while sexual excitement is being induced via exposure to pornography works, if at all, only for a very short time and in a limited fashion. Laughably trivial aversive behavioral modification techniques such as snapping a rubber band on one's wrist every time one thinks homosexual thoughts—a supposed mental discipline technique—produce no lasting positive results, as well as creating a few negative ones, such as disdain for one's therapist.

INDUCING/ENHANCING SPECIFIC PREEXISTING EMOTIONAL ILLNESS

OCD

Reparative therapists who recommend praying the gay away can foster a patient's OCD by fostering guilt and enhancing scrupulosity and ritualism.

Histrionic Personality Disorder

The use of trivial therapeutic interventions such as the self-hypnotic suggestion of Emile Coué[3] (consisting in the reparative realm of telling oneself the equivalent of "every day in every way, I am going straight") risks advocating/enhancing histrionic suggestibility.

Avoidance/AvPD

Reparative therapists can induce avoidance by promoting celibacy or by recommending isolation from a world that triggers gay fantasies, effectively advocating isolation from life. Patients can develop avoidant personality disorder (AvPD) when they come to believe, because they are told, that it's better to be isolated than to be gay, and that there are great rewards in the future/afterlife (however illusive these may be) for those who do not live sinfully in the present. Patients who give up on love completely wind up living a limited, agoraphobic life, avoiding forbidden places (such as a certain beach) to avoid having their environment relight their gay fantasies. Gays told that for therapeutic reasons they must stay away from other gays, predictably abandon their friends and lovers in order to get better through avoiding having their homosexuality fired up by contiguity. Only then, especially as they get older, they generally find it harder and harder, after reparative therapy fails, to go back to resume their old gay friendships, and just as hard to go forward and make new gay friends. The gay community responds with hostility to those who abandon it once by exiling them forever. Some gays compensate for the loss by gaining strength from their faith and joining a religious community that now better accepts them as "a good person, one of their own." Other gays are left totally isolated and lonely, with nowhere to go and no one to love, stuck somewhere between being recent (not yet integrated) entrants to a (reluctant) straight community, and recent dropouts from an angry (gay) community. Too many gays told that as a condition of their renaissance they must sever all ties with other gays, drop gays who love them (and whom they should, and do, love in turn) just because these others are gay. Some go on to abuse other

gays as part of this cure—in order to more fully separate themselves from gays and move on to become straight. These other individuals suffer a great deal from being dropped with little to no explanation offered. As victims they feel mystified as to why they were dropped and react to the cruelty involved by becoming depressed. Gays who try to act straight when they are still gay often enter marriages that turn out to be a sham: disastrous for the adults involved and possibly harmful to any children born should the marriage dissolve, leaving the children to grow up in a broken home.

"As-if" Personality Disorder

Too often therapists encourage patients to act "as if," an emotional state which Helene Deutsch[4] has described and which some view as part of the *borderline* condition. Therapists who recommend living "as-if" straight may in fact be inducing more general "as-if" psychopathology character-ized by pathological identifications (with straights) strengthened by not-me ("I'm not gay") dissociative defenses that can even lead to the development of a full multiple personality disorder (*DSM-5* Dissociative Identity Disorder).[5]

A homosexual man thus encouraged said he had become straight when in fact he had only become "as-if straight." His new identity consisted mainly of showily carrying his books like a boy not a girl hoping that "since sissies carry their books like girls, carrying my books like a man will make me hetero." He hung out with other men who called him "dude" as his way of dealing with his being effeminate, doing so at the suggestion of a therapist who in effect was telling him that he should see himself (not as gay) but as others at least pretended to see him (as straight).

Anxiety and Depression

Patients say, "Tell me what I need to know about myself to get better." But they really only want to say, "love me as I am, and don't try to make me over. Stop telling me bad things about myself. For that makes me extremely anxious and very, very sad."

Reparative therapists who are excessively critical/rejecting can make their patients anxious and depressed simply by telling them that they are in fact broken. Promises to fix gay patients when broken predictably make patients angry and depressed because these patients come to blame them-selves for not changing over to become someone else as requested—as if their recalcitrance is the problem, when it is the therapists' excessive expectations that are creating much of the difficulty. Patients who please

not themselves but their therapists by getting married heterosexually often end up in failed marriages, not only depressing themselves but also doing considerable harm to their spouses.

Patients who enter therapy with preexisting emotional problems are particularly vulnerable to becoming depressed as a side effect or direct result of their treatment. This is especially true for patients with obsessive-compulsive disorder (OCD) who are likely to feel even dirtier than before, and for paranoid patients who are already suspicious that the whole world is against, and punishing, them because they are gay.

Paranoia

Even patients who voluntarily apply for reparative therapy may come to feel if not entirely forced into treatment then controlled and coerced by it. They feel misunderstood and badly treated, and that shades over into their feeling persecuted.

Identity Problems/Gender Dysphoria

Gay patients already in conflict about their identity/their sexuality can respond to challenges to their core (gay) identity with destabilizing identity crises. In my experience, it is particularly destructive to demand effeminate gay men to forcibly enhance a masculine identity they choose to disavow, threatening them to the point that they feel as if they are being not changed but mutilated.

PTSD

Messages of disapproval drummed into patients can induce posttraumatic symptoms including such anxiety equivalents commonly associated with post-traumatic stress disorder (PTSD) as gastric distress, headaches, muscle tension/back pain, and hypertension where patients, instead of feeling uncomfortable and openly anxious, express their anxiety indirectly in the form of bodily (conversion) symptoms. Many gays whose reparative therapy has induced PTSD complain of constant nightmares, particularly ones involving being separated from their partner, or from their family, and not being able to contact the person—generally (these days) because a cell phone doesn't work, or works until a crucial moment at which time the connection becomes lost: clear expressions of the separation anxiety inherent in feeling forced to walk a new and lonely path—fearful of being traumatized, abandoned, and hence being alone forever.

Masochism

High-risk, self-punitive sexual behaviors such as barebacking or compulsive sex-club clubbing, even when identified as self-realizing, self-fulfilling, and self-affirming, in fact may instead represent self-destructive masochistic enactments. Those therapists who recommend mortification may be openly advocating what some consider to be masochistic self-abuse.

Dependency

Reparative therapy like any form of therapy that is excessively prolonged or intense can create dependency on the therapist and the therapy. Patients in a prolonged reparative psychoanalysis can take up living a pseudo-life characterized by a full focus on their therapy and their therapist. As a result, they spend too much time in and between sessions wondering whether they are improving. They obsess about whether or not they have fully digested what their therapist said. They become overly involved in wondering whether what they told their therapist was right and complete. They brood about what their therapists think of them and whether or not they are loved. Eager to get not a life but to their next session to determine how they are doing and correct, complete, or expand upon what they said, to work out old problems in their pasts that they think might be contributing to their being homosexual in the present, and to look for signs that their therapist does or doesn't like them, they fail to live a full life outside of treatment, doing so precisely so that they can more fully focus on what goes on in therapy. Abandoning day-to-day functionality and the pleasure principle for a future orientation founded in too much reality principle, sacrificing now completely for some of tomorrow, they focus on future goals over present pleasures, only to ultimately abandon their current humanity in favor of improving their future identity.

Confusional States

Reparative therapists quite likely confuse their patients about what is right and what is wrong (e.g., about what is moral and what is sinful). Their patients, feeling mentally destabilized, start questioning themselves (one form of the "Q" of LBGTQ = lesbian, bisexual, gay, transsexual, questioning). Then they attempt, usually without success and only adding to their confusion, to reorganize by doing exactly what their therapists advocate for them. Should they become isolated from their peer groups at their therapist's prompting, they can experience additional sensory deprivation, and that causes them to become even more perplexed than formerly.

Impotence/Asexuality

Sexual impotence can be the result of SOCE's creating an unconscious link, which may be semipermanent or permanent, between "sex" and "bad," leading a man to lose his ability to get an erection and/or develop ejaculatio tarda, or a woman to lose her desire and experience reduced sensation. Too often while sexual performance deteriorates, sexual urges prove tenacious and even paradoxically increase, leading to strong desire corrupted by the inability to gratify it. Both sexual intimacy with other gays and nonsexual intimacy with straights can be affected, sometimes to the extent that patients, feeling, and being, alone and isolated, resort to compulsive Internet dating and street pickups to replace the personal relationships that they no longer have but still feel they desperately need.

Alcoholism/Multiple Substance Abuse

Alcoholism and substance abuse can represent an attempt to feel less lonely by substituting (often multiple) addictions for (forbidden) relationships. Both alcoholism and substance abuse can represent an attempt to deal chemically with emotional disappointment about not getting better from a promised cure clearly not forthcoming.

EXCESSIVELY FOCUSING ON GAY ISSUES

Reparative therapists overly focused on sexual orientation neglect other aspects of the patient—his life, and her well-being. Blaming all of a patient's problems on being gay, these therapists convince themselves that if a gay individual is depressed or has interpersonal conflicts, it must be entirely due to his or her being gay. Focusing on reparation, as the *Psychodynamic Diagnostic Manual* (PDM) puts it, the therapist "fails to explore and expose those aspects of [a gay man's] personality that [a gay man] spends inordinate energy trying to subdue."[6] Moreover, involvement in reparative therapy too often deters a patient from seeking help elsewhere and assistance for pressing problems that may have nothing to do with being gay.

BLAMING THE PATIENT

Reparative therapists say that patients who attempt/commit suicide during or after treatment do so "not because I told them that they were broken and needed to be fixed" but "because (part of their already being

broken) they entered therapy already suicidal, and were in their suicidality just following a predetermined downhill path."

BEING HOMOPHOBIC

As discussed in Chapter 12 and throughout this book, many reparative therapists have negative (homophobic) countertransference (highly personal) responses to their patients. Nicolosi even openly states "that it is legitimate to place higher worth on heterosexuality within the framework of one's value system."[7] Reparative therapists like him, by viewing gays as defective and as second-class citizens, routinely intensify the therapist-patient power imbalance that already exists in all doctor-patient relationships, further poisoning the therapeutic atmosphere by hurtfully demeaning their patients through putting them down simply by not treating them as equals—thus causing/exacerbating the patients' internalized and internal homophobia, leaving the patients feeling as if they are unworthy, misguided, ill, and amoral. Thinking that they are thus being helpful by firming up their patients' desirable straight identity, reparative therapists instead weaken their patients' legitimate gay identity, which for gays is the one that is authentic and the one that carries the most weight. Simultaneously, by sending negative therapeutic messages to their patients, reparative therapists predictably worsen gays' preexisting emotional symptoms, especially any anxiety and depression to which as gay patients they may be prone.

Some reparative therapists act homonegatively out of the conviction that they can best motivate their patients to change by shaming them into it. But it is equally true that being shamed seriously interferes with the very change making that the therapists are presumably trying to promote. Instead of favoring progress it stalls forward momentum by bringing on iatrogenic (doctor-induced) depression, alcoholism, and self-destructive behaviors including suicidality, all of which impede the cure that therapists are presumably trying to bring about.

Characteristic of homophobic countertransference is the view of homosexuality as both a mental illness and as sinful. Both views can be a product of the therapists' own mental activity, the first of the therapists' self-hatred, the second of the therapists' self-shaming. In the first case, prejudice reigns, in the second case scrupulosity is what is ascendant. In some cases an assessment of the "Deity as He is" is but a concrete representation of the therapists' own morbid psychic condition as it is. In such instances religious beliefs lead to practices that are symptomatic of the therapist's need to be rigid and harshly controlling often to the point that

they disregard their patients' own needs and disavow their patients' rights to freely express themselves, thus making such therapists coercive individuals who force gays to share in the dictates of their own harsh consciences of whose tenets they are convinced represent the only valid and authentic possibilities. When therapists' religious edicts represent a product of this sort of projection, the therapists demand that their patients be as utterly free of sin and immorality as they themselves want to be and feel they should be. This demand leads to recriminating against their gay patients exactly as they recriminate against themselves—their self-hatred projected onto hating gays, their negative ideas about gays a reflection of their negative view of themselves displaced outwardly onto their patients, as these therapists cleanse themselves of the bad inside by cleansing those outside of themselves, their patients, of evil, the same evil they perceive to be within, as a result abandoning their personal humanity in favor of widespread cruelty to those whom they call (with a great deal of irony) "my patients."

Homophobic reparative therapists often deaffirm gays not by what they, their therapists, do but by what they, their therapists, omit doing. They also do so nonverbally. Because most patients assume that nonverbal messages, coming as they do out of the therapists' unconscious, better reflect what their therapists really think than what their therapists actually say, nonverbal (negative) messages are especially potent and as such increasingly likely to become part of the patients' (negative) self-identity.

Homophobic reparative therapists routinely go on to minimize their negativity by excusing it as a reflection of personal opinion and belief to which they are, like everyone else, entitled. They also spin their homophobia as being a product of admirable allegiance to the conservative, especially to the evangelical conservative, movement.

A reparative therapist's homophobia took the form of his viewing being straight as mainstream and gay as off-kilter. He decried how gay cultures inappropriately elevated homosexuality to the same level as heterosexuality, with gays foolishly assuming that both were morally and functionally equivalent; denied that homosexuals were psychologically and emotionally identical to (the equal of) heterosexuals; and additionally insisted that gays suffered from a uniquely wide variety of psychiatric disorders, in particular those marked by suicidality and substance abuse. He routinely affirmed such stereotypes as homosexuality's equivalency to effeminacy. To him, homosexuality was triggered not by physical appeal but by chronic apprehensions going back to childhood, particularly those arising out of a fear of one's father leading to the adult gay man's desire to

have many men to prove that he has no reason to be afraid of any one man. That is, while this therapist viewed heterosexuality as being a product of sexual *instinct*, he viewed homosexuality as a sexual *symptom*—the product of the acting out of an unhealthy conflict-laden state ending up in a yielding to various fears such as a fear of disapproval and a fear of being shamed. He routinely defamed the homosexual lifestyle based on his own heterosexual biases, as he made his own, not his patients', life values the only ones worth approving of. He brooked no disagreement about his own values. Though they all were heterosexually imbued, he considered them to be the only ones worth embracing.

EXCESSIVE COMPLIANCE WITH PARENTS

Reparative therapists harm their patients by siding with parents who, feeling guilty about making a child gay, live out their guilt by sending their children off to therapy to "help" them. Parents mostly do this less for their children's sake than for their own benefit—to relieve the guilty feeling that their children became gay because as parents they raised them up badly, for example, "because I as his mother was remote my son never knew how to relate to women," or "because I as his mother was engulfing I turned my son off women and onto men." For obvious reasons, reparative therapists support the parental (guilt-inducing) belief that family interactions caused a child to become gay. Though being gay is a biological event that generally no one can influence, and so for which no parent should ever be held fully accountable, reparative therapists blame parents almost entirely for making their child gay when instead they should shift from blaming to ascertaining: helping the parents determine what their child might need as a human being in preparation for providing him or her with that exactly. Reparative therapists preoccupied with changing a sexual orientation that can't and shouldn't be touched fail to encourage parents to try to deal with any covalent (additional) problems a gay child might have, in preparation for handling their child, not mishandling his or her homosexuality.

THE HARM MY REPARATIVE THERAPY DID TO ME

Aspects of my reparative analysis offer a cautionary tale for others tempted/attempting to go straight by analyzing their gay away.

My "reparative analyst," failing to turn me straight, instead kept me living for five years "as-if"—as-if I were a straight man. I was integrated into the community, dating women occasionally, and, by living

a lie, prospering financially in private practice. But I was nevertheless constantly lonely and longing to return to being unashamedly gay—even at the risk of being exiled by my professional community or, worse, losing my medical license and perhaps even my physical freedom, after being caught in a raid on a gay bar. During this time I substituted alcoholism for homosexuality, with drinking heavily the only extramarital indulgence, so to speak, the medical culture of the time permitted me. For alcohol provided me with a degree of relief from the always building sexual tensions pressing in on me in the absence of any opportunity for real, satisfying, physical discharge. Not surprisingly, while my outward behavior had changed considerably, my sexual fantasies remained steadfastly unaffected.

My actual "psychoanalysis" consisted of my lying on the couch four days a week talking about random matters seemingly without a defined goal/objective/purpose, hoping that somehow I would convert, as if reborn, through that always forthcoming (but never actually occurring) definitive "aha moment." I went along, having felt a strong sense of guilt about being gay. That, along with the medical community's almost official condemnations for what I was, reined me in sexually somewhat throughout most of my analysis. I say "somewhat" for as I look back, on occasion I must have relapsed into my old, gay ways. For example, after years of celibacy I got drunk and had sex with a friend. My analyst's response wasn't "No surprise, after all it's been over four years and for someone only in his twenties that's a long time," but "Of course that happened because you were as usual acting out, and as is your wont, opening yourself up by getting drunk around someone gay." That is, her therapy wasn't working not because it was ineffective but because I was defective, and my being defective was the reason why I wasn't fully cooperating in the taking of her cure. This attitude on her part made me feel as bad about myself as I gradually came to feel about my future prospects. For I began to see that what I was doing was sacrificing my personal life for my professional life, preparing for a future that would never come by making my present one that could never satisfy.

Although my professional life had improved in many respects, in others it actually suffered greatly. I was accepted as variously straight and ex-gay by my medical establishment. But in addition to my regular job I had to work at night just to pay for my analysis. As a result, I didn't do my regular job as well as I otherwise might have: because I was always borrowing time from my regular job to get to and attend my night job, the one that was paying for my analysis; because I was constantly running off from work for my therapy hour; because I was constantly exhausted from working two

jobs; and because I was constantly depressed since I was working all the time and not doing much else, leaving me feeling not only fatigued and exhausted but also alone and isolated.

The beginning of the end of my analysis came when my analyst, after having waited for four years to make it clear that I was persona non grata as far as being accepted by the psychoanalytic institute was concerned, told me, in certain terms, that she would block my being accepted as a trainee in her Freudian Institute. I could never become an orthodox psychoanalyst. Had she been leading me on all this time because she needed a warm body to fill her practice, only to reject me when, as I later found out, someone else, someone "more analyzable" (according to how she thought and worked) came along? Though I knew I couldn't realize my dream of becoming an orthodox Freudian psychoanalyst, at least in Boston, yet, being very passive and hooked on the transference, my view of Freudian analysis as nirvana assuming primacy, instead of moving on and perhaps seeking a non-Freudian non-Boston institute that might have welcomed me, I continued with my Freudian analyst, preferring to fail at something first rate than to succeed at what I perceived to be second rate. Clearly for me coming in second signified not being a runner-up but coming in first, only now on a list of losers. So I never pursued other equivalent pathways. Instead of asserting my true value elsewhere I accepted being devalued by officialdom at home. Thinking it was me, I never challenged those in power, though it was them, not me. Thinking I failed to match up to their ideals, I went along completely, even though theirs were not so much ideals as they were ideology: of an inflexible, immutable, excessively scrupulous sort formulated and lived out in a kind of bathysphere where certain characteristics that supposedly described which doctors were unsuitable as orthodox analysts ruled, pigheadedly. As it happened, the characteristics they deemed undesirable were those that exactly matched the characteristics I possessed.

Hearing that there was no future for me as an analyst I deliberately chose to go "from the cloth to the blacklist." I got a boyfriend. As a result I was given the choice: give up my new boyfriend or give up being analyzed. I chose the boyfriend, and my analysis was over. I was asked to leave. Finally I realized that by sacrificing living for pseudo-living for a higher cause I had become the undesirable equivalent of an acolyte who had taken the vows of poverty (because of the high analytic fee); chastity (because I was following the rule of abstinence); and obedience (because I was faithfully attending sessions four times a week, living in an enclosed bathysphere where the only acceptable action was one that, having a life of its own, retained little relevance to the wider world out there—taking

place as it did in a micro-parallel universe with little to no contact with, or serious meaning for, anything resembling real life.)

I was being analyzed to become straight when I should have been analyzed for wanting to become someone I wasn't and for having become self-sacrificial, without protesting, even to the point of having developed what amounted to Stockholm syndrome—consisting of excessive and self-brutalizing passive conformity, as if I were a kidnapped child who could have escaped but was held there by an invisible emotional thread. For years, though knowing things were not working, I had chosen not to leave but to stay—out of fear, hope, misplaced devotion, and masochistic abrogation to a more powerful person, a False God with bad ideas infused by a Good Higher Cause based however entirely on shaky flawed earthly premises. So, though I should long ago have terminated the analysis myself and found another analyst, I had instead waited until my analyst terminated me—abruptly, as it turned out, because I wasn't up to her standards. Even so, afterward, for years, instead of feeling rescued from, I felt thrown to, the lions.

Recovery after my analysis was over was slow and painful. Not surprisingly, after months and months of having no meaningful sexual life I turned from being virtually celibate to being actually promiscuous, mostly seeking partners after getting drunk. Alcohol released a sexual attraction that, though after having been stymied as forbidden for so long and ultimately weakened, was now beginning to once again forcefully reemerge.

I spent my next years lonely because by now I had separated not only from my friends but also from the gay culture to which I, of course, belonged. I now longed to rejoin the gay subculture to which I had an affinity. But that reality wasn't so simply achieved. My postanalytic life was characterized by a stifling chronic regret over time and money wasted and energy misspent. I also had no way to easily make up for the five years lost. I was in a persistent state of grief about my having failed at going straight. I realized that the reality is that it's one thing to look for a partner when you are 21 and another to seek one when you are closer to 30.

As I look back at this postanalytic phase, I think I was suffering from chronic PTSD-like symptoms due to having been rejected personally and professionally. I could have used help with this and my concomitant depression; but I didn't seek it because I thought that I had simply failed miserably, and for that there was neither excuse nor remedy.

I had developed was what I now recognize to have been an ex-gay treatment syndrome. For me this consisted of residual social problems accompanied by depression and serious characterological difficulties associated with persistent problematic behavior in part driven by a persistent

lack of self-confidence that led me to reject the obvious need to put my failed reparative experience completely behind me to seek a new life, one that was both real and vibrant, and especially one where I avoided multiple sadomasochistic relationships characterized by my picking people who rejected and punished me. It was as if by my being so constituted I was reenacting my failed analysis, this time with new people, doing the same thing over and over again, now hoping that the outcome would somehow be different. I was repeating that rejecting, dismissive time over and over in an attempt, generally failed, to come to some form of peaceful closure—running through one relationship after another consisting of repetitive submission to, alternating with tyranny toward, a series of men I now both idealized and disdained—much as I had once come to both idolize and distain my analyst. My malformed personal relationships with others occurred in a setting of severe alcoholism, and recurrent explosive temper tantrums when things didn't go my way. I was pretty clearly in a state of rebound hypomania, one that overstated "At least I'm through with that," and consisted of a prolonged struggle to rid myself of the poisonous introject of my analyst by relating to, then rejecting, one person after another, hoping this time to find someone different, someone new, someone less traumatic, someone better to help me get going again. Instead, secondary self-blame for my wild behavior was the order of my day, making me even more depressed. Depressing too was how I blamed others for my failures and how that, along with being wary about letting others into my life, kept me from finding a worthy partner.

I finally entered therapy with an affirmative therapist: someone who urged me to give myself not a hard time but a vote of confidence, to put my bad experience behind me and get on with my life, grieving once and for all for what was not to be and getting it over with, without more regrets and further delay, without continuing to be as self-punitive as was presently my wont.

Fortunately, a new era in psychiatry was dawning where psychiatrists had begun to recognize that it was the reparative analyst not the to-be-repaired patient who had the most problems. Fortunately my new therapist helped me finally close my jaundiced superego's eye previously always open to the evil that I had supposedly done. Fortunately, too, I started living in the present, refusing to yield to orgiastic self-condemnation/torture, no longer desiring to constantly work on problems I didn't have for some better, greater cause, working toward some more gallant future—only to destroy my here-and-now functionality for a tomorrow that would never, and didn't actually need to, come.

6

The Role Played by Religion and Religious Beliefs

In this chapter I discuss the interfaces between homosexuality, homo-negativity, and religion, noting that these constitute trouble spots marked by conflicts—particularly those involving the difficulty aligning the vengeful, punitive religious view that gays are evil people who do bad things with the more sanguine, more compassionate, more sympathetic religious view of gays as human beings who, like everyone else, are neither saints nor sinners, but simply people who rightfully can, and deserve to, give themselves a vote of confidence and in turn most certainly get one back, especially from those who claim to be in the business of trying to help them.

A uniquely difficult task involves reconciling the positive, loving therapeutic messages in the Bible with its negative pronouncements about homosexuality to create a model for helping gays reconcile their sexual orientation/sexual orientation identity with their religious orientation/religious identity so that they no longer have to feel, as they often do now, so hopelessly torn between two seemingly incompatible, diametrically opposed philosophical extremes.

Mainstream therapists do not generally view homosexuality from a moral perspective. They see it as being neither bad nor good but as value free. In contrast, reparative therapists often view homosexuality as a moral failing. They devalue it as a sin that must be forgiven and as a crime that

must be punished. Mainstream therapists view homosexuality as a variant of normal, and as such as a moral nonevent. In contrast, reparative therapists view homosexuality as an abnormal form of heterosexuality, existing on a lower moral and spiritual plane. Mainstream therapists often condemn religious reparative practitioners as bigots. Reparative therapists often condemn mainstream therapists as unacceptably activist emancipators. Peace seemingly cannot be forged between the two so very different positions and concomitant practices so that homosexual individuals can become fully comfortable with both their sexuality and their religious beliefs. It can seem as if contrary viewpoints can never be incorporated into an overall benign rational comprehensive view that reconciles all sides, resolving conflicts between turfs so that all concerned can both live free of conflict within themselves and make peace with others around them. Those who on religious grounds stigmatize gays for their sexual orientation and atheistic tendencies need to reconcile with those who, homosexual or not, condemn the negative aspects of religiosity as anathema, antithetical to what they stand for, irrelevant to the positions they embrace, and completely antagonistic to what they believe.

Unfortunately, in my experience at least, conflicts between being gay and holding orthodox religious views about homosexuality are basically irreconcilable. I personally believe that for many gay people the best and only way to "resolve" these conflicts is through a process akin to *dissociation*. Here differences are not fully reconciled, or even diminished, but ignored by simply "agreeing to disagree"—in a polite way, stepping aside from factional fighting (with others and within oneself) to simply allow differing viewpoints to coexist in close approximation to one another—to dwell side by side *without* the feeling that one has to make a full attempt to reconcile them. Dissociation doesn't involve compromising either one's faith or one's sexual identity by disavowing one or another side of the argument, thus disavowing some important and personally meaningful principle or discriminating against some significant aspect of oneself. It simply involves maintaining extreme opposite views intact, though irreconcilable, after abandoning any attempt to finally bring them fully into line, thus keeping both one's faith and one's sexuality, *and* respecting both one's religious and sexual sides—instead of trying to make the two somehow jibe, but instead simply accepting that one in fact has to be hypocritical about this matter just to function effectively and even solely to keep one's sanity.

Few serious observers suggest and embrace such a solution. They claim that that involves using the pathological defense mechanisms of denial and splitting, and doing so in an unsavory way just to force some

semblance of healing. To them, while the end is vital, it doesn't justify that means. They say that as a solution this particular one is inauthentic. Then they stigmatize inconsistency even more than they stigmatize homosexuality.

But I believe that there is no other choice but to live with mixed views about one's homosexuality, without trying to be rigidly, 100 percent, consistent. Certainly gays themselves must live with the opposing views, often as much within themselves as on others' parts, that homosexuality is okay *and* that homosexuality is a spiritual/moral failing, a sin, and/or a crime.

To some this dissociative mindset is sophistic. To others, including myself, it is both unavoidable and acceptable, for those who hold such views do not manifest an imperfection of mind wracked by uncertainty, but simply reflect how impossible it is to make one's mind up in the face of poignant conflicts like this.

I believe that all gays should actually read the Bible. If they do, they will find in it examples of that very inconsistency that I describe. To me that gives gays permission to themselves to be inconsistent—to accept contradictions like the ones in the Bible without ever having to profitably address and completely bridge over the disconnect between gay love and gay shame, between gay satiation and punishment for being gay—instead modeling themself on the inconsistent Great Book thereby allowing themselves, as the Great Book does, to hold opposing views while letting each contrasting view survive comfortably alongside its obverse, so that gays can simultaneously take the good things in the Bible to heart and hear what it says in its less than sanguine sections, as they simply accept the good-bad dichotomous view gays sometimes have of themselves, much as others have of them.

There is simply no purpose in arguing whether being gay is a sin against God, a natural/God-created condition, or is without any relevance whatsoever to issues involving an Almighty. One's faith is, according to Freud, writing in the *Future of an Illusion*, an "illusion" because its tenets can neither be proven nor disproven.[1] Recognizing this helps one embrace both religious and secular approaches to oneself.

This said, all concerned have to deal with a painful reality: that the minister who prays the gay away has as his or her true rival, the psychiatrist who prays the minister away.

Nonreparative and reparative religious psychiatrists tend not to accept each other. Mainstream psychiatrists who overlook that for many gays God has significance for, and a special input into, their daily lives, fail to offer religious gay men and women the possibility of receiving some of the benefits of their beliefs and so the potential benefits of therapy with

faith-based features. Such therapists try to keep religious beliefs entirely out of their therapy even when they know that religion can heal. Mainstream therapists who do this can effectively deprive gays of the beneficial aspects of religion should they suggest that gays submerge their religious beliefs entirely to resolve the struggle between their religion and their gay identity, and especially if they recommend doing so because they believe that religious faith is pathological, as if talking to God, with, or without, God talking back, is always symptomatic, that is, must belong to the category of delusional and/or hallucinatory.

In contrast, some religiously oriented therapists insist that their religious beliefs are authentic and primary constructs and can be effectively employed as therapeutic tools founded in reality. These religious therapists as exemplified by those who advocate praying the gay away may believe that religious approaches are effective for reasons that nobody fully understands. Or they may feel that the healing is God's work, and as such is not the result of suggestion or autosuggestion processes, that is, closely allied to the placebo effect. A positive effect indisputably exists and that it is God's work is not fully deniable. So people like Robertson urge that homosexuals turn to Jesus, for "Jesus will take sins away, if you're a homosexual he'll take it away."[2]

Other religious therapists, not going quite so far, cite an author such as Cameron, according to whom "studies showed that not only was a belief in God strongly correlated with positive treatment outcomes, but the intensity of the belief in God paralleled the degree of hopeful expectations placed in therapeutic interventions." According to Cameron, "this observation accords well with William James support of the 'mind cure' movement which allowed for thinking yourself to better health. The effects of transcendental meditation on physiology, e.g., blood pressure, are well known eliciting the 'relaxation response,' a response also elicited by visualizing a soothing image or memory or repeating something [one believes] in, perhaps a spiritually significant, religiously specific word or short phrase like 'Hail Mary, full of grace.' Some observers see at least an aspect of the positive results of religious therapy as due to the placebo effect, which they cite as a real reason religious, or any, therapy works. They don't deny that the effect is as Cameron says, a "part of the power of positive thinking and autosuggestion . . . a poorly understood aspect of the mind body connection consisting of latent variables that we don't measure though we harness [them] for sanguine therapeutic purposes."[3]

Too often religiously infused therapists suggest gay men submerge their gay identity entirely in favor of their religious identity even if that means hiding who they are from others and from themselves, even when doing

so out of guilt and shame. They might even recommend that gays not come out at all but rather stay in for religious reasons—at least until they first explore how their being out can, and routinely does, clash with their religious convictions and even their life objectives. They recommend gays consider the possibility that they would be better off staying in, in order to fully retain their religious affiliations, as if coming out necessarily involves renouncing aspects of a desirable, religious self, the one right, and moral, persona.

Too many religiously oriented therapists not only aver that their beliefs are not simply the product of their own mental activity and claim a direct connection to God but also formulate this God as a punitive Being who originates/shares their view of homosexuality as "dirty and diseased" and who asks His worldly representatives to do His job for him: cleansing homosexuals of their sins. Those with a narcissistic mindset fed by omniscience and omnipotence even make therapy more into a sermon than what it should be, a patient-therapist joint endeavor, demanding not health through growth but redemption through prayer and faith. These religious therapists too often act as if they hate not only the sin but the sinner as well.

RECOMMENDATIONS

Faith-based therapists can best help their patients by using religiously based *affirmative* techniques that guide their patients to become not more but less self-punitive to the point of being able to count their homosexuality among their blessings. A first step involves therapists freeing themselves up from noxious, assumptive, rigid, projective, punitive value systems and judgmentalism. Idiosyncratic belief systems and prior foregone conclusions about what is good, right, and appropriate for gay patients—especially beliefs such as the one that as reparative therapists they have ascended to a higher moral and spiritual plane than the one their gay patients are on—should instead be diluted by therapeutic neutrality where therapists stop forcing their patients to accept the therapist's own personal, speculative faith-based concepts of what is and what is not The Almighty's design. Such therapists need to avoid conflating what is moral with what is normal and instead accept their patients' (differing) approaches to/views of normalcy as being as acceptable as their own. While these therapists are unlikely to subject their belief that one can pray the gay away to a procedurally correct controlled scientific study (that like any other well-done study requires careful selection of subjects and the use of rigorous control groups), they can at least accept that some gay

patients choose to be philosophically agnostic and religiously atheistic, and so where these patients are concerned modify their faith-based principles and techniques in the direction of encompassing more traditional psychotherapeutic tenets and practices, ones that avoid at least some of the more rigidly controlling, idiosyncratically religiously based interventions that some reparative therapists find so tempting to advance and put into action.

Scrupulously religiously oriented gays might themselves benefit from becoming more flexible in their religious outlook. To do so they might seek personal relationships with the more liberal among any of their very conservative contacts—perhaps those who while still religiously orthodox nevertheless encourage gays, as they do all people, to accept themselves as they are. Such gays might even consider attending a less conservative church—one that advocates being not what they should, but what they truly aspire to, be. They might modify their view of God to consist of fewer negative projections of personal animosity and more projections of personal positive self-acceptance, leading to a conviction that their God loves gays, not that they are unworthy of His adoration.

Gays *must* filter out personal negative overly zealous external religious views that assess them as sinful, criminal, diseased, and unconscionable unless they become heterosexual or celibate. They need to accept themselves more as they are, after developing less of a self-directed savior complex of their own, one that demands obeisance to the dictates of some higher homonegative sadistic power that longs to affirm their own punitive conscience-driven hypermorality however much the morality comes out of a fire and brimstone value, moral, and ethical system that demands that gays find themselves wanting because as godless reprobates they need a religiously oriented therapist to offer them not help and succor for their problems, but comeuppance for their crimes and punishment for their sins.

7

Personal Problems of Reparative
Therapists/Homophobia

In this chapter I supplement my discussion of some of the more academic shortcomings of reparative therapy with a discussion of some of the more personal and professional shortcomings of reparative therapists. My goal is to help gays who are either considering or actually in reparative therapy evaluate not only their therapy but also their therapists—to evaluate the treatment they are about to get/currently being given so that they don't take everything that their therapists say as gospel, and to try to get as much that is positive as they possibly can out of a treatment experience that is likely to be overall more of a negative involvement than a positive encounter. Too many reparative therapists commit the following offenses.

They

- ## MANIFEST A TENDENCY TO NAIVELY BUY INTO
  AND ACT UPON FAULTY RESEARCH

Some reparative therapists naively accept as gospel, and act upon the tenets of, poorly designed studies in their field, as if they must do precisely what these studies imply/indicate/recommend. For example, many reparative therapists make it their mission to turn gays straight after reading the Regnerus study's supposed proof that children of homosexual parents are significantly more likely to have problems than children of opposite-sex

parents. They use the study findings to support SOCE based upon the belief that children raised in homes where both parents are of the same sex differ (in negative ways) from those raised in traditional homes in such parameters as level of education, impulsivity, and suicidality, so that "at least modest benefits . . . accrue to the children of married biological parents."[1] However, according to other observers, this study has too many serious methodological and conceptual flaws for it to serve as a proper motivator for doing SOCE. For example, Regnerus mainly observes not children of same-sex couples but children who were raised in households where one member had a relationship with someone else of the same-sex (and so had been unfaithful to his or her spouse). To me, the clear existence in these particular households of relational problems of a general nature makes the issue of homosexuality moot. For in my experience, in heterosexual-style marriages where one spouse is cheating on the other with a same-sex partner the damage is done less by the spouse's "homosexuality" than by his or her infidelity. The children know that one parent is being unfaithful to the other, and it is that that most affects these children in negative ways. Loyal to the "victimized" parent, the children resent the unfaithful parent's actions and additionally feel threatened because they fear that they will ultimately be living in a broken household. Being miserable affects their entire outlook on life and so warps their ongoing development. So, for me, the call in such situations is not to do SOCE but to do marital therapy—to solve not sexual orientation issues but marital problems, whatever the sexual activities/preferences of the participants in the marriage under study happen to be.

Another example of a study with serious procedural flaws creating clinical problems for those who believe in and act upon its results is Spitzer's study of reparative therapy.[2] Ultimately Spitzer apologized to the gay community for the results of the study. There was no "scientific misconduct," only a problematic study design mainly involving Spitzer's poorly conceived method of case selection, as well as, according to Besen, distractive affiliations and the personal ambitions of "an over-the-hill stage horse."[3] (This study is discussed further in Chapter 11.)

Still another example of a poorly designed study that reparative therapists buy into and live by is Bieber's study[4] where only psychoanalysts (likely with preconceived notions) treating gay patients, and not the patients themselves, were interviewed. Bieber didn't report what he heard directly from gays who had applied for treatment to change. Rather he essentially recycled what other therapists (psychoanalysts) treating gays (or possibly just thinking about treating gays) told him. I had some professional contact with Bieber and worked with many of his associates. I also

personally knew many of their analysands. To date, I am still amazed that these personable and intelligent analysts, men and women who treated many of my colleagues effectively in nonreparative therapy, participated in Bieber's study instead of detecting (and reporting on) its flaws. Perhaps this is because these analysts themselves sometimes used reparative procedures, without that amounting to their doing full reparative therapy. This is not uncommon or surprising, since reparative procedures, if not full therapy, are, if only secretly, a cornerstone of many a therapist's armamentarium/practices. That is, some mainstream psychotherapists/psychoanalysts in effect subspecialize in reparative therapy for a select group of their patients. Clearly this makes completely banning reparative therapy not only undesirable but also impossible, for to do so one would have to ban all therapy that incorporates any reparative procedures into its protocol, even when that therapy is overall nonreparative.

• ARE INADEQUATELY TRAINED

Many problems with reparative therapy are not unique to this form of treatment. Rather they occur in all forms of psychotherapy. In particular, though psychotherapy is a "discipline" that requires study and practice, many of its practitioners, from all schools, have nevertheless undergone inadequate (poor or essentially nonexistent) training. Poorly trained therapists, simply making things up as they go along, get things wrong both in theory and in practice. As do psychotherapists from other persuasions who are legitimately licensed, some reparative therapists, especially those with minimal training that is additionally of a low quality, for obvious reasons tend to inflate the worth of their training to make it seem as if they have had adequate, extensive, advanced schooling and supervised experience. Not a few reparative therapists with a strictly religious affiliation substitute their religion for scientific training, and their faith in God for their knowledge of correct procedure, in effect making God their professor. NARTH's own claim to be a "scientific and professional organization that *includes* (italics added) highly qualified academics and fully licensed mental health professionals"[5] says it all, for NARTH is unwittingly confessing that some unqualified academics and some unlicensed mental health professionals are doing reparative therapy under its rubric (and likely citing their affiliation with NARTH as an indicator of their reliability and official status).

Even with legitimately licensed MHTs (mental health therapists), not all their licenses are created equal. Just having a license doesn't assure an acceptable level of competence. Some licenses assess and assure medical

competence better than others. Some licenses are fraudulent or fraudulently obtained, in some cases the licensing exam questions having been somehow pilloried, even purchased by those who choose to spend more money to bribe others than to give of their time and money to learn things themselves. Many licenses are inapplicable to the procedure being performed (e.g., not every "Doctor" is a true Healer). Almost anyone can call himself or herself a therapist (e.g., a "mental health worker"), hang up a shingle, go into practice, and sometimes even qualify for third-party payments. Although some think that anyone with a desire to help people with emotional problems can do psychotherapy, I strongly believe that proper intensive training is necessary before taking on the considerable responsibility of working with patients, gay or straight, who are emotionally troubled in any way. The fact that psychotherapy is "all talk" doesn't excuse one from having to be officially prepared to say, and do, things right. Too often, once a therapist proclaims "I am a reparative therapist" the focus shifts from "what training and experience have you had?" to questions relating to the value and legitimacy of this particular form of therapy, obscuring an equally important issue—that poorly trained therapists, reparative therapists or no, by definition are incompetent.

In conclusion, while imperfect, at least licensing offers the patient some form of protection from outright quackery. While unlicensed therapists are sometimes sincere and natively talented people, the very fact that they have made no effort to obtain a license raises questions not only about their capability but also about their honesty, a quality to be treasured in therapists. The question then becomes not only "should reparative therapy be banned?" but "is this always psychotherapy in the first place?" and, if it is, "shouldn't everyone be required to have adequate training/a license before they are allowed to do it?"

Many therapists in the ex-gay movement are not only not LMHPs, some have a shady past—a few of whom have even been imprisoned for doing harm to others—often the kind of harm to others that makes it likely that they will also do harm to their patients.

That reparative therapy sometimes works even in the hands of untrained therapists is in part attributable to the power of suggestion and the working of the placebo effect: (overlapping) powerful forces that can be harnessed by many different therapists of different persuasions, varying technical abilities, various degrees of training (if they have any meaningful training at all); and various types and qualities of licensure. Also therapeutic "success" may simply reflect natural progression—the natural progression from gay to straight that is especially common in bisexuals (and many of those who apply for SOCE do so because they are troubled

bisexual in the first place). Too, even badly done therapy can have unintended good consequences, and that is often the case with those suffering from sexual compulsivity, relational challenges, and drug addiction. While therapists doing wild therapy/analysis should be reminded that "the road to hell is paved with good intentions," some therapists discover, serendipitously, that good intentions can be associated with, or by themselves be enough to pave the way to, a degree of therapeutic nirvana.

• ARE SELF-SERVING

As for reparative therapists who are themselves "ex-gay," there is no shame in turning one's life around by helping others—unless one is not helping, but sacrificing them as a way to do something for oneself. Reparative therapy seems to be a field that attracts those with a psychopathic bent—ranging from those simply motivated to deliberately defy the (traditional) establishment, to those who after serving a sentence for one crime defiantly go on to commit another—less to intellectually smash icons and more to sadistically hurt people, in this case, patients. Certainly, being ex-anything does not by itself confer the ability to help others change along similar lines. In fact *touting* being ex-anything generally speaking represents more of a countertransference problem than a psychotherapeutic solution, because "touting" represents a position that is antithetical to the neutral "blank-screen" stance I believe to be an essential for all therapists attempting to help other people realize who *they* are. In my opinion, extreme activism can become pathological, and pathological activism can interfere with the therapeutic passivity necessary to be maximally effective with patients who, as some gays are, are struggling inwardly with their own identities, and longing to find out, and realize, not someone else's ambitions, but their own.

• ALLOW THEIR COUNTERTRANSFERENCE TO PREVAIL/UNDULY INFLUENCE

Many therapists doing reparative therapy display a variety of personal psychological problems, subtle or overt, conscious or unconscious, that influence how they view and treat their patients. Some reparative therapists are doubly challenged due to being poorly trained as reparative therapists. (All purely reparative therapists are by definition effectively poorly trained because their theoretical positions, the cornerstone of training, are flawed, and for some their judgment is flawed by being pathologically infused by homonegativity or other forms of bias.) Many reparative therapists then

go on to excuse (rationalize) their psychopathology by giving sophistic explanations for their shaky positions. Often they aver that gays *are*(and are not merely socioculturally "perceived to be") second-rate citizens who to become first rate members of society first need to be made straight. They then call this psychopathological belief "a scientific theory," when in fact such a theory originates in personal problems of an idiosyncratic, and often extremist, nature.

For example, one reparative therapist was personally erotophobic due to his fixation on unresolved childhood guilt-laden incestuous feelings directed toward his mother. He dealt with his resulting personal pathology by embarking on a course of helping gays overcome *their* own sexual psychopathology—much as he hoped to overcome his, but could not do so successfully. This process, a "projection" and "reaction formation" combined, led him to develop a widespread ostensible (but not actual) (over)concern for gays and lesbians whom he said he wanted to "help avoid living the painful life that I believe that all homosexuals (like myself) predictably lead." As a result, he became a reparative therapist whose platform was that gays should all go straight, but (he would insincerely insist) "not for my own, but for their own good." He had become not only antigay but antifreedom—that is, it was obvious to his patients that he felt that they should go straight "not because you want to" but because "I want (order) you to, as I want (order) myself to do exactly the same thing."

Developmentally, his controlling attitude harked back to power struggles he had with his own parents, for he carried on the tradition of his parents' authoritarian relationship with him with his patients, repeating with them his own early traumatic submissive/rebellious relationship with his Mom and Dad. Predictably, with new candidates who consulted him he would spin his reparative approach to help assure that their "informed consent" was in fact "informed"—by him and his own psychopathology—the latter leading him to paint a distorted picture of gay life ("take it from me, it's a horrible way to live") and an equally distorted picture of himself: as a caring concerned individual whose main goal in life was to help gays turn their lives around—starting with their sexual orientation and ending with their not having to suffer through the gay lifestyle he himself was living, one which he so abhorred in others, and to such an excessive degree.

This reparative therapist frequently cited the Regnerus study that supposedly proved that same-sex parenting produced children who were broken compared to children who were raised in traditional households. He resonated with that study and made it his own because he himself had severe problems with his own parents, parents who, though straight, had

a troubled relationship with each other (they had divorced when he was very young and subsequently constantly fought over his custody). He blamed his parents for the homosexuality he so abhorred in himself and so deeply wanted to cleanse others of.

For this reparative therapist, much of the therapy he did was in fact an assault on gays disguised as a wish to help them. Helping gays represented a reaction formation against his hating them. His hating them originated to a great extent in self-hatred and self-disgust projected outward, then defensively transformed into misplaced empathy.

Serendipitously there were unintended positive consequences that arose out of his misplaced empathy. He was actually able to convince some of his patients that he really cared for them. And to some extent he actually did, and he even came across as being a loving person, and that more than anything else made him in some ways an effective therapist.

Here are some specific examples of therapist psychopathology that make up psychopathological countertransference responses to patients in reparative therapy and which manifest, often subtly, in the therapist's attitude toward gays and being gay.

Narcissistic reparative therapists attempt to cure homosexuality less for their patients' benefit and more for their own. They almost always claim to be more knowledgeable than mainstream therapists, as if they, but not those following traditional practices, possess special insight into the causes of, and methods for curing, homosexuality.

Paranoid reparative therapists treat others' differences *with* them as attacks *upon* them—as attacks not only on their expertise but also on their basic identity. They routinely view their patients' challenges to them not as self-realizing (for their patients) but as antagonistic (to their therapists)—as anathema to what they are trying to do, almost as if patients (and professionals) who have minds of their own are not merely challenging but are in fact persecuting, them, though the therapists are simply trying to help. Their reliance upon "faith" can represent more than a simple ideology. It can also be a way to keep challengers at bay, for faith by its very nature is not something that can or should be challenged, or if challenged, is something that inherently suggests ready defenses along the lines of "I am entitled to my belief, and yours is an anti-religious bias."

Histrionic therapists see their patients as rivals and compete with them as to who is in charge and who will triumph; for example whose sexuality is more and whose sexuality is less valid, acceptable, and enjoyable.

Masochistic reparative therapists pick fights with traditional therapists predominantly to bring down the wrath of the scientific establishment upon their own heads so that they *can* wallow in seeing themselves as

"poor me." They enjoy being seriously misunderstood and gravely wounded, yet, through it all, being, and being seen as, strong, unyielding defenders of their beliefs and guardians of their faith.

Hypomanic reparative therapists tend to be activists who see themselves not only as saving the world from homosexuality but also as affirming their own "unalloyed goodness," congratulating themselves that they, unlike gays, uphold family values—and by so doing are being both admirably right and highly moral—out of both respect and love for their Savior and because they are fully supportive not only of their own Mom and Dad but also of all the Moms and Dads of this world.

Depressive reparative therapists have a bleak worldview, which they convey to their patients by emphasizing the downsides of being gay, as they emphasize the downsides of most other things. They also operate in blaming terms as they see homosexuality as a personal downfall, one for which their patients, having chosen it, are entirely responsible. They blame all of their patients' life problems on their patients being gay, even though the thing at fault may not be "living while gay" but "being treated homophobically while living." Depressed therapists thus subtly stifle their patients' attempts to thrive and their wish to have a wonderful gay life (one that in reality they could much more easily fulfill if they didn't affiliate themselves with abusive therapists, who are the very thing that makes their life unenjoyable and unrewarding). Such therapists' "insightful interpretations" are in fact often veiled accusations which depress their patients, leading their patients to restitutive homosexual enactments that in turn move their therapists to recommend that their patients undergo even more of their treatment.

OCD reparative therapists' magical thinking, at times amounting to delusions of omnipotence, goes along with excessive perfectionism where they pride themselves on scrupulously and consistently following the rules, though these rules generally hold only in the bathysphere in which they operate, which means that the rules they go by have little to no connection to real life, rather being distinguished more by the quality of internal consistency than by the virtue of substantial relevance. That is, what OCD therapists say, and do, has less connection to who and what gays are than that it has a life of its own wherein therapists have become intellectuals with an impractical bent who concoct elaborate esoteric psychodynamic schematic patterns of developmental causation based on supposed meaningful early developmental events, especially parent-child interactions, that although in theory these postulates are literally accurate within the closed operating system they espouse, the theories are inapplicable outside of that system—having little or nothing to do with the

development or sustainability of homosexuality in real life. As a result, the reparative theory they hold and the reparative things they practice thrive without relevancy to or impact on the realities involved in becoming, growing up, and getting over being gay. Theirs has become an idiosyncratic ideology that suits some particular esoteric personal bent and satisfies some immediate and long-term inner personal and institutional need, but, thriving only in some parallel universe of theirs, meets few of the immediate and long-term needs of their patients.

Additionally, many reparative therapists with OCD tend to be overly controlling individuals who feel justified in ordering others about "for their own good." (They then attribute this practice not to their own controlling "anal" nature but to the need to impose controls on people like gays because, as all-knowing therapists, they are in the best position to tell gays what to do and encourage them to do it.)

They assure themselves that they have the mandate to demand that gays go along with them because, being right, powerful, and on top, they are entitled to assume command. In particular, issues of control contaminate OCD therapist's transactions involving informed consent. On the surface these therapists, thinking they are being admirably permissive, say they bend over backward to let their patients decide what they want for themselves. But though they speak of "informed consent" they actually obtain "uninformed, coerced submission." They have ways to seem to be letting their patients decide what they want while they subtly undermine aspects of their patients' free will that countermands their own (the therapist's) ideology. They influence their patients covertly, often by the use of such nonverbals as the eyebrow raised in a way that contradicts their own ostensible neutrality, of course in the direction of where they want their patients to go. Typically theirs is an ultimately readily discovered hidden agenda whose goal is to remake gays whose sexuality needs to be revised because gays are by nature abnormal, disgusting, criminal, and sinful (especially when it comes to having gay sex and refusing to stop it).

Their bullying of gay men and women into complying with their personal ideology can amount to a form of child abuse, both of the actual gay children and adolescents they treat, and of the child that exists within any person, gay or straight, man or woman, of any age, who becomes a patient. For all patients come to therapists with a childlike neediness and anticipation that makes them vulnerable to the ministrations of the "powerful expert"—with adults as vulnerable as they would be if they actually were children—leaving even grownup patients to respond to therapy not with an adult's skepticism but with a child's dependent, overly trusting, acquiescence.

Secular obsessiveness often shades over into religiously oriented obsessiveness leading to attempts at reformation along overly scrupulous nonsecular lines. Overly scrupulous religious therapists become not merely religious, but excessively, morbidly so, as they rigidly focus on issues involving morality (good versus bad) and contamination (clean versus dirty), so that they come to view homosexuality not as different from, but as immoral and filthy as compared to, heterosexuality. Some seem to be saying that while all sex is uniquely dirty, homosexuality is more uniquely dirty than heterosexuality.

Psychopathic reparative practitioners in a guilt-free manner vocally promise their patients more than they can ever deliver just so that they can encourage their patients to readily accede to their ministrations and do their bidding. They don't actually consciously lie. Rather they lie unconsciously, for lying for them is not their métier, but their symptom. Many deliberately, again unconsciously, justify the work they already plan to do by exaggerating the problems associated with being gay. They advance the (homonegative) hypothesis that being gay is a lifetime disadvantage just so that they can step in to help their gay patients have a better, more advantaged life. They are often persuasive enough to actually inspire their patients to have heterosexual relationships, often not to please themselves, the patients, but their healers. The heterosexual relationships their patients attempt under such circumstances tend to be forced, joyless, and hurtful to others, especially to heterosexual partners to whom these patients ultimately promise more than they can deliver, leaving their partners ungratified and disappointed when, as can happen, these patients ultimately cast these heterosexual partners aside for homosexual liaisons. I know of many unhappy spouses with unhappy children who have entered family therapy because of the tensions at home between a still gay man attempting to act straight and his straight wife attempting to act accommodating. The negative effect on such children was far graver than any negative consequences there may have been of being a child in a stable gay marriage raised by a stable gay couple. Children of unhappy partners where the partners are in a sham marriage can become children brought up not by a loving couple but by a newly divorced, single, struggling, unhappy mother or father.

Excessively *sadistic* therapists are fundamentally hurtful individuals who enjoy making others feel bad, so they comfortably abuse their patients (like they abuse everyone else) by denigrating them. They do so almost personally—by calling them names—typically pejoratives masquerading as scientific formulations concocted to prove an unscientific

but seriously hurtful point: that being gay routinely makes one (and others) unhappy.

Sadistic therapists emphasize negative aspects about gays and homosexuality, especially the minor negative things that in this imperfect world always regularly exist in association with things that are positive. (Something "shameful," in this case "the unacceptable gay lifestyle," can, by those who want to see things that way, always be found in association with something admirable, in this case, the satisfactory gay lifestyle.) To read sadistic therapists' descriptions of gays one would think that all gays are narcissistic, mean, vulgar, unrelated, distanced, hypersexual sadomasochists devoted and doomed to a life of unhappiness, disease, and loneliness predictably leading directly to their early death. Thus van den Aardweg sadistically proclaims that "homosexual behavior is . . . one of the sexual behaviors that inspire abhorrence in other people."[6] Sadistic therapists according to Besen are "heartless and lack even a hint of decency and humanity,"[7] for they claim that all homosexuals live a lurid lifestyle, since they are all promiscuous as well as all diseased and all heading for premature death while spreading disease to others, all the while recruiting children in the cause, and for sex. (Because some gays are pedophiles and some pedophilia is gay pedophilia, supposedly therefore all gays are pedophiles, and all pedophilia is gay pedophilia.) Again according to Besen, Cameron even concluded that gays should all be quarantined so that they don't "commit sexual mass murder."[8] Such therapists spiritually put a scarlet A on gays' foreheads, effectively isolating them from decent society, or even secretly fantasize putting in place a "final solution," one presumably involving starting a gay holocaust where straights build concentration camps to exterminate all those who are gay, lesbian, bisexual, and transsexual, and perhaps even those who, being bisexual, are merely questioning who they are.

Another sadistically pejorative criticism of gays making the rounds from time to time is the pervasive ideas many reparative therapists cling to: that all gays are socially withdrawn and that their loving relationships are all regressive reenactments of child-parent relationships, that is, theirs are not relationships at all but throwbacks, so that interactions between two gay adults are in fact interactions between two children both of whom are fighting to see not who is going to love and be loved the most, but who is going to mother whom, and who is going to whine the loudest when the teat, as it predictably will be, is actively pulled away, and/or turns spontaneously cold. According to a number of observers in the sadistic mode homosexuals are supposedly all immature, immoral, physically and emotionally diseased exhibitionists, fetishists, and voyeurs. As well they are

all paranoid, for they all supposedly imagine that they are being persecuted simply to gain sympathy as a way to be embraced socially, and by
those who are fair and just. Thus when speaking of gay love, van den
Aardweg hurtfully says that true gay love doesn't exist, only "erotic craving"[9] which can never be satisfied anyway because of a masochistic need
to be "the rejected one."[10] Too, he portrays gay love as consisting of a
"need to fulfill oneself and enhance one's self importance, a childish merging fraught with self-pity," a wallowing in the perversion of not finding
"the obvious natural objects of the propagation-directed sex instinct
attractive."[11]

Therapists who believe such things, who picture all gays so negatively,
often act out by discharging their sadistic fantasies toward gays not openly
but passive-aggressively, that is, in the guise of "sexually reorienting gay
patients *for their own good.*" They act out this way simply in order to convince themselves and others that they are not haters but lovers, and that
they do reparative therapy not because they dislike gays, but because they
like them so much that their only wish is to help them get better.

• MANIFEST HOMONEGATIVITY (HOMOPHOBIA)

Among reparative therapists who hold homophobic views about gays—
the very people they claim to want to help/cure, is Nicolosi, who criticizes
homosexuality (as well as God who presumably made men and women as
homosexuals) by asking, "How could I have been designed by the Creator
for anal sex?" Nicolosi "scoff[ing] at the American Psychological
Association's idea that homosexuality is equivalent to heterosexuality"
notes that "anal sex is damaging to the body. . . . It's demeaning to a man's
dignity; it's unhealthy. I couldn't have been created for a same-sex relationship whose *very design* makes biological parenthood impossible. . . . So
I was designed this way? Then I have been created by an absurd God."[12]

Reparative therapists who claim to be traditional mental health practitioners simply specializing in converting gays to straights can in fact be
seriously biased individuals with a gay negative mindset which, no matter
how they attempt to hide it, is readily observable and capable of being
conveyed to their gay patients, predictably leading their patients to view
themselves as negatively as others view them, thus enhancing patients'
self-questioning and intensifying their self-abusive, self-defeating self-
views, often to the point that these patients become so depressed that
they begin to literally view themselves as the second-rate citizens, sinners,
and criminals that their reparative therapists are virtually accusing them
of being. Now gays who yearn to change want to do so less to become the

normal heterosexuals that they say they long to be and more to become the less despicable people that their reparative therapists hint that they should admire.

Reparative therapists hold their negative views of gays in part because they lack complete knowledge of the subject of which they speak with such authority. This is partly because they rarely deal with a representative group of gays. Rather they tend to formulate their opinions about all gays from some gays—specifically those who, seeking treatment, are by definition of necessity somewhat unhappy about who and what they are—often because they are depressed, or in the throes of adolescent turmoil, or are borderline individuals struggling with identity confusion crises. Many such individuals, though ostensibly applying for therapy to go straight, are hoping to rid themselves of an emotional disorder that alarms them even more than does their being homosexual.

Reparative therapists tend to overlook the many well-adjusted homosexuals who exist, but, rarely coming to clinical attention, are generally invisible to therapists and so unavailable to present them with a well-rounded view of gays and gay life—that is, what a vigorous gay life is really all about and what it is actually like to be healthily gay. Reparative therapists formulate their theory and practice based on ideas that, almost understandably, seem to them to be rational and comprehensive. Only they are irrational and limited simply because, inapplicable to the population of gays as a whole, they only apply to those relatively few gays that populate the treatment rooms of reparative therapists.

The upshot is that homonegative reparative therapists tend to view homosexuality as an *emotional disorder in its own right* and/or as routinely leading to, and being *associated with, symptoms of another emotional disorder*, and one averred to be characteristic for all gays. Too, they view the *homosexual impulse itself* as abnormal, as van den Aardweg says, as "a craving of the infantile ego."[13] Reparative therapists couch such homonegative formulations in scientific terms so that they can claim that their goal is science, while in fact they make the connections they make between being gay and being emotionally disturbed for entirely unscientific reasons: because they wish to devalue gays, which they do by being thoroughly, if not always ostensibly, critical of them.

In the realm of homosexuality routinely *leading to and being associated with other symptoms of an emotional disorder*, homosexuals are in addition to manifesting their perverse sexuality supposedly all more likely to be suicidal, substance abusers, and sufferers from depression than on average. Though homosexuals proclaim they are "gay and proud," their pride is merely a hypomanic denial of/compensation for the shame they feel, and

not necessarily unreasonably, about themselves. Hypomanic gay pride merely maniacally covers up depressive gay shame.

Being gay is supposedly routinely associated with severe *personality problems*: As *paranoid* individuals gays blame others for things that they themselves are, do, and ought to alter; they are hypersensitive individuals readily hurt by others as they too quickly take offense at what everyone says about, and does to, them; in particular they blame others for outlandishly criticizing gays—for acting outrageously.

For most gays, homophobia is imagined, a projection of their own misanthropy. Thus others are not deaffirming gays as much as gays, being paranoid, are deaffirming others (for deaffirming them).

As *depressives*, gays poor-me mouth. They feel unacceptable; then, hoping to be convinced that they are acceptable, and to be accepted as they are, seek others' reassurances that they do, after all, match up. That is in part why they are so promiscuous. They are seeking out one sex partner after another hoping to be admired and embraced if not fully loved. Only, once again, failing to achieve their goal of being loved and to be validated along the way, they have to repeat the process until someone actually does fully embrace them, which, however, never actually happens because they predictably *imagine* that others don't like and don't accept them, and they believe such a thing even when it's not true.

Gay men closet themselves due to depressive guilt. As such they become ego-dystonic homosexuals, or what Nicolosi calls "*non-gay homosexual*[s]," which Nicolosi defines as same-sex, attracted men who "experience . . . a split between [their] value system and [their] sexual orientation."[14] Hating themselves for being gay makes them incapable of embracing themselves overall. That keeps them from achieving their true potential and frequently causes them to become suicidal.

As *passive dependent* individuals, gays have an excessive need for reassurance, the goal of which is to have others make them feel special. Additionally, their love affairs are marked by clinging. This readily yields to rage reactions when they feel that they are not the full object of others' attention.

As individuals with a *(bipolar) affective personality disorder*, gays employ denial (hypomanic) defenses to help them be sufficiently happy to get through their sad, miserable little lives. Denial makes them prone to evaluate their "condition" not merely as normal but as better than normal, even something to be proud of. Though heterosexuality is God's design, gays, maniacally and so foolishly, declare that homosexuality is the equal of heterosexuality, if not an actual improvement on it.

As *narcissists*, homosexuals are prone to display a false self, a self oriented to, and preoccupied with, theatrical image making, making them excessively concerned with how they look, and outrageous in the ways they make themselves up, and over. Many are self-preoccupied to the point of being so excessively concerned with external appearances that they choose image over substance, rendering them likely to be easily hurt and offended by anyone's being even superficially critical of them.

As *masochists*, gays provoke homonegativity toward themselves because they need to be, and so are virtually begging to get, punished—usually because they, quite properly, hate themselves for being sick and for destroying society. They often act out this masochism by voluntarily getting involved in a long painful treatment process (reparative therapy) that they already know will change little to nothing. According to van den Aardweg they are also (*sexual*) *sadists*, as particularly manifest in some of their more "underground" sexual proclivities (e.g., they are "underwear fixat[ed, and involved in] urinary and fecal sex").[15]

As individuals with a *gender identity disorder* (*gender dysphoria*), unlike straight men who are real men, gay men, though men, are never comfortable with being truly masculine. Basically effeminate men, they can be made straight by prohibiting such sissylike behaviors as sewing, by forcing them to indulge in masculine behaviors like playing sports, or even just by calling them real-guy names especially the transformative "dude."

Most gays are *infantile* overall, that is, globally *immature*—in their thoughts, their behaviors, and their relationships. In general, they manifest a childlike primitiveness that they subsequently deny by falsely accusing others of being infantile, that is, of being homophobic.

In conclusion, reparative therapists mischaracterize gays as diseased to paint them as individuals who are personally inept. They view them as being without moral fiber, as lacking in solidity, and so as of less value than straights and therefore of no substantial consequence to a world that properly devalues them as narcissistic, selfish, and self-aggrandizing.

Additionally, homohaters routinely specifically disparage *gay love*, *gay relationships*, *gay sex*, and *gay professionalism*, viewing them all as necessarily less healthy than their heterosexual counterparts. Also, they do this in such an effective way that gays buy into the picture painted of them, leaving gays feeling devalued, disgraceful, and dirty, and so in need of the radical change and the complete overhaul that reparative therapy promises them.

In the realm of *disparaging gay love*, reparative therapists hold the homophobic belief that gay love supposedly doesn't actually = real love.

Rather gay love is a manifestation of some emotional problem making it no better than a mere neurotic infatuation. True gay love doesn't exist. It is composed entirely of neurotic parts. These are primitive longings; anxiety about heterosexuality due to a fear of women leading a man to feel inadequate performing as a heterosexual; hatred of members of the opposite sex; and a hatred of members of one's own sex (compensated for by loving them to excess). Narcissism is an important driving force of gay love: a need to be filled up leading to attention seeking, making this love more like egoism. Moreover, gay love is self-love because gays' partners are the gay persons themselves, with gay individuals falling in love with others only as a way to love themselves/satisfy their own needs. As a result, gay love is never altruistic, only selfish. All attempts to satisfy a partner are really attempts to satisfy oneself. And no matter how much others love gays, gays always feel unloved, for, as van den Aardweg suggests, sadly homosexuals cannot use true affection coming from others because they are "obsessed with the tragic idea of being the rejected one."[16]

In reality, homosexual love is viable as long as individual personality problems don't interfere with its full expression, much as they can also do for straights. For example, so often it is not one's homosexuality but one's lack of self-affirmation that interferes with the capacity to affirm, and so to love, another. If gays are unduly critical of one another, and see their partners as defective, and cheat (more than on average), they generally do so not because they are gay but because they are neurotic, that is, depressed, because for example they believe "How desirable could anyone be who chooses me?" And so they seek the next partner as part of their need to improve upon their negative experience with their last one.

In the realm of *disparaging gay relationships*, reparative therapists see all gay relationships as self-absorbed affairs where closeness is really dependency, the expression of a need to control others, and a seeking to satisfy one's ambition to feel relationally successful in order to feel less like a personal failure.

In reality, gay relationships are not all of a lesser sexual and personal value. Gays, in spite of what many reparative therapists say, can be truly happy in their homosexual relationships. Just because a relationship is a same-sex relationship doesn't change its basic nature. Relationship problems of a loving and sexual nature do exist for gay men. True, developing a calm, peaceful relational and sexual life can sometimes require that gays work out a neurotic inability to relate healthily. But the answer is not to go straight; it's for loving (sexual) partners to become more attuned to each other. Each must make their relationship of primary importance as one relinquishes personal needs to avoid depriving the other of what he or

she needs, wants, and ought to have, with each moreover giving up personal habits that displease others, and doing so appropriately in a way that assures that each person involved in a meaningful relationship makes certain that empathy guides the interactions. Doing thusly can reduce relational volatility, eliminate much partner conflict, and reduce interpersonal violence thus preventing physical injury. Making sexuality as spiritual as it is physical thusly allows physical attraction to grow as the spiritual connection leads to relational peace characterized by the absence of interpersonal conflict. Some homosexual relationships are marked by conflict, but conflict in a homosexual relationship is not necessarily due, and/or in any meaningful way related to, the homosexuality. Gays have the same relational conflicts as straights. If there are differences between gay and straight relationships these are only in the nature of a thin patina that covers the basic identicalness that deep down exists between the relationships of people who are gay and people who are straight.

Not surprisingly, according to reparative therapists, all gays also supposedly make bad relationship choices. Or if they have made good ones, they aren't able to see them through very effectively. To me this would be, if true, a reason to help gays make better relational choices and develop them creatively. But to reparative therapists it is a reason for gays to change their sexual orientation from gay to straight, as if this would of necessity not only improve their sexuality, but would also enhance their relational ability.

Reparative therapists who thereby disparage gay relationships hurt gays' chances of developing happy, healthy, more satisfying, rewarding, perhaps monogamous relationships. They overlook that while in reality some gay relationships are problematical, many times the problems with gay relationships can be resolved. Relationship *problems* are not "intrinsically gay." Rather they are intrinsically neurotic, characterized, for example, by a lack of adequate motivation to make a given relationship work, where partners become fully dedicated to being loving and to, if they so choose, remaining partnered forever. Just because some homosexual relationships have problems doesn't necessarily mean that all homosexual relationships are problematical and so therefore ought to be devalued as compared to straight relationships. It rather means "we should just redouble our efforts to make this work." Many times relationship problems lessen on their own. And if not, it is *affirmative* therapy that can best help gays develop their relationships along more sanguine lines.

Gays involved in excessively *narcissistic* relationships can learn to make their partners' needs take precedence over their own, and to stop using their partners (and other people) as an opportunity to satisfy their own

needs and longings, and instead to start making their relationships less of an opportunity for self-expression and more of an opportunity for altruistic enactment, sometimes including self-sacrifice, as gays start acting with more sensitivity to their partners in a way that precludes an excessively selfish involvement with themselves. Sensitivity training can help enhance love by enhancing altruism and so the pleasure not of taking but of giving, which goes along with the ability to identify with a partner and to give to him or her what one might want for oneself, even though that giving happens to be at one's own expense. Narcissistic gays usually also benefit from being helped to deal with their superiority complexes and so to instead accept equality as the right and proper basis for a relationship.

When gay relationships are excessively *dependent*—gays become aggressive when their dependency needs are not met—gays can learn to get beyond the need to cling and merge, and to become angry when thwarted over nothing, to instead develop a personal independence that avoids overall divisiveness while yet encompassing a reasonable degree of closeness.

When an excess of *competition* mars a gay relationship, gays can learn not to view the other person as defective and themselves as flawless, and not to force one partner to be submissive so that they as the other partner can be triumphant, but to emphasize the seeking of cooperation between two over the exercise of power of the one.

Gays with *avoidant* tendencies that create connection difficulties can learn to accept more in the way of anxiety-free closeness and commitment. Gays with *phobic* tendencies that make relationships volatile and unstable due to the anxiety can beneficially uncover the exact meaning of their concerns and so reduce their need for fearful withdrawal. *Borderline* gays whose excessive sensitivity, perceptivity, and fluidity on the one hand cause them to have relational problems can on the other hand draw upon these very qualities to relate better by minimizing anger outbursts and reducing the number of crises du jour.

In conclusion, many homosexual relationships are conflict free. The belief propagated by reparative therapists that this is not so is symptomatic of the homophobic subtext of all reparative therapy. Since to reparative therapists being gay is sinful, criminal, and dysfunctional, and the reality is in its perception, everything about gays, from the relationships they form to the sex they have, reflects that dark view, and that darkness can, as reparative therapists see it, only be lifted when these reparative therapists shine their heterosexual light upon the black void. Though many reparative therapists say that no gays are happy, many gays are extremely so, and the thing that most threatens to make them unhappy is

reparation. And for those gays who are unhappy, and certainly they exist, therapy for their despair wouldn't be reparative. It would not be with a doctor who wants to change them to straight. It would be with a healer who wants to help them improve their lives: affirmatively showing them the way, changing not their troubled sexuality, but their troubled attitude, in the direction of an untroubled one toward their sexual orientation.

In the realm of *disparaging gay sex*, sexual difficulties supposedly characterize, and are an integral aspect of, all gay relationships. Many reparative therapists, speaking of the neuroticism of homosexuality, believe with van den Aardweg that "the vast majority of active homosexuals are promiscuous."[17] Suggesting that gay individuals cannot form stable ongoing committed peer/loving relationships or have real families that resemble those of heterosexuals, he says that "the fairy tale of faithful homosexual 'unions' (with its slogan, 'What is the difference from heterosexual marriage apart from the sex of the partner?') is a propaganda item, to win privileges from the law and acceptance within Christian churches."[18] But "partner variability is inherent in [gay] sexuality" and "the opportunistic marriage concept [just] serves emancipatory purposes, [and] the legalization of adoption by homosexual couples."[19] Additionally, all gays are childish, with their sexual loving inherently adolescent in nature (and adolescents, all being immature, hardly know of what true love actually consists).

Supposedly gay sex is much less gratifying than heterosexual sex because the former is inherently too evil to be at all good (enjoyable). Gay sex is also too cumbersome to be enjoyable, for two people who are homosexual necessarily experience problems due to those physical limitations that make mutuality impossible, assuring that only one partner can be satisfied at a time. But in fact the same challenges can also be found in heterosexuals, and so are not characteristic of one or another sexual orientation but of all sexual physicalities and possibly representative of neuroses gays and straights happen to share. The therapeutic goal then would be not sexual reorientation therapy for homosexuals (to change the sex from homo- to hetero-), but sex therapy (to make the gay sex better).

As for infidelity being a unique symptom of homosexuality, infidelity is generally a psychological not a homosexual event: a reparative attempt to fill a void or to replace feelings of loss with chimerical gain, or a vengeful attempt to retaliate for harm one person believes was done to him or her by another. Conversely, relationally sound gays experience the same normal feelings of sexual ownership (fidelity) as do straights, and if they don't they can be helped to come around, for example, by opening up to their partners after relinquishing their need to hold back a part of themselves from those they love.

Finally, in the realm of disparaging gay *professionalism*, as a rule all gays are supposedly unprofessional or, if professional, can only really thrive in a profession that is itself "very gay" (which can but mean that that whole profession is itself "very sick"). In reality, professional ability and sexual orientation basically come from different parts of the brain. And while some professions are more welcoming of gays than others, meaning that while gays can relax and be less closeted in some than in other professions, there is no one profession where gays are so constituted that they predictably always do a better or worse job overall than straights.

DIFFERENT MODELS OF HOMOPHOBIA ESPOUSED BY REPARATIVE THERAPISTS

I believe that it is homophobia not homosexuality that can be an emotional symptom, and moreover one with a differential diagnosis—that is, there are different discernible psychopathological reasons for being homophobic.

I divide homophobia into two categories. In the first group of homophobias, the homohating is mainly of *conscious* origin. Here homophobia may be not symptomatic but rather a legitimate philosophical stance at least seemingly fueled by reason and guided by faith.

Overly Scrupulous Religious Homophobia

Scrupulously religious homophobes corrupt religious beliefs/tenets purposely so that the homophobic individual can retain/develop a condemnatory view of homosexuality as being a sin and as constituting immorality.

Overly Scrupulous Psychoanalytically Oriented Homophobia

Scrupulously psychoanalytically oriented reparative therapists apply psychoanalytic principles to help explain the development of homosexuality, even though applying psychoanalytic principles of causality to an event that is not psychological in its causation is a scientifically flawed approach meant to better condemn homosexuality by harnessing the "scientific terminology" of abnormal psychology. In the psychoanalytic view, homosexuality is backward-looking in the sense of being incompatible with psychological maturity. Thus being gay is a regressive form of being straight: a fixation then a fallback, the result

of being thwarted somewhere along the developmental line in one's attempts to master being heterosexual, or the result of having been traumatized in one's early relationships, especially in those with one's parents.

Socially Attuned Homophobia

In the condemnatory *social* view of homosexuality, homosexuality is supposedly dangerous both to the homosexual, for almost all homosexuals risk getting a sexually transmitted disease, one reason they all have a short life expectancy, and to others, the victims of homosexuals, especially children, as well as the societal institutions, particularly the institution of marriage, that homosexuals corrupt.

Politically Attuned Homophobia

In the condemnatory *political* view of homosexuality, gay rights (emancipation) campaigns are nothing more than gays living out their victimhood, which they do in order to blame and attack others for being insufficiently understanding and excessively unjust. Gays lie that being gay is biologically determined because if it were otherwise, if being gay were actually a choice, gays would have to take responsibility for being gay and so could not be excused for being homosexual on the grounds that "they can't help themselves." They would have to take the consequences for something they deliberately brought on themselves, both as a practice they consciously condone and an activity they intend to continue. Gays who manipulate their reputation this way are supposedly taking advantage of others' all-too-human need to support the perceived underdog.

In the second group of homophobias the homophobia is predominantly unconscious, that is, it is a psychological symptom/illness just like other neurotic or psychotic symptoms/illnesses, such as a phobia or an obsession. While homohaters view homosexuality as an illness, it is homohating that is the illness, and as such that which is symptomatic, however much homophobes claim that their negative beliefs represent a conscious preference, a true assessment of reality, and/or a socially validated mindset. That is, homophobia, not homosexuality, is the disease, and it is the homophobe, not the gay individual, who suffers from the malady and needs the cure. The following categories describe both homophobia in laypersons and homophobia found in reparative therapists.

Paranoid

Homohaters are often *paranoid* individuals. Like other paranoids they have a "me-against-them, I am good you are bad, and I am a victim and you are persecuting me, so I persecute you to get back at you for persecuting me" mentality, and thus an overall "It's you not me" mindset. This persecutory mindset is exemplified by the dedication in van den Aardweg's book *The Battle for Normality*: "The ideology of 'normality' and 'unchangeability'. . . discriminates against those who know or feel that that is a sad lie."[20]

Dynamically speaking, paranoid homophobes project their own self-hatred/self-disgust onto gays and condemn gays as they are wont to condemn themselves along the familiar lines of "it takes one to know one" as "it takes a self-hater to hate others for exactly the same thing."

Obsessive-Compulsive

Obsessive-compulsive homophobes like other obsessive-compulsives are preoccupied with good versus bad, as they are with cleanliness versus dirtiness, and sexuality that is virtuous and godly versus sexuality that is amoral and ungodly.

Histrionic

Histrionic homophobes like other histrionics are excessively, outrageously, pessimistic about gays and their prospects in life and so are excessive in their condemnations of (often expressed in the Bowdlerized terms of sanctimonious worry about) the gay lifestyle.

Avoidant

Avoidant homophobes like other avoidants are isolated people, removed from others, reality, and the world, who seize upon others being gay as an excuse to drift away and move out of the orbit of homosexuals much as they look for an excuse to move away from the orbits of everybody else.

Bipolar

Bipolar homophobes are alternatively excessively loving toward and full of hatred toward homosexuals, shifting as a compensatory mechanism to

make themselves feel better as they cycle from loving in order to feel less unacceptably hateful, over to hating in order to feel less unacceptably loving.

Sadistic

Sadistic homophobes are equal opportunity haters, as full of hatred for gays as they are for everyone else. As they often do, here too they bend "scientific" theory according to their own ends: to make it seem as if their hatred is legitimate, as when they call homosexuality a choice so that they can hurt those who make such an (outrageous, defiant) selection.

SOPHISTIC DEFENSES OF HOW THEY THINK AND WHAT THEY DO

Reparative therapists defend themselves, their practices, and their philosophy using a number of psychological defense mechanisms which they install to enhance the seeming validity of their homohating and so to increase its sphere of influence.

Those who are primarily manipulative paranoid individuals defend themselves against criticism using the *projective paranoid defensive misattribution* of "it's not me, it's you," that is, "my therapy is not misguided, the problem is your bias against reparative therapists, along with your irrational defense of homosexuals, one that buys into the typical homosexual's tendency to resist anyone who tries to help them go straight by condemning these others for being controlling and coercive." Also, *identifying with the aggressor*, reparative therapists justify themselves and their activities by not answering asked questions and not responding in an explanatory fashion to legitimate challenges but by instead savaging those who ask and challenging them in turn, their way to avoid answering and responding by instead devaluing and demonizing others who question them, silencing them and any of their viewpoints that differ significantly from, and are critical of, their own. Such individuals condemn those who challenge their reparative beliefs and practices as being politically motivated, or as being ignorant, antiscience, creators of myths, individuals who spread lies about honest, thoughtful, reparative practitioners. Further identifying with the aggressor they launch vilifying frontal attacks on the studies that suggest that reparative therapy doesn't work, citing antireparative bias originating with an invidious sampling of the individuals who are doing the studies—calling all such participant individuals "anti-reparative ideologues primarily out to determine that reparative therapists are religious

nuts, all of whom practice a form of witchcraft whose so-called science is really the work of Satan."

Using the defense of *selective attention*, reparative therapists deny the harmful effects of reparative therapy by citing its helpful aspects only, though its (occasional) helpfulness is at best spotty and often merely entails unintended (and unanticipated) positive consequences.

Using *rationalization* and *displacement*, reparative therapists defend what they do by deflecting blame from the therapy itself onto its ("merely occasional") misuse by untutored practitioners. They say that if reparative therapy is ineffective, as its detractors claim it to be, this would not be because the therapy itself doesn't work, but because those doing it are doing it incorrectly, as well as applying it to patients who are unsuitable candidates, for example due to being inadequately motivated for this particular form of treatment. They further rationalize their approaches as being unchallengeable because they come from on high, from God, meaning that how they think and what they say and do is by definition being faith-based, intellectually irreproachable. They use the defense of *sublimation* to turn their negative into positive motivation, as they attempt to morph their reforming tendencies into a true, admirable desire to help.

Using a *passive-aggressive* defense, they devalue being gay but they only do so "indirectly"—by praising the value of being straight.

Using *intellectual* defenses, they distract others from their true motivations by selectively citing the supposedly irreproachable reality of their statistics: statistics that purportedly prove (however sophistically) that most homosexuals experience earlier deaths than heterosexuals and raise children poorly, compared with how children of straight couples are raised.

Instead of documenting his results for all to see, a reparative therapist in true paranoid fashion took umbrage at and devalued those whom he saw as devaluing him and his therapy. He cited the need to maintain appropriate secrecy about his practices as merely his way to keep others from stealing his original therapeutic methods. He simply denied using the extreme behavioral remedies he in fact did employ, such as aversion behavioral therapy that employed painful electric shocks, at most admitting to using some of the "milder" aversive therapies such as inducing vomiting while exposing his patients to sexual stimuli. He further defended his practices with very personal, ad hominem, attacks on his critics—blaming them for inexcusably being atheists who don't believe in God and ridicule Christians, as well as for being self-serving—criticizing his practice so that they could steal his patients. He accused those straights who defended and protected being gay as being themselves latent, closeted "non-gay" homosexuals. At

times he simply shouted down those who seemed to disagree with him. According to him, he wasn't depriving them of their right to their opinions about the value of reparative therapy or stifling their protests about the undoubted harm it could do, just challenging them to a purely intellectual discussion that for obvious reasons they had to have if they really sought the truth.

Justifying his bad behavior by comparing it favorably to another, worse behavior he would deny his practices were harmful by comparing them favorably to practices that were even less sanguine, as in "I am as of recently no longer using painful electric shocks."

A central tenet of his was his belief that patients who listened only to him and did what he knew was in their immediate best interests and for their ultimate good were thereby becoming less confused because they weren't being misled by the bad information given to them by others.

In truth those who condemn reparative practices often don't know exactly what reparative therapists say and do because reparative therapists shut them out from their inner circle, thus preventing them from attending key meetings, many of which take place behind doors closed to keep outsiders out. And in truth, as this defender never tired of pointing out, those who attack reparative therapy *are* often somewhat biased themselves, making them at best unfair and at worst self-serving. In fact, some of those who attack reparative therapy aren't aware that it is as much a product of the practitioner as it is a practice of a defined group, meaning that there are gray zone therapists who as honest practitioners are working in a mode that is somewhere in between affirmative and reparative, where the patient's well-being is in fact their main concern and focus, and the reparative approaches they use are only a minor part of what they actually do.

8

❖

Cognitive Errors (Erroneous Beliefs) of Reparative Therapists

This chapter discusses how reparative therapists often act ineffectively and/or inappropriately because they think illogically, that is, because they make cognitive errors that lead them to develop seriously idiosyncratic notions about how to view, cope with, and handle their gay patients. These distortions can have a chilling effect on their patients in part because therapists who make these errors misunderstand, feel negatively toward, and even bully those they set out to help. And rarely do their patients, most of whom are in awe of an authority that they assume has impeccable medical credentials and/or religious imprimatur and/or knowledge and experience, understand that they ought to challenge their therapists' misguided notions, or at least compensate for them so that they discount them as potentially misguided and wrongheaded. Instead, patients, buying into their therapists' illogic, become even more anxious and depressed than they were previously. Then they cannot fight back effectively. Instead, tolerating their mistreatment and cowed into inaction, they remain quiet and suffer in silence. Or they do fight back but in self-destructive ways, such as by becoming counterproductively stubborn in treatment, but without quitting their therapy altogether.

Cognitive errors do not always lead therapists to have *negative* notions about their patients. Sometimes they instead create false *positive* notions that lead therapists to love their patients to excess, so that they try too

hard to help people who are okay as is, or even come to believe that they are capable of accomplishing miracles to the point that, in a state of misplaced "loving empathy," wanting not to hurt but to help gays, they persist in what they are doing and go too far in it, intervening with their patients to keep them from destroying lives that are not now, and never will be, broken. In a common scenario, they overeagerly tell gays that to get better you have to "stay out of gay-imbued places and away from all your old gayfriends." As a result, gays, following this misguided advice, come to feel lonely, then depressed, and perhaps even suicidal. In one respect at least this excessive illogical "helpfulness" is as, or even more, destructive than excessive illogical negativity. For at least gays with excessively negative therapists can properly rebel against a clearly defined enemy. But gays with excessively positive therapists have a hard time mustering enough anger to defend themselves against their therapy, so that now they cannot protect themselves from undue mistreatment by becoming antagonistic to their therapists.

Identifying and eliminating distortive reparative therapeutic concepts leading to unhelpful or harmful therapeutic interventions starts with understanding the structural and dynamic, often highly emotional, origins of therapists' cognitive errors, as described in the following section.

DEFENSIVE ASPECTS

Projection. Cognitive errors with an excessively negative caste, where therapists start with and retain the premise that gays are bad, sinful, disgusting, criminal, godless, child-abusing perverts, and even mass murderers, often originate in self-hatred based in guilt about and revulsion toward one's own homosexual tendencies kept in check by condemning others for the same thing, thus deflecting personal criticism, and dealing with one's conflicts by resolving them not in oneself but, vicariously, in others. To illustrate, in male therapists the mistaken notion that all, not just some, gays are promiscuous often expresses the projection of a personal erotophobia, leading therapists to criticize gay men for being sexually wild in order, by demanding another's complete virtuousness, to proclaim one's own, and by externalizing and condemning one's own similar, similarly forbidden, and similarly wild sexual fantasies/wishes in others keeping them under control in oneself. A hope is that if as therapists they view gays in a sufficiently condemnatory fashion, then God will forgive them (their therapists) for their own transgressions in thought and deed. Not surprisingly, therapists who think this way feel most righteous when they flay others the hardest, hoping for maximum personal redemption for

themselves. This externalized self-hatred often takes the form of excessively harsh "therapeutic" meanness deliberately (if unconsciously) planned and introduced as a way to be deliberately (if unconsciously) abrasive and even punitive in order to prove to oneself, and the world, that "you should admire me for thinking that people like this shouldn't be let off lightly, and, as you can clearly see, for a therapeutic orientation along with good intensions that are proof that I, distinguishing myself from my gay patients, am not at all like them in that way."

The best defense for gays subjected to this "projective therapeutic abuse" is to retain their pride by countering the proffered shame by not taking it personally, reducing its personal impact by recognizing the projection (of the personal shame/self-hatred) involved—how, in other words, "it is taking one to bash one," and "one is bashing another to deal with/avoid bashing oneself." Unfortunately, too often therapists brought up short by patients complaining that they, the therapists, are projecting their own meanness onto them manage to counter their patients effectively by claiming that they are not projecting self-blame, just properly affixing blame where it belongs. Therapists furthermore can effectively rationalize their blaming of others (gays) as being determined not internally but externally at the urging of (a harsh) God whose bidding they are merely doing. And they can also effectively counter criticism for their blaming nature by saying that no matter how hurtful they seem to be now, in the larger scheme of things they only hurt others for a specific ultimate good: to set limits on patients who, out of control, are hurting *themselves* both personally and professionally (and are at the same time hurting others as well and in a way that they as therapists have a responsibility to stop).

Projection also extends, after the fact, to disavowing culpability when therapy fails, as it usually does. If gays fail to get better from the therapy offered it presumably isn't because the therapy is no good, it's clearly because the patients, for reasons of their own, are not responding as they should. Thus my analyst "threw me out of therapy" when being able to stand being celibate no longer I rebelled and started what was to be a long-term gay relationship. She told me (wounding me severely) that I was no longer a suitable candidate for therapists like her. What she ought to have said was that her form of therapy was not now, nor was it ever, a suitable approach for patients like me.

In short, paranoid therapists disown who they are by foisting their guilt about their own "badness" onto their patients. They rage at gay transgressions as they rage at, and instead of raging at, their own.

Reaction formation. Reparative therapists often compensate for their anger at gays by becoming excessively loving to them.

Denial. Poorly trained unknowledgeable therapists often deal with their self-perception of their lacks by convincing themselves that they know more than they do, and even know it all.

Control. Reparative healers formulate their therapeutic interventions and overall plans less to help their patients than to demand that their patients see things their way, and go along.

Introjection/identification. Without necessarily realizing it, many reparative therapists are self-questioning individuals who tend to prop themselves up by identifying, merging, and so becoming one with, the homonegative thinking of gurus they admire—which are as often as not those gurus currently in the public eye.

Passive acceptance. The more *passive* reparative therapists join homonegative groups and buy into a group mentality that grows exponentially due to the mutual respect and support that the group members give to each other.

Rationalization. All reparative therapists tend to secondarily rationalize what they think, say, and do by spinning the cognitive errors they make in a way that appears to render them legitimate, that is, not extremely distortive but highly logical. Their spinning tries to convince themselves and others that their homonegativity is realistic, along the lines of "I don't have a problem thinking that homosexuality is sinful, it is my observations of the gay world that have judiciously, but inevitably, led me to conclude that being gay does definitely = a sin." Theirs then readily becomes a "blame-the-victim" mentality where reparative therapists see themselves as being "entirely reasonable, for they are not being homophobic, just decrying gays after accurately perceiving how, realistically speaking, all homosexual behavior is sufficiently bad for all gays to deserve being thusly criticized."

Narcissism and grandiosity. Many reparative therapists who think that being gay = being bad, sinful, disgusting, criminal, and godless owe the origin, persistence, and strength of their false beliefs to the narcissistic view of themselves as Saviors. Now they can see themselves not so much as mere therapists but as full missionaries and, as such, as emissaries of a harsh (but still loving) God.

EGO-IDEAL PRESSURES

Structurally speaking, cognitive errors are like symptoms in that they originate in pressure from the "ego ideal," that is, in that "part of the mind" that tells one how as therapists they ideally *should* think and what as therapists they ideally *should* believe—regardless of the reality they should be consciously and dispassionately assessing.

MOOD (AFFECT) DRIVEN

Mood or affect spins off erroneous cognitions, particularly pessimistic versus optimistic thinking that leads depressed and/or hypomanic therapists to feelthink their way into being gay as either bad or good according to the therapist's ascendant affect/mood/mood disorder, high or low, of the moment.

In conclusion, it follows that therapeutically speaking gay patients will feel less despair and can even avoid months or years of misguided and likely harmful therapy if they not only recognize that their therapists' negative (and also positive) "logical" formulations about them are descriptively cognitively flawed, but also understand why therapists are that way, so that they can better fathom why their therapists think as they do—in preparation for considering the possibility that what the so-called experts tell them is true therapeutically involves in fact significant "therapeutic misconceptions."

SPECIFIC COGNITIVE ERRORS

What follows are some specific characteristic (overlapping) distortive thoughts (cognitive errors) that color/contaminate reparative therapy, with correctives (for therapists and patients alike) either stated or implied.

Now = Forever

Reparative therapists often assess an issue's/event's durability according to its present strength/impact then conclude that if the moment is meaningful then by that criterion alone it means that it will endure for a lifetime, as if "all is now lost" means "all will be lost forever." As a result, when adolescents currently display homosexual tendencies their therapists conclude that radical treatment is necessary lest they become permanently afflicted (by their homosexuality). If a child's lifestyle is for now demonstrably "sexually problematical," concerned parents convince themselves, and some reparative therapists, that that is not related to transient adolescent turmoil but is a "certain portent of things to come." As a consequence, parents and therapists alike panic and take desperate, often irreversible, sometimes harmful, "corrective" action due to having given nonexistent, trivial, and transient problems an imprimatur that imparts a significance to them that they don't have/warrant and lends them a quality of permanency that they don't actually merit. What they forget is that

the gay "indulgences" of the young are not necessarily the gay "sins" of the mature, if only because most young men and women go through a temporarily wild polymorphous perverse phase that sooner or later blows over, so that they "settle down" on their own. Still, therapists and parents alike predictably make things worse by overreacting and becoming so alarmist that they make unnecessary contingency plans and take irrational, excessive, often destructive precautions, thus intervening in a controlling straightjacketing way that risks perpetuating the very lifestyle/misbehavior that they originally sought to eliminate. In many cases, gay adolescents' gayness should neither be encouraged nor discouraged but viewed dispassionately and from afar putting the often lurid adolescent lifestyle into perspective as something that will ordinarily spontaneously simmer down as the young person grows up and settles in. What is going to be will in any event be, for not only can spontaneous progress not be readily influenced, but in so many cases attempts to do so by intervening to try to protect adolescents from themselves will not only be of no use, but will likely, having an unintended reverse effect—be somewhat, or seriously, counterproductive, and possibly even harmful.

In my experience, those gays who stay gay from adolescence on tend to alter their gay *lifestyle* over time, and many, changing for what might be considered to be "for the better," become, after their adolescence (actual or prolonged) is over, less wild/promiscuous as their relationships broaden and deepen and they focus down on one particularly meaningful, one especially important relationship. So, if not a certainty, it is at least highly likely that a hyperactive gay youth will eventually turn into an older, more stable, more mature gay adult, perhaps one who dedicates himself or herself to one partner, where each enriches the life of the other and also the lives of others in their orbit. Conversely, radical interventions by ex-gay therapists doing ex-gay therapy for the young (often egged on by panicky parents) are quite likely to interfere with the natural course of things in an unpredictable way, and just as likely, having a paradoxical effect, to do a lot of harm, and very little good.

Common = Normal, Rare = Abnormal, Different = Bad/Worse

This erroneous belief equates exceptional in the good sense of "unusual or extraordinary" with exceptional in the bad sense of "deviant," that is, "off-kilter," so that all concerned—therapists, gays, and their parents—view that which is not statistically common not as incomparable and thus as outstanding in a good way, but as "anomalous" and thus as standing out in a bad way. For such people common/unexceptional (heterosexuality)

= normal (good), and less common/exceptional (homosexuality) = deviant (bad) on that evaluative basis alone. Many reparative therapists believe God shares this erroneous belief, so that unthinkingly and unaccountably they assume that people of His creation are innately heterosexual with homosexuality representing not a difference but a deviation, a kind of occasional (if not entirely rare) mutation from what He intended and tried His best to accomplish. That is, God didn't create two things, he created one thing, with the other an infernally, unfortunately, anomalous version of the first.

In fact, being different merely = being dissimilar, without implying "better" or "worse." Thus being untraditional is different from, not better or worse than, being traditional, just as being gay is, at least in some ways, different from, not better or worse than, being straight.

Therapists doing reparative therapy premised along such distortive lines naturally make unscientific, often pejorative, value judgments due to assigning a negative value to being gay and to the gay "lifestyle." Using this distortive reasoning they conclude that being gay is a mental illness because it is less common than being straight, which, simply by being common, represents the standard of "wellness." (They go on to equate mental illness with a sin or crime along the lines of the similar = the same thing error described below. Next they view mentally ill patients as sinful criminal people who, being out of control, have become antiestablishment due to their rebelliousness and so, rebellious like all mentally ill patients, have ultimately become and are now dangerously subversive.)

All-or-None (Dichotomous) Thinking

Reparative therapists thinking dichotomously divide the gay world into black and white, good and bad, then, too readily pigeonholing gays, compartmentalize them as losers not as winners and as devils not as angels, although most gays are mere mortals somewhere on the continuum between these two extremes. Reparative therapists who think dichotomously see gays as fallen straights much as they might see the devil as a fallen angel (with the fallen ones requiring above all a therapeutic, and somewhat punitive, exorcism). Fear and dislike compound their negatively perceived view of the gay object not as a variant of normal, but as inherently completely abnormal. For such therapists, individuals who believe that anything less than 100 percent = zero, gays are not merely humans with human flaws mixed in with virtues, but inhuman in the sense of being *completely* inferior (to their straight counterparts) and as such *thoroughly* despicable. (Critics of reparative therapy, also thinking dichotomously,

tending to conclude that *not* all = nothing, dismiss faith-based treatment as completely unscientific and therefore as always and totally completely ineffective. Yet faith-based treatments can have positive effects, and these can be understood scientifically through a dispassionate, however rarely performed, exploration and elucidation of the nature of the power of suggestion and, what is related to that: the existence and nature of the placebo effect. Both are real effects that can be understood and quantified, and so should therefore never be automatically dismissed as hokum.)

Similar = The Same Thing

There are some superficial similarities between homosexuality and emotional disorder (for example, both are relatively speaking uncommon and both involve the passions). But to reparative therapists that means that homosexuality is per se an illness and thus, like any other psychiatric illness, can be understood by invoking causative, psychopathological developmental and psychodynamic events, such as fixation (e.g., an anal fixation coloring sexual style) and/or regression (from heterosexuality to homosexuality). Thus homosexuality can be cured, as can other psychiatric symptoms, by identifying, analyzing, and so reversing and thereby undoing what is causing the developmental obstruction and/or backsliding. (An especially bizarre example of this similar = the same thing thinking is the theory that says that it helps gays to go straight if caffeine is reduced or eliminated, for caffeine is a stimulant and sex is stimulating, therefore reducing or cutting out caffeine reduces [homo]sexuality.) The common belief that homosexuality = an addiction (and so can be cured with a kind of 12-step program) relies on the making of this cognitive error, where similarities between "gay" and "addiction" are emphasized (both do, after all, have compulsive qualities), but (major) differences are overlooked/ excluded/ignored, leading to the grossly oversimplified conclusion that having homosexual sex bears some meaningful fundamental equivalency to having too many martinis.

Personalization

A cadre of ex-gay therapists believes that one becomes gay simply in order to initiate a personal attack on straights. Thus some reparative therapists, taking being gay personally, see it as an assault on them because it is an assault on their personal morality/deeply held values. So they treat gays psychotherapeutically not so much to help them go straight as to fend off/ get retribution for a personal attack from without, in other words to

retaliate after the fact for having been personally assaulted. When they cite "defending tradition," as they often do when speaking of gay marriage, they are likely to just be defending themselves—against an *attack* on "what I traditionally stand for," therefore "on my identity" and so "on me" (and on that "me" who has to stand up against the "dangerous armed gay troglodytes of this world"). This self-defensive element of reparative therapy helps explain why some reparative therapists are so irrationally devoted to the ex-gay therapeutic process. It is not only as if being gay is a sin and a crime, the sin and crime of being gay necessarily involves another sin and crime: an emotional and almost physical attack on straights, and so a personal assault on their straight therapists.

Reparative therapists who hold the belief that gays don't want to change are substantially denouncing gays personally, indulging in a narcissistic somewhat paranoid view of gays' motivation that involves stereotypically viewing gays as adversaries by design. It's as if being gay is a deviation deliberately created to dislodge straights from their smug heterosexual perch. Therefore, heterosexuals can but be expected to defend their territory, and so think themselves justified, by a legitimate desire for self-defense, in putting sensible pressure on gays to stop doing what they are doing and at least starting to go straight.

Parents of a gay child often assume that everything gay that their child says and does is somehow relevant to them and thus their business, their concern, and a proper source of their anxiety. And what is worse is that they often come to believe that a child has become gay in order to spite them—having become purposely, thoroughly, and deliberately disagreeably gay in order to attack them by launching an assault on their straight values.

Comparative Thinking

Reparative therapists who think comparatively fail to assess the value of a thing on its own merits. Instead they look only to how something is aligned with, and so how it does, or does not, measure up to something else. For such people nothing has an individual identity; and all things have to be evaluated in comparison to something else. Thus these therapists view being gay predominantly not as different from, but in comparison to, being straight. Reparative therapists imbued in the sanctity of heterosexuality routinely manage to deny the existence of successful happy homosexuals by comparing negative gay stereotypes to positive heterosexual stereotypes (then establishing therapeutic objectives accordingly). Allegedly they do so simply to keep gays from failing both

personally and in life as compared to straights. (For according to them being gay inevitably causes one to become emotionally disturbed both now and in the future.) But in attempting to keep gays from being themselves and who they are simply because gays are comparatively inferior to straights, they overlook that gay life is unique, so that comparisons to straight life are apple-to-orange syllogisms (of an invidious nature). Gays compared to straights are in fact not comparatively flawed and broken but, unless emotional problems exist or supervene, are living a life that, though different in some respects from the straight life, has its own unique joys, as well as, admittedly, its own equally unique challenges.

Sequential = Causal

For reasons that are often more emotional than intellectual, reparative therapists seeking to uncover the secret of homosexuality (and to reverse homosexuality through an uncovering form of therapy) tend to force arbitrary, approximate, and hence false linkages between randomly occurring events in the past and present lives of homosexuals according to their belief that if one thing occurred before another then these two things must of necessity have been causally related. For example, reparative therapists convince themselves that a son's parents caused him to be gay by what they did to him when he was growing up—first by identifying parent-son problems, then by blaming their son's becoming gay on these problems simply because the problems occurred first and their son "became gay next." In an extreme example of how this reasoning can be faulty, a therapist claimed that a schizophrenic mother who heard voices that directly involved her child caused her child to become a lesbian. Yet her child in growing up had ignored her mother's problems as being "just the way she is" and, taking what was good out of the parent-child relationship, loved her mother a great deal and became a lesbian for other reasons entirely. To a large extent what happens to children depends not only on their environment but also on their temperament (which is in great measure inborn). For temperament influences how children view and interpret their parents' personal issues/problems. Thus a temperamentally passive son can view his father's passivity as welcome ("he leaves me alone") not as an undesirable absence/abandonment, and a son with a sunny temperament can view his father's intrusiveness not as undesirably enveloping but as an expression of the father's loving feelings toward his son. The child so temperamentally constituted as to welcome an in fact engulfing mother as being close and warm has as a counterpart the child temperamentally prone to viewing a distant, rejecting mother's

remoteness not as cold but as encouraging independence. At any rate, so many developmental issues occur in everybody in a jumble that there is rarely, if ever, a provable linear relationship between "parental behavioral characteristics and actions" and homosexual outcome. Looking for a relationship between parental upbringing and homosexual outcome is too often the province of those therapists who so disapprove of being gay that they seek any negative thing they can uncover to use to "explain" homosexuality via the use of psychodynamic developmental child-parent theories that have little to nothing to do with the actual development of homosexuality.

In my experience, reparative therapists who blame parents for making children gay generally have the very same problems in their own parental upbringing that they attribute to their patients—because these therapists view their own meaningful parent-child interactions as being entirely applicable to their patients and definitively causative of their patients' homosexuality.

Covalent = Causal (Guilt by Association)

Reparative therapists who view covalent problems as causative unwisely connect dots that should remain separate. In particular, they commonly dub covalent narcissism as causative of homosexuality. For example, if a man is both narcissistic and gay they claim that it is his narcissism that both convinces him that he can redo nature along his own desired lines and shapes his sexuality by causing him to love only people like himself (so that he seeks men as part of his idolizing himself, then falls in love with this idealized self). This theorizing stops short of explaining being gay in people who are not, or who are even the opposite of, narcissistic (such as the many altruistic self-sacrificial charitable gays out there). It also fails to explain why most very narcissistic individuals are completely straight.

Ad Hominem Reasoning

In ad hominem reasoning irrelevant linkage is established between the so-called quality of the product and the assessed value, generally based on a distortive *negative* valuation of, the producer. Gays therefore must go straight because one cannot be both gay and good at . . . (fill in the blank).

In fairness, ad hominem reasoning can be, and often is, unfairly employed in order to criticize not gays but their reparative therapists. Not all reparative therapists who have themselves been gay and given it up out of

a sense of guilt are simply neurotics dealing with their own low self-esteem by curing others the way they feel that they themselves should have been, or had actually been, cured. Not all reparative therapists apply their distortive/negative self-evaluations to others and so attempt to better themselves by bettering (curing) others of similar or the same "flaws." Not all reparative therapists predictably judge their patients as a way to judge themselves. Some reparative therapists are sincere and smart. However misguided they may be in their sincerity, ineffective, and likely to do more harm than good, they are at least honestly trying to help.

False Equivalencies (So That Ego = Id, and Instinct = Self)

In psychoanalytic terms, reparative therapists equate ego with id—their patients' "higher" intellectual functions with their "lower" instinctual behavior—then judge their patients accordingly, along the lines of "no gay person (instinct) can, or should, be a psychoanalyst (intellect)."

Selective Attention/Abstraction

Reparative therapists who make this error register only the troublesome (outlandish and self-destructive) "gay behaviors" that they themselves observe in and hear about secondhand from their (more troubled) patients, and that do admittedly exist in some gays and in some parts of gay society. Then, taking these "facts" out of statistical context, they assume the worst and view all gay life in the darkest possible, least acceptable, light, further allowing these therapists to retroactively justify some of their pre-installed negative assertions about gays. They convince themselves through information gleaned solely from gays seeking psychiatric help that few gays are in fact happy and successful, so that by definition all gays are going to be failures regardless of who they are and what they do, simply because they are gay, so that only being straight can possibly lead to being happy. Those with a religious bent in effect using the Bible as their *DSM-5* overlook statements in the Bible that could help them feel more loving toward gays. Instead they focus on the biblical statements that express and justify negative feelings about homosexuality—fertilizing notions into convictions by giving the former a fertile field in which to grow. (An important one of their motives is to avoid having to take responsibility for their own negative feelings toward gays by shifting blame—by assigning their negativity to an external source, the Bible, thus allowing them to whitewash their own negative attitudes toward gays as originating with and validated by others thus to instead steadfastly

maintain their self-view as one of personal innocence—of anything even slightly resembling irrational homohatred.)

The cognitive error of selective inattention/abstraction characteristically leads reparative theorists to overlook how sexuality is not only a psychological (supratentorial) but also a physical (infratentorial) phenomenon (the tentorium is a structure that divides the higher, cortical from the lower, subcortical brain). They thus focus on the psychological aspects of the moments that occur just before (mindful) gay fantasies erupt (and attempt to cure homosexuality by having their patients avoid bestirring, as well as having them actually attempt to garble/suppress, these fantasies). What they forget is that when it comes to sex it's both "mind over matter" and "matter over mind," because it generally takes both fantasy (stimulating the new brain) and looks (stimulating the old brain) to make for a full sexual response.

Selective abstraction also accounts for reparative therapists' undue fondness for using simplistic behavioral methods to treat homosexuality—simplistic-to-absurd interventions that emphasize pure sexuality in (and so overlook the uplifting aspects of) homosexual relationships, such as those involving closeness, bonding, altruism, and heroism. These behavioral methods often measure outcome (gay responses) entirely by using the crude penile plethysmograph (that measures tumescence by measuring changes in penile circumference), thereby simplistically measuring sexual arousal by a response to porn, overlooking such elements as love, hero worship, and positive identification that gives gay sex its full measure. Thus, they simplistically recommend that since some gays cruise in parks to go straight, all gays should stay away from parks, and/or keep out of men's rooms, for, as they believe, doing so will make gays less gay by reducing/eliminating presumed specific onsite temptations (triggers). But where one goes to have sex doesn't explain everything about sexual arousal. It just means that these places are the ones some gay men go to have sex. The implication is that the place where gay men go to have sex makes gay men feel sexual. But in many respects, it's not the meat rack that catalyzes feeling sexual, but feeling sexual that catalyzes going to the meat rack.

Reasoning according to the principles of selective attention, reparative therapists cite only the studies that bear them out, not those that weaken their positions. They omit inconvenient truths such as the one that patients rarely tell their therapists the full story about themselves and their lives, and that many patients claim that they have been cured although they (conveniently forget that they) continue to have occasional ongoing gay relationships, which they sometimes indulge in in a dissociative-like

state. Such therapists typically selectively protect their results as positive by conveniently avoiding doing long-term follow-ups. They frequently claim, "Our therapy works because it works sometimes," as, overlooking their failures, they selectively cite only instances where their therapy helped, if only a little bit.

Part = Whole

Reparative therapists who reason according to part = whole make and promote unwarranted extensions from the specific to the general so that they typically define the whole class of gays according to one or a few of certain gays' assumed unsavory aspects, then accordingly create a false picture of all gays from one or a few examples of gay men and women, making one or a few aspects of the thing into the entire matter. Turning the partial into the whole (entirely negative) view, they estimate all gays as unworthy and devalue all gays pejoratively, striking fear into gays' hearts about their current and future prospects—not, however, prospects they determine by a fair assessment of what gays are and predictably will certainly become, but instead prospects determined according to how the reparative therapists have inaccurately evaluated (and devalued) gays' potential in life based on one or a few of some gays' life problems, potential or actual. The research studies that this error can produce are fertile examples of such selectivity, based as they are on the small sample of gay men and lesbians who are out there acting out, while overlooking the myriads of out gay men and lesbians who stay close to home, living quiet lives, never feeling that they need to apply for psychotherapy. (Part = whole thinking is related to the some = all thinking and selective attention/abstraction thinking discussed elsewhere in this chapter.)

Part = whole thinking leads reparative therapists to focus on developmental causes of homosexuality while completely ignoring its genetic components. Also, forgetting how most things being imperfect have both a good and a bad side, reparative therapists who think this way overlook that there are merits comingled with the so-called demerits of being gay. Instead, after learning about one or two negative aspects of the reality of gay existence, they view the negative side of being gay as its entirety, as they come to view rare, uncharacteristic, and exotic behaviors on the part of gays both as more significant and more widespread than they actually are and as (stereo)typical not just of a few gays but of all members of the class.

Such therapists devalue diversity after overlooking the disadvantages of cookie-cutter sameness. They feel alienated from or antagonistic toward gays because they believe that *all* (not just some) gays, being

different, are outsiders, and as outsiders are thereby by definition entirely antiestablishment. Global negative judgments based on using the sexual yardstick as the sole measuring instrument by which they assess their gay patients both personally and professionally lead them to judge gays as thoroughly disabled due to being sexually compulsive (and they simplistically define "sexually compulsive" as having more than one partner at one time, or in rapid succession). Simultaneously, having concluded that a "disability" in this one area is a disability in all, they fail to recognize that gays, like anyone else, compartmentalize, so that gay sexuality, coming from a different part of the brain from gay professionalism, is unrelated to gays' ability to do their work/value as a worker. Then in therapy these therapists work on their patients' being gay to improve their functionality at work, instead of working directly on any work disability their patients might have to help their patients improve their functionality at being gay.

Making the same cognitive error they note that many gays on some level wish they weren't gay, then go on to cite not the normalcy of ambivalence about being gay (like about most other things) but the supposed truth that fundamentally all gay men and lesbians only want to become straight.

Selective abstraction, also discussed above, is related to part = whole thinking. For some therapists, supposedly gays constitute a collective to be typified by its unsavory members so that all gays are what a few actually are: promiscuous, impulsive, and unreliable. And all gays, not just a few, supposedly have developmental lags and unresolved dynamic conflicts, which are causal not incidental and significant not trivial. The disadvantages of being gay make being gay overall disadvantageous, this even though homosexuality can sometimes be an advantage, as it can be for psychotherapists treating gay men, or for activists hoping to help minority groups avoid/cope with prejudice and discrimination. To evaluate gays in this limited, pejorative way reparative therapists of necessity have to fail to recognize that much so-called across-the-board "typical negative/undesirable" gay behavior is in fact situation specific—thus representing an acquisition of the moment—according to whim and passing need, existing in the first place due to having been specifically provoked, and continuing, if at all, due to having become a learned behavior.

Overgeneralizing this way, reparative therapists now come to make predictions and draw conclusions after tarring all gays with the same negative brush, and making the whole picture out of one of its brushstrokes only as they conveniently forget that out there raunchy underworld gays who recruit and molest children are very few compared to the (many) gays out there who have children and love them well.

Reparative therapists who see only the baser aspects of gay sex up close view being gay as resolutely unsoftened by the loving, erotic bath that covers more primitive/animalistic behavior with positive, even ethereal, emotions.

Thinking = Doing

Reparative therapists who make this error believe that if a gay individual can "think straight" then he or she will "go straight." This—an example of the belief in the omnipotence of thought also found in shamans—in the magical thinking characteristic of obsessive-compulsive patients (who for example, equate naming something with having an influence upon it) and in patients who are overtly psychotic and suffering from delusions of omnipotence, creates therapists who believing in their own special mental abilities come to feel that they can defy natural law and heal others through using the power of the mind. Such therapists believe that they can simply make being gay go away by having a patient not think about (ignore) the facts of the situation (as if it will thereby somehow completely cease to exist); by having the patient actively, often ritualistically, and/or prayerfully, wish the gay gone; or by having a still gay patient thoughtfully act as if he or she isn't gay, for example by stopping crossing his legs. This aspect of reparative therapy often amounts to little more than an ongoing, prolonged ritualism as typified by the recommendation to get cured by obtaining supposedly life-changing, gay-reducing gender-identity feedback/support from others (e.g., "you will become an okay straight man if you can only find other men to call you 'dude'"). In some cases this magical thinking incorporates such mind over matter induction techniques as purging, fasting, and self-flagellation accompanied by continuous prayerfulness likely amounting to autosuggestion by mantra, along the lines of Coué's presumably curative reparative chant, "every day in every way I am getting better and better."[1]

Along similar lines, ex-gay therapists too often confound wishing with accomplishing, trying with succeeding, the value/intensity of their objectives with their actual viability, and even merely having objectives with actually achieving them. Faith-based omnipotent/magical thinking supposedly confers Godlike power and capacity for influence upon such therapists, leading them to imagine that if they wish something to be so, then so it will be, and their patients will change—along the lines of "because I desire it strongly enough, that will certainly make it happen."

It is simply illusory to believe that there is a relationship between thinking of oneself/seeing oneself not as a real man and becoming homosexual

and, as Nicolosi suggests, curing one's homosexuality by being thought of/seen as a real man by a real man (the therapist) leading to one actually becoming "a real man" and thus a heterosexual. (Quoting Nicolosi, "when a real man sees me as a real man, then I become a real man.")[2]

The supposedly magical powerful effect of wishing and believing, along the lines of "if you wish hard enough you can make it happen" only drives the plots of fairy tales. Sometimes patients can pray gay into seeming, usually temporary, submission, but I personally know of no cases where a gay person went straight by praying that that would happen. Prayer can enhance general health; but becoming healthier physically and mentally in a general way, while an admirable objective, certainly won't bring about a specific change in sexual orientation.

Catastrophizing

Reparative therapists who catastrophize turn the reality of being gay from an issue of modest life-influencing significance into one of life-destroying proportions. Thus being gay is not something their patients can integrate and coexist with, but a complete disaster for all concerned: not only gays, but also their parents, their coworkers, and their society. Unable and unwilling to help their patients take being gay in stride, and at times virtually delusional about gays' negative prospects, catastrophizers view everything about being gay, and about the patient who is gay, in hysterical fashion, as being of calamitous import and thus a potential debacle. Then they respond accordingly—in extreme cases going so far as to lock gay patients up and force them to consent to SOCE done under the hollow rubric of being loving and for the good of those being so totally loved.

A component of catastrophic thinking is selective abstraction (as discussed above) where therapists see only the negative aspects of being gay after obliterating from their consciousness such positive elements as the upsides of diversity and the joy of homosexuality itself, both of which considerations could potentially lead to one's having pride in, not being alarmed about (and ashamed of) being gay.

Catastrophizing on the part of reparative therapists can but make gays sheepish about being gay, furthering a patient's negative self-attitude, ultimately strengthening the therapist's own self-created untenable position that "gays have to change for their own good, and for the good of everyone else around them."

An important component of catastrophizing consists of the flattening of the response curve to the point that even trivial minor meager stimuli are believed to be sufficient to create major reactive responses. Therapists

who believe thusly use trivial therapeutic interventions based on the misconception that these are of sufficiently momentous impact to have a lasting effect.

Catastrophizing therapists suspend the sense of humor of all concerned. Therapists and patients alike become excessively gloomy, and progressively lose more and more perspective about "the situation." Depressive feelthink takes over to replace neutral/optimistic logical thinking. Therapists fail to staunch unfavorable assessments of theirs that lead their negative responses to completely take over and seriously influence their views of gays based on the therapists' own failure to make subtle or gross distinctions that would have permitted them to appreciate the extant differences between being gay and being an invalid. As a sideline, catastrophizing therapists miss the opportunity to learn exactly what their patients are like. Instead they feelthink their way into a gay world they create out of the stuff of their own fearful fantasies/nightmares, making it difficult for them to perceive their patients as they actually are and hard to see their patients' future life as it will likely turn out. So they miss the opportunity to shelve the gay issue in favor of satisfying what *is* often a highly poignant need of gays who apply for therapy: not to change to straight to become less troubled, but to get over their troubles to do away with their need to turn straight.

Ex-gay activists typically deliberately harness catastrophic thinking for their own social/antisocial purposes, consciously employing it not only to justify their own views but also to create a specific advantage for themselves and get a competitive edge over others, as well as to impress and convince colleagues and the public to change over to their way of thinking. Controlling narcissism ultimately attaches to their catastrophic reasoning along the lines of "because I think this, it is so" (e.g., "because I think you need to change *completely* you must do so, abjectly, and for me, by agreeing with my extreme position, however autistic it is, about what direction you definitely should take").

For those reparative therapists prone to being near psychotic, catastrophic thinking often involves a projection. Feeling gay-imbued chaos within, they project this chaos outwardly, leading them to evaluate gays, and gay life, as necessarily being entirely chaotic.

Inexact Labeling

Inexact labeling, a characteristic manifestation of feelthink, leads to ignoring the facts in favor of an emotional interpretation of them, in this case predictably leading to the taking of unnecessary and often

inappropriate antigay (ex-gay) therapeutic actions. Reparative therapists often also actively *encourage* their patients to label themselves inexactly in order to develop a false (straight) identity. They urge gay men to re-label their gender identity (to masculine) as if doing so along "butcher" (more macho) lines will, perhaps through a process akin to self-hypnosis, or involving some other form of magic, walk back to change their actual gender identity preference (from effeminate to masculine), and thence to alter their innate sexual orientation (from gay to straight).

Tangential (Slippery Slope) Thinking

Reparative therapists think tangentially in order to convince themselves and others that their patients have to change lest one thing lead to an-other and their patients go from being gay to being promiscuous, to be-coming diseased, to dying of an STD, to killing others off by contaminating them, to destroying the whole world. Their thought processes run along the distortive intellectual path of chain thinking characterized by slippery slope ideation marked by slow-but-sure off-course, deviate mentation that by taking small intermediate steps entirely ignores overall the common-sense consideration that because something can conceivably occur doesn't mean that it is likely to or will actually happen; for common things are common and rare things are rare and commonly things somehow work themselves out before "it comes to that," so that only rarely does a series of unstoppable intermediate steps lead inevitably downward to Armageddon—for most times most things stop somewhere short of the bottom of the so-called downward-spiraling slippery slope, so that just because there is want of a nail doesn't necessarily mean that a whole ship is going to be lost, thus so that "if we allow gay marriage we would also have to allow a man to marry his turtle."

Blaming/Scapegoating

Reparative therapists can be blaming individuals who as such believe that natural occurrences like being gay cannot and do not occur without some-one having to be responsible for what has happened. Thus, under the guise of learning about the developmental psychodynamics of homosexu-ality, they blame gays' parents for making their children gay. They alter-nate between the two extremes of blaming Satan and blaming God for homosexuality itself. Conversely, scapegoating gays is often a reparative therapist's convenient way to *avoid self*-blaming—by other-bashing. We hear, "The reason I am not a good therapist is that I have a typical

difficult, recalcitrant gay patient," when in fact it is equally or entirely true that the reason patients aren't good patients is that they are involved in bad therapy.

Hypocrisy

Hypocritical therapists overlooking, or dissociating away, glaring inconsistencies that subvert their own fondest contentions, condemn others for doing or being something that they themselves would like to, or actually do, do. Often enough, gay activists espy reparative therapists (and their patients who give ex-gay testimonials) at gay bars or rest stops on the highway. This suggests that some reparative therapists "cure" themselves through suppression of and reaction formation directed toward their homosexuality—defenses which, not surprisingly, readily crumble due to the inability to sustain them, for it is difficult to disavow oneself and one's sexuality for more than a short period of time.

Sophistry

Webster defines sophistry as "reasoning that is superficially plausible but actually fallacious."[3] Sophistry thus involves thoughtful spinning. Sophists intellectually justify emotionally determined (even delusional) premises after the fact, often for personal, typically utilitarian, purposes. Reparative therapists sophistically create skewed (positive) studies of reparative therapy by selecting the subjects of the study so that they have the opportunity to spin off research papers that prove what the therapist wishes to prove, which involves either a distortive personal belief (bias) or reflects a (biased) affiliated group's shared thinking. Rationalization fixes the original sophistic premise, as exemplified by how one reparative therapist saw himself as fair because though negative he treated everyone the same way (badly). Rationalization often also takes the form of citing how approving one happens to be of something in one gay person as if this, being the exception, disproves the rule that one happens to disapprove of everything related to being gay in almost everybody else. Such rationalizers typically point to their selective tolerance, their acceptance of a few things gays and lesbians "do well" to deflect from their overall unselective intolerance of gays and lesbians, as they give "evidence" that they are not part of a hate group by saying something along the lines of "some of my best friends are gay" and "some of the members of our inner circle have been (!) homosexuals themselves." Patently absurd cover-ups fly by amounting to "I justify my devaluations of gay men and lesbians simply by

calling my negativity a sign that I am just being painfully honest." Such individuals claim "We have a right to our opinion," which is true, but that is a truth that should not cover a greater and more significant truth, that theirs may not be a legitimate opinion to which they have a right, but a bigoted formulation masquerading as a neutral opinion, leading to junk science which amounts to malpractice, which is not protected constitutionally, and to which they have no right. They typically rationalize their frequent reparative failures by saying that no therapy is always effective (disregarding how most time-honored forms of therapy are at least minimally helpful to some extent, or at least are not so regularly harmful as is reparative therapy). They confess to a smaller flaw to obscure a larger one as they admit "we can't cure everybody" to deflect from a greater truth— that they can't cure anybody. Admitting to a degree of fallibility in order to appear to be overall infallible, they add that all forms of therapy have a significant failure rate, and that it's not true that reparative therapy never works just because it doesn't work for everybody all the time. They agree that not everyone changes and that there are relapses—just so that they can claim that their frequent failures are not special but actually expected, that is, part of and built into the process. In noting that the failure/relapse rate of reparative therapy while great is not much greater than that which occurs in any other form of psychological treatment, they overlook that failures/relapses do not occur in the absence of cures.

In short, gays have a right to be protected from practitioners who hold opinions that are in fact prejudicially opinionated—in the sense that they consist almost entirely of antihomosexual attitudes and beliefs. Free speech may protect the expression of some opinions. But the free speech doctrine doesn't hold for psychotherapists doing psychotherapy. They are expected to speak not freely, but accurately.

Post Hoc Reasoning

Reparative therapists reasoning post hoc retroactively create a framework that supports their theories then claim that it's their theories that support their framework. They alter and interpret facts to fit their theory (after formulating their theory in order to fit their models). To conclude that homosexuality is acquired, through psychological mechanisms—for example, developmentally—they cite the absence of definitive evidence for a homosexual gene, even though many biologically determined behaviors, such as some components of schizophrenia, are, in the present state of our knowledge, not traceable to a specific single gene/multiple genes. As mentioned throughout, they view homosexuality not as inborn but as

acquired because they must do so if they are to remain committed to reversing it through psychotherapy. Psychotherapy can do little to nothing to reverse that which is inborn, like homosexuality, for verbal psychotherapeutic interventions cannot effectively change that which is innate. If being gay is a biological/genetic development, then it is unlikely to be curable through verbal psychotherapeutic interventions. But if being gay is acquired along identifiable psychodynamic lines, analyzing its origins can potentially undo being gay much as analyzing a personality disorder psychoanalytically can potentially cure one's psychological problems. (Reparative therapists often complain about hypocrisy in GLBTQ individuals, saying that they are being manipulative when they claim that being gay is not acquired but inborn just so that they can be classified as belonging to a minority group, and thus be awarded legal protection against homohaters.)

Displacement

Due to displacement many therapeutic formulations about gays are really less formulations about being gay than they are, though couched in gay-specific, often gay-unfavorable, terms, more relevant to, and so translatable back into, important but nongay issues. So often the term "gay" is itself really a displacement from "untraditional," "lacking control," "unacceptable," and antiestablishment"—pejorative terms all reflecting mindsets originating in the generally scrupulously family-oriented values almost all reparative therapists highly embrace themselves and recommend to others.

One therapist expressed *control* issues of his own by displacing them onto "all gays are passive," while another expressed control issues of his own by displacing them onto seeing gays as all hopelessly stubborn, along the lines of "you can't tell them anything, especially to get the help they need to get over being gay and return to some sort of sanity." Another therapist expressed *competitive* fears and wishes of his own by displacing them onto value judgments meant to invalidate gays whom he saw as a competitive threat to him because they were "all about to take over the true meaning of marriage, and so my marriage." As another example of displacement from a competitive to a gay issue, a reparative therapist expressed fears about her own not winning/losing out in life as the reassuring (for her) belief that the "gay competition" in fact consisted of a bunch of inferior people—for gays were less worthy than straights like her, and the gay way of life, as compared to her straight way of life, was necessarily one that scored second best to her own style of living.

Sublimation

Sublimation consists of rechanneling socially unacceptable ("dirty/contemptible") into socially acceptable ("clean/acceptable") ideation. Therapists often sublimate (rechannel) their personal concerns about gay sex ("contemptible, disgusting") into what they consider to be positive, helpful therapeutic interventions ostensibly meant to help gays become ex-gay. The therapeutic goal of celibacy can represent, among other things, an extreme sublimation involving turning "bad sex" into "good social behavior."

Primary (Delusional) Thinking

Primary delusional thinking, related to, but the next step beyond, being merely opinionated, is typically installed to prepare a therapist to be intimidating along the lines of "you better believe it, for I know it is so because I know it is so, unquestioningly and exactly." Because religious precepts are in fact illusions (in the sense that the reality can neither be proved nor disproved beyond doubt) it can at times be difficult to distinguish faith-based ideation from primary delusion.

Paralogical Predicative Thinking (von Domarus Illogic; Similar = The Same Thing)

Here connecting links are forged to become strict identities. The (false) reasoning is this: if A can be meaningfully equated with B in any respect, and C can be meaningfully equated with B in any respect, than A = C. The oft-cited example of this kind of thinking is, "I (A) am a virgin (B). The Virgin Mary (C) is a virgin (B). Therefore I (A) am the Virgin Mary (C)." Therapists think this way about their gay patients for a reason: so that they can create favored stereotypes that, however irrational these are, they can subsequently use to condemn a gay individual for being gay. Many reparative therapists determine that being gay is a disease by reasoning: being gay = a developmental lag; being neurotic = a developmental lag; therefore being gay = being neurotic (and being neurotic= being diseased). Therapists who claim gay = addiction are in effect saying the equivalent of "gay = pressured; addiction = pressured; therefore gay = addiction." They also reason that "sex involves seduction, homosexuals are sexual people, and therefore homosexuals routinely seduce other people," and, what is more, are all by nature prodigious seducers, especially of children. And they further reason that "homosexuality is unique, sick people

are unique, therefore homosexual = sick." Reparative therapists come to judge their gay patients' overall professional potential on the basis of their (devaluing) assessment of gays' sexuality/sexual performances when they reason: work is a performance, sex is a performance, therefore sex = work (therefore bad gay sex = bad inferior work—and so all gay men lack professionalism, and will predictably be professionally ineffective, and especially in professions where it's required that they be forceful and masculine enough to fully take charge).

Reparative therapists assume that gay men have many (both personal and sexual) features in common with girls and so just by virtue of being gay = girly boys or effeminate, as well as personally passive to the point that they all allow themselves to be bullied without complaining or resisting, but instead swooning in the face of all serious onslaughts to their pride and position.

Reparative therapists seizing upon the occasional shared characteristic make the point that there is a supposedly meaningful relationship between homosexuality and character disorder/defect, so that homosexuality must of necessity be in its turn accompanied by specific characterological problems/relational misbehaviors such as shyness/promiscuity. This kind of thinking uses Von Domarus elaborations to "prove" that gays are all narcissistic or are all aggressive (bitchy). Those gays who are passive sexually must by definition necessarily be passive relationally as well (subservient), and those gays who are active sexually must by definition necessarily be active relationally as well ("butch")—"tops" who must of necessity actively victimize others who are bottoms—the stereotype of the bitchy gay man/the aggressive lesbian.

(Altering) Effect Can Influence Cause (If A Leads to B Then Altering B Will Retroactively Alter A)

Reparative therapists aver that since all homosexuality leads to promiscuity, treating promiscuity, perhaps with an antidepressant like Prozac, will cure homosexuality. Presumably since masculine men say "Hey Dude" to each other, saying "Hey Dude" to a gay man will walk back to make him masculine, and therefore heterosexual. Many reparative therapeutic approaches arise out of this thinking: heterosexual men are he-men; homosexual men are she-men; therefore acting like/becoming a he-man will make the gay she-man into a heterosexual he-man. According to such beliefs, since gay men sew, you can cure homosexuality by requiring them to stop stitching, and gays can go straight by forming same-sex *platonic* relationships with straights. Because a heterosexual man's relationship

with (straight) men is nonsexual (compared to a homosexual man's relationship with straight or gay men) developing nonsexual same-sex relationships with straight men will hopefully walk back to the homosexual's becoming heterosexual. Or, since homosexuality is mysterious, demystifying other men will presumably reduce or eliminate the mystery of homosexual sexual attraction.

Because gay = nonassertive (a false equivalency in and of itself) it therefore follows that increasing assertion will decrease being gay. Because gays are ashamed of themselves, therefore enhancing self-esteem will walk back to change gays over to straights. Because being shamed by a man leads to being gay, being approved of by a man, as a man, will lead to his becoming straight. Because repression of one's true self leads to homosexuality, a false self, decreasing repression presumably leads to heterosexuality by releasing one's true self, the straight self.

According to many observers since gay men develop a sexual attraction to men as a compensation for having, historically speaking, been deprived of a man's warmth and love, making up for what has been lost by giving a gay man nonsexual warmth and love will of necessity reverse homosexuality. Therefore, some therapists recommend supplying nonsexual warmth and love through touch and chaste hugging offered by a same-sex mentor. The belief is that this will allow the he-man inside the she-gay man to burst forth and return to the heterosexuality that was presumably thwarted early on due to the long-term absence of sufficient nonsexual warmth coming from another man.

Unfortunately, psychic life is too complex to assure that that which is done can be undone by reversing course along some (unfortunate) path once taken. Aside from the psychodynamic inaccuracies involved, as we know from emotional disorders such as PTSD, traumatic psychological events often leave lasting posttraumatic scars that do not simply heal with the application of any psychotherapeutic procedures, which too often merely act as Band-Aids placed on top of large sealed-over but still festering wounds.

Underlying this form of thinking is the belief that underneath it all gay people are really closeted straight people, that is, straights in a gayjacket. They not only really want to be straight but actually are straight, no matter how gay they at first appear to be. Altering cause by inputting into effect is then supposedly an efficient and rapid way to bring this changeover about.

Overcompensation

Reparative therapists typically use the concept of overcompensation/reaction formation to explain facts that don't fit their theories. Thus if gay

men appear to be loving they are merely overcompensating for their underlying hatred of all men, including other gay men, a fact that helps explain why gay love is a charade. And, of course, homosexuality, being innately shame based, is itself a compensatory pride mechanism, one installed to deal with shame over heterosexuality. Hence reparative therapists cure by restoring the opposite of A back to A itself. When it comes to homosexuality, the so-called fact that it purportedly defends against heterosexuality (that is, is the opposite of heterosexuality) becomes a license to see sissified behavior as a protest against the underlying masculinity gays hide due to shame about being a real man. Then one can cure gays by taking them back to the straights they really are.

Stereotyping

Stereotyping is a complex cognitive error consisting of a number of contributory suberrors, some of which have already been discussed.

Therapists who stereotype homogenize gays simplistically even though gays are in fact a highly diverse population/group. They then evaluate and judge gays, rarely positively and mostly negatively, not as individuals but as members of (some) fantasized cohort with supposedly overarching distinguishing (again, generally negative) characteristics. Some stereotypes hold simply because the ones doing the stereotyping subsequently actually validate the venal stereotype by selective observation (e.g., using case samples from STD clinics to represent all gays), or because the stereotyping subtly or actively influences its victims in the direction of fulfilling the stereotype. Thus reparative therapists stereotype all gays as depressed after providing them with therapy whose premise, that they are broken, is actually what is depressing them.

There are a number of reasons for stereotyping. In *parroting/echoing*, as if by osmosis reparative therapists, including ex-gay reparative therapists, *learn* stereotypes unique to an inner circle of affiliated therapists, such as "you can spot gays by their effeminate ways and unmanly interests." Intelligence does not always protect therapists from such learning, for they also "learn" for political, monetary, and unconscious reasons, the latter serving a psychological function, particularly the need to belong to, to be an official member of, a homogenous group all of whose members think the same way about gays and about ways to make gays ex-gay.

In *condensation/symbolization* the stereotypes are codes whose creation is generally inspired by personal bias. For example, a reparative therapist's seeing all gay men as symbols of evil incarnate was his shorthand way to express, in a nutshell, reassuring thoughts about his own sexuality (e.g.,

"here's what all gays are and do, and how it differs considerably from what I am, and do").

In *assuming guilt by association*, homonegative therapists stereotype all gay men by categorizing all of them in an excessively negative category based on grading them according to the company they keep—the people who do, or presumably will, surround them: friends who are wanting, part-ners who are cheating, parents who are being troublesome, and the like. Therapists who think this way (believing in human osmosis) assume that if a son is gay he will necessarily absorb the tenets of, and so be an integral part of, the gay underground and therefore already has, or soon will have, pierced nipples and eyebrows and a preoccupying underwear and foot fetish, just like all his friends. (And even when that is the case, catastrophizing, they go on to see this likely trivial problem as a serious one without recognizing how often such things are temporary, and/or if permanent, accompanied and softened by other neutral, nonexploitative, more loving behaviors.)

In *circular reasoning* therapists stereotype homosexuality as an illness because it is not mainstream, and as not mainstream because it is an illness.

In *zero-sum thinking* reparative therapists stereotype gays as immoral because as therapists they presume that there is a finite amount of moral-ity in the world, and it is quantifiable, so they conclude that if they accept gays as moral that will simultaneously reduce the quantity of morality left over for straights. The most familiar example of this is the belief that gay marriage doesn't simply stand side by side with, but somehow, in ways that nobody can definitively define/prove, diminishes straight marriage.

In *ignorance*, which can be either innate or motivated, the stereotyping involves getting the facts wrong. Ignorance differs from psychopathic dis-simulation for effect, the latter involving the deliberate manipulative cre-ation of illogic to gain some specific practical advantage—as when psychopathic ex-gay ministers stereotype all gays as reprobates in order to keep their own ex-gay ministries alive.

All these forms of stereotyping lead to flat assertions both too broad and too narrow, where opinion supplants facts and irrationality supplants rationality leading to the reparative theme with variations that being gay is a very difficult, harsh, unrewarding experience. By implication, thera-pists instead of simplistic stereotyping where "it becomes obvious," and "everyone knows that" should focus on rational multifaceted nuanced views about being gay to the point that they drop their irrational one-dimensional opinions, having remained sufficiently flexible to stay open

to arguments to the contrary, leading to their developing ideas that are different from those they currently favor, and resulting in their relinquishing procedures they too often use, exclusively and excessively, to harm people they are supposedly in the business of helping.

9

Positive Aspects of Reparative Therapy

This chapter attempts to evaluate reparative therapy not only according to its failures, of which there are, unfortunately, many, but also according to its successes, of which there are, fortunately, a few, in order to determine if any real, lasting therapeutic assistance can ever result from this form of treatment, and if the patients undergoing it can ever find a way to extract a measure of a successful outcome from its ministrations.

We should ask: If, when it comes to reparative therapy, the notoriety is so much more pervasive, the criticism so much more bitter, and the opprobrium so much more intense than that meted out to non-ex-gay therapists and their therapies—even though the latter also have their own, well-known, yet somehow more acceptable, disadvantages, including low-cure rates and high incidences of complications? Are there biases involved against reparative therapy that lead those who evaluate reparative therapy to overstate its negative effects while downplaying the good that it can (however occasionally and sporadically) do? While the conservative religious right overestimates reparative therapy's positive features while downplaying the harm that it routinely does, too many traditional therapists become even more negative about reparative therapy than its actual theory and practice necessarily warrant. Would reparative therapy survive as a technique unless it actually helped some gay people—if only through its unintended positive consequences? Is the downside of receiving re-

parative help always greater than the upside of the assistance some patients do get out of select aspects of their reparative encounters?

True, reparative therapy mostly wastes time and money and hardly ever changes a person's sexual orientation. True, at best the only anticipatable change is in the direction of celibacy, and that involves the renunciation of a natural function. For celibacy is not a sign of a new and superior form of healthfulness, but a manifestation of a new sort of disease, one characterized in great measure by the unhealthy defense mechanisms of denial of who one actually is and suppression of what one actually wants. For celibacy, instead of curing mental disorder, constitutes one of its symptoms, as in some ways it resembles a psychiatric disorder of its own: OCD with its ritualistic behaviors (saying mantras to ward off temptation); borderline personality disorder with its "as-if" characteristics (acting "as if" one is asexual instead of gay when that is not basically true); self-punitive, self-destructive, depressive isolation (as one comes out of the closet only to abjectly, guiltily retreat back into its lonely walls); and/or masochism as displayed in one's involvement in what is at least that spiritual self-mortification characteristic of asexuality.

I have spoken to some patients who although unable to resist having enrolled in a program of reparative therapy were nevertheless able to disregard its negative aspects and withstand its deaffirmative messages, at least enough to be able to grasp the wisdom that sometimes exists in rudimentary form even in this most overarchingly misguided type of treatment. These patients actually managed to benefit from the supportive elements in their reparative therapy. As Besen says, "There are grains of truth to some of the . . . stories, and we ought to listen to what they are saying [that some gays are] treated poorly, cast aside . . . not getting their legitimate needs fulfilled by the GLBT community because of how they look, or [because] they are poor, or strange. For gays like this, reparative therapy can help detoxify the negative experiences they are accustomed to getting within the gay community [because] therapists at least offer them warm promises of unconditional love."[1] Unkindness, with its accompanying gay-on-gay bashing, will always exist within the gay, as it does within the straight, community, and for those gays who feel guilty about being gay and get depressed and want to change at least in part as a result of being ignored and bullied, reparative therapy can help—if only indirectly, and only those individual patients who can learn to work the reparative system to make the best out of a very bad situation. While this form of therapy is never to be too highly recommended or too actively sought or undergone without much thought for one's ultimate well-being, there are times when a patient already in this form of treatment, for whatever reason, can turn

even this method of therapy to his or her advantage: not to be cured of being gay, but to at least get somewhat better in areas other than those directly involving his or her basic sexual orientation.

Reparative therapists can do less harm than good when patients learn how to turn their reparative experience from repair to affirmation and hence from psychic brutalization to psychic mending. (Those therapists who participate in bringing about this sanguine outcome often argue that banning reparative therapy outright, especially by politicians ill equipped to understand its nuances, should not be an option.) There are some reparative therapists who in practice, if not in theory, have been able to modify their approach and their therapeutic techniques enough to reduce their therapy's harmful, and maximize its helpful, effects.

Reparative therapists who do *short-term* therapy can at the very least offer their patients a mode of treatment that being time-limited doesn't sap their energy, waste too many of their hours, squander too much of their money, or demand a life-stifling involvement—involvement that is, like some prolonged mainstream psychoanalyses, so engulfing that it keeps the patients from living a normal life for a long time, and when it is over, often after months or years of hope for change, and nothing actually happens, leaves gays devastated to discover that they are no different from what they were when they first began treatment and are possibly actually worse off than they were before they started trying to get help.

A possible line of defense of reparative therapists (and an elucidation of the good they sometimes do) lies in comparing them directly to mainstream therapists. Many therapists doing ex-gay therapy use the same (mainstream) methods other therapists use: methods such as inculcating insightful understanding, correcting misguided thinking (via cognitive therapy), behavioral modification (as part of their cognitive-behavioral approach), "relaxation therapy" that ranges from meditation to dealing with anxiety by consulting a wallet-sized card that suggests "think of peaceful green meadows"; exhortation; environmental manipulation; suggestion; autosuggestion; control/mind control; modified brainwashing; and promoting group affiliation (especially promoting healthy intragroup identifications). That is, some reparative therapists use techniques that are, though often not called that, structurally and conceptually not so different from the techniques mainstream therapists use, so that at least some of their patients can harness the mainstream aspects of their treatment to undergo beneficial change. To illustrate, mainstream behavioral therapy consisting of issuing a patient a phobic pass (a piece of paper that says, "You will be okay, you don't have to be afraid, I guarantee it"); supportively supplying a service animal to reduce anxiety; or advocating

defensive denial, reaction formation, and suppression rather than deep understanding (which can make patients anxious) uses techniques that don't, in their essentials, differ a great deal in content and structure from those reparative therapists use when they suggest a gay person employ autosuggestion to go from gay to straight by not calling himself gay, or, as Nicolosi suggests, by actually calling himself "non-gay."[2] (While "non-gay" individuals have same-sex attractions they don't think of themselves as homosexual.) Is praying the gay away that much different from, and any more or less effective than, the traditional behavioral technique called "stop thought" where patients with OCD are told to tell ("yell out to") themselves to "*stop*" brooding and so to chase the brooding away by refusing to let their repetitive thinking get out of hand?

Psychoanalytically oriented reparative therapists misapply what are valid techniques to understanding and curing a phenomenon (homosexuality) to which these techniques are inapplicable. Yet in having a discussion of the so-called individual dynamics of homosexuality with their gay patients therapists are perhaps, unbeknownst even to themselves, giving their patients a method that they can use to change for the better, not straight though once gay, but straightened out though once troubled.

Reparative therapists often help gay men with *compulsive* sexual problems reduce their sexual compulsivity by helping them diminish their pressure to perform so frequently simply by offering their patients structure, which is a helpful aspect of almost any type of regular therapy. Just being in therapy is organizing if only because patients now have less time and energy for involving themselves in sexual activities to excess.

Questioning(ambivalent) gays whose ego-ideal/identity remains strongly as straight in spite of clear tendencies to be gay (some of Nicolosi's "non-gay homosexuals")[3] can feel that just by being in reparative therapy they are somehow affirming the nongay identity part of themselves. These individuals can point to ongoing therapy as a cover, one that enhances the reputation they have with others who decry being gay, while enhancing the reputation they have with themselves for "at least doing all I can to go straight." Patients with OCD who see being gay as a sin fear that if they don't change they will be damned and believe that they are at least symbolically and ritualistically warding off damnation by trying to go straight; they also care less if therapy works as promised than that they are at least going through the motions of seeking help—a fantasy of undoing who and what they are, one that is often enough to make OCD gays feel more good than evil because at a minimum they see themselves as participating in something angelic, and so for that reason alone they should be forgiven for all their prior and current "gay" mistakes, sins, and crimes. Not a few

OCD patients favorably (if misguidedly) experience their therapists' ministrations as exorcisms whose admirable objective is to expel sin through prayer, backed up through a climactic cleansing achieved by the performance, as advocated, of daily kindnesses and gratitudes. Almost any form of therapy will offer OCD patients the illusion that they are at least trying to defy evil and beat Satan by not yielding up to temptation, and thus escaping damnation for being gay and recovering from having fallen.

Patients who are *delusional* and *hallucinating* (such as those who are thinking paranoid thoughts/hearing voices) can replace hostile accusatory delusions and hallucinations with kinder gentler ones, for example those no longer calling them to task but to order. Just being in treatment can help those who are additionally *decompensating* emotionally, with ex-gay therapy constituting at the very least a kind of occupational therapy in the sense that it functions as a diversion from the patient's troubles.

Gays prone to using *dissociative* defenses (e.g., patients with multiple personality tendencies) can gain some benefit, however superficial, temporary, and subject to relapse, by shifting into a new personality where they act *as if* they are heterosexual and even get heterosexually married. The more they act as if they are straight the more they actually begin to see themselves as, and in rare cases actually become, that way. (Typically, however, when their therapy either fails from the start or wears off, as it usually does after a short period of time, these individuals relapse, only to find themselves in a marriage that now seems utterly foreign and straight-jacketing to them.) And gays with *as-if* (*borderline*) personality features often buy into the therapist's promises that just acting straight takes away the gay, so that they can become heterosexual simply by naming, or in this case renaming, who and what they are.

Depressed gays welcome the unconditional offer of help which they feel they get from a reparative therapist who shows care and concern. Therapists who merely tell patients that they, their therapists, want them to keep coming for sessions can help sad, lonely patients feel wanted/un-rejected and as a result less depressed. Depressed patients who feel isolated often respond favorably to just being connected with their therapist. They also respond favorably to others in their lives liking them better because, while they may still be gay, at least they are taking the necessary steps to go straight.

Denial that the therapist and patient share ("if someone calls you 'dude,' you turn straight") can lead to a *hypomanic* shift toward fantasies of therapeutic productivity that can help gays feel that they are, for sure, heading in the right direction.

Gays whose *masochistic* self-loathing leads them to have frequent anonymous sexual encounters meant to humiliate and shame them can get a measure of at least temporary relief out of the demeaning aspects of reparative therapy. They feel that their therapist wants to punish them, and paradoxically that means, to them at least, that their therapist cares enough about them to give them what they at least feel that they deserve.

Gay individuals who graduate to become ex-gay therapists can improve their own functionality by going about helping others become ex-gay. Gays with a checkered past can help heal themselves by loving and caring for someone else with a checkered present. Those who are ex-criminals can help other individuals with a criminal bent much as former drug addicts by becoming drug counselors can help current addicts. Still, it remains true that those with a shady past may in attempting to heal others damage them instead.

Also counting in their favor is that some reparative therapists *avoid* doing some harmful things that mainstream therapists do. Unless they are physicians, and not too many are, they don't prescribe masculinizing hormones for sissy boys, don't give excessive amounts of addicting anxiolytics to reduce the emotional pressures that supposedly lead to homosexual cruising as well as to homosexuality itself, and don't prescribe antidepressants in the belief that homosexuality is a depressive equivalent that can be cured by manipulating the brain's serotonin levels.

Though it is wrong to excuse bad behavior by citing even worse behavior, it remains true for some patients, particularly those with an intellectual bent, that the reparative therapist who starts with the premise "you as a gay man are broken" is possibly no more deaffirmative than the cognitive therapist who starts with the premise that "your thinking is all off."

Perhaps other than cognitive therapy and pharmacotherapy there are as few creditable controlled studies of the therapeutic effectiveness of mainstream therapy as there are of the therapeutic effectiveness of reparative therapy. In all cases therapeutic value is too often assessed through citing anecdotal information (such as testimonials) to evaluate how worthwhile a given treatment might be and to determine which form of treatment to select.

Ex-gay therapists are not the only therapists who are inadequately trained. Not all "mainstream" therapists are LMHPs, as not all LMHPs are mainstream therapists. Too many LMHPs do wild analyse because they are self-proclaimed psychoanalysts who, even though creditably licensed in some venue, have in actuality not attended creditable training programs or have attended creditable training programs but not the ones that are applicable here.

All forms of therapy can be sexualized in the transference. Probably as many straight nonreparative therapists try to seduce their patients as ex-gay reparative therapists try to seduce theirs. (I personally know of several examples of straight therapists seducing straight patients, with one occupying a prominent position as head of a major mainstream psychiatric training program and another who, though in training to become a mainstream psychiatrist in a major mainstream psychiatric training program, was never disciplined for exactly such unethical, distinctly untraditional, actions.) Nor are mainstream group therapies exempt from the problems of reparative groups—where an undercurrent of sexual attraction often exists between the group members, sometimes involving the group leader. In all forms of therapy there is an inherently thin line between caring, supporting, and coming across, intentionally or not, as seductive, and an equally thin line between helping someone and actually falling in love with and seducing that person "in order to do him or her the most good."

Finally, reparative therapy is done no more or less often than is traditional therapy because the therapist needs the money, or has political reasons, or wants to get into, or remain in, the limelight.

In conclusion, though reparative therapy attracts more notoriety and criticism than mainstream therapy, a reasonable defense of reparative therapists is that such criticisms being also applicable to mainstream therapy therefore say as much about psychotherapy in general as reparative therapy in specifically. Thus if reparative therapy doesn't work it may not be so much due to the problems associated with the impotence/harmfulness of reparative therapy as the problems associated with the impotence/harmfulness of all forms of (verbal) psychotherapy, with some mainstream therapies, even those of the most sophisticated, best researched nature, in some cases working no better (and doing no less harm) than some reparative therapies. To cite two examples, reparative therapy is not unique in promoting symptom substitution, and mainstream therapy that doesn't work right can lead to the same depression that reparative therapy is often cited as inducing because of *its* supposedly unique ineffectualness.

As a result, aspects of reparative therapy can be seen as no more misguided than, and even misguided in the same ways as, other therapies and therapists—even therapy done by the best trained therapists possessed of the most highly developed skills and impressive academic credentials. Therefore, in my opinion, the criticism that reparative therapy in many respects certainly merits ultimately should be leveled not only at its methods but also at the use to which these methods are put. The best answers to the questions "Does reparative therapy work?" or "Does it mainly do harm?" lie not in global condemnation of the method, but in specific

point-by-point contact with, understanding of, and ultimately supervisory control/condemnation of the individual using any given method including, but not limited to, reparative techniques.

Part II

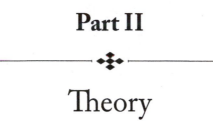

Theory

10

---◆⬧◆---

Speculative Causal Theories

This chapter outlines some reparative theories about the supposed psychological causes of homosexuality. Although I believe that these theories are all wrong, at one time or another they have almost all nevertheless actually formed the basis of one or more (ineffective) reparative treatment approaches.

AN OVERVIEW

Serious disagreement about what causes homosexuality exists between reparative therapists who claim that emotional factors are almost entirely responsible for same-sex attraction and more traditional therapists who, disagreeing, believe that one is born gay and that the so-called psychodynamics of homosexuality are not causative of, but consequential to, being homosexual. According to the latter group, how children develop, including what their parents and society says and does to them, doesn't cause homosexuality. Rather, the developmental events described in the reparative literature as the *cause* of, are in fact, if at all related to a *response* to, being gay, so that a homosexuality already in place might cause a mother and father to respond in a unique way to a child they sense is different and perhaps, as they see it, inferior, defective, or especially needy. Does the mother's overprotection of the son cause him to be timid and perhaps effeminate, or does his timidity and effeminacy cause the mother to overprotect her son? Is a son or daughter gay because the father

hates him or her, or does his father hate him or her because he or she is gay?

I believe that homosexuality can no more be meaningfully understood psychodynamically than a composer's compositional ability can be explained conceptually. For musical ability is, like homosexuality, innate, so that the homosexual is no more aware of what is causing homosexual attraction than, as Ned Rorem says, referring to the process of composing in *The Later Diaries*, composers are aware of how they compose, for when it comes to composing, "one doesn't know how, one does."[1] Writing about the composer Benjamin Britten, Adams put it this way (paraphrasing a sentiment usually attributed to Saint-Saëns) about Britten's compositional ability: "He produce[s] music like an apple tree produces apples."[2] By analogy, gays produce their homosexuality just as an apple tree produces its fruit, and, like Rorem's composer and, unfortunately, the reparative therapist, know little about why, how, or what they do.

Van den Aardweg, in a (surprising for him) partial agreement with traditionalists, notes that "homosexuality is not adequately explained by a disturbed or detached relationship with the same-sex parent, and/or an overattachment to the opposite-sex parent, no matter how frequently these are associated with it. For one thing, such relationships are often seen in pedophiliacs as well, and in other sexual neurotics. . . . Moreover, there are normal heterosexuals with similar parent-child interactions."[3]

It is conceivable that there are many possible theoretical developmental pathways each producing homosexuality, or even each producing a *different type* of homosexual/homosexuality with differing characteristics, so that there are "different homosexuals" who can perhaps be differentiated by their varying personalities (passive or active, etc.) and varying sexual proclivities (femme or macho, bottom or top, etc.). This is unlikely, however, because many (most) homosexuals' personality characteristics/sexual preferences, however strong they might be at any given time, generally speaking are at least somewhat mutable over one's lifetime. That is, the personality characteristics of many a given gay person shift over the months and years, and sexual preferences are, as is the case with heterosexuals, not necessarily limited to one specific desire or always triggered by one specific type (e.g., only macho) or a specific set of circumstances (e.g., one is only attracted to a man who comes on strong, or only to a man who is disinterested/overtly rejecting).

Recognizing that homosexuality is *inborn*, not acquired, has academic, personal, political, and occupational consequences.

In the realm of the *personal*, gays who recognize that they didn't choose homosexuality but were born that way can simply accept that they are gay,

and are gay due to forces beyond their, or anyone else's, control. Now they no longer have a reason, as many still do these days, to feel guilty about being immoral, damned, disreputable reprobates to be properly censured and correctly punished as sinners and criminals. And gays who (rightfully) discount the psychodynamic theories about what caused their homosexuality can stop trying to analyze their gay away, and instead accept themselves as they are, no (psychodynamic) questions asked, or if asked, satisfactorily answered as being irrelevant, of no legitimate interest whatsoever, and of no consequence at all.

In the realm of the *political*, if being gay is not chosen by, but for, the individual, then the same equal rights apply to gays as to any individuals who, as the saying might go, "can't help themselves"—who "can't help who and what they are," like "those who are disabled, or of color, or belong to a certain race."

In the realm of the *occupational*, many psychoanalysts believing (as some still do) that being gay involves a regression from being straight proclaim that no gay man, being regressed in one area and so likely to be regressed in all, should be allowed to enter certain professions, such as be admitted to be schooled in the practice of psychoanalysis. But gay men who refuse to accept the fact that they are regressed straights will see themselves as well, not sick, and fight, perhaps aggressively, for their rights—one of which is to apply to join a psychoanalytic institute of their choosing: to be considered for admission without being blackballed, and so to be able to practice psychoanalysis (or whatever medical specialty they choose) without being excluded. More specifically, all theories that supposedly tell us exactly what causes homosexuality are wanting for the following reasons.

SEPARATE CENTERS EXIST IN THE BRAIN FOR SEX AND FOR PSYCHODYNAMICS

The centers of the brain that control sexual attraction are anatomically rather separate from the centers of the brain that control cognition. Therefore, analyzing psychodynamics won't likely alter sexual orientation.

THE THEORIES ARE INTRINSICALLY CORRECT BUT IRRELEVANT (INAPPLICABLE TO HOMOSEXUALITY)

Gays like anyone else can have developmental problems. There is evidence, for example, through psychoanalysis, including dream analysis, linking developmental issues to personality difficulties/severe emotional

problems such as a major depression in gays as well as in straights. But there is no evidence meaningfully linking any "psychodynamics of homosexuality" to homosexuality itself. In other words, theories that are dynamically correct in their own right don't, as intended, explain what they set out to explain, which is homosexuality. Many of the psychoanalytic theories put forth by Christian psychoanalysts to explain homosexuality do reveal some meaningful things about a gay patient's psychodynamics. So they can be helpful in resolving gay patients' conflicts, thus helping gay patients improve their overall psychological state. These theories raise issues worth considering. But they apply only to someone who is "neurotic" not to someone who is "merely" homosexual. The explanations these theories proffer can be helpful to understand gays who are anxious and depressed, but not to explain their being gay per se. No theory convincingly connects causal hypothesis with homosexual outcome. Rather all extant hypotheses fail to bridge the gap between theory and outcome, process and product, leaving the gaping chasm between mind (psychodynamics) and body (sexuality) not bridged, and so incapable of being crossed.

Applying (or misapplying) classical orthodox psychoanalytic theories of causation to what is in fact remotely if at all a psychodynamically infused event reminds me of what a patient of mine who had a reparative therapist enamored of early trauma theory told me her therapist had told her: "my viewing the primal scene (my parents having sexual intercourse) was what made me cross-eyed."

A MULTIPLICITY OF THEORIES SUGGESTS THAT NO ONE THEORY IS ASCENDANT AND SO NO ONE THEORY IS CENTRAL

There are too many theories for anyone serious about explaining homosexuality psychodynamically to be able to comfortably and finally chose and legitimize one theory over the others.

Opposite theories exist/rule so that the theories I have come across to explain homosexuality not only contradict themselves internally but also disagree with one another.

Ultimately, the theories taken together remind me of the formulation of one of my mentors discussing how birth position determines the development of emotional disorder: "Those who become neurotic are generally the first born, the last born, the middle child, or a child who occupies some other unique position within the family and with his or her siblings."

Theorists attempting to explain homosexuality identify some parents who are one way and other parents who are another way. Unfortunately,

the characteristics of the parent in one study can be exactly the opposite of the characteristics of the parent in another. Parental upbringing theories either note that the fathers of homosexuals are too passive or too castrative. Many of the gay men I worked with had mothers who were absent when they were theoretically supposed to be engulfing and fathers who were absent when they were theoretically supposed to be overwhelming. Gay men with classical passive fathers no boy can, or should, identify with nevertheless turn out, when they grow up, to be quite masculine after they mature, having *compensated* for their supposedly resultant weak masculine identities—often by becoming hypermasculine in protest. Few of my own personal parental relationships seemed to trace any sort of classical developmental patterns or convincingly lead to my specific gender identity/ sexual preferences. My mother, for example, was absent while my father was intrusive, not what one would expect from reading the many studies about how a heterosexual boy supposedly gets turned into a homosexual man.

Sometimes, when all siblings are treated equally by their parents, only one goes on to become homosexual, and in these cases when looking back one cannot find a convincing link between how the parents treated a given child and what the child's sexual orientation turned out to be. (In many such cases a child's temperament, which is inborn, is what makes the difference.)

Different contradictory explanatory theories seem to produce the same outcome. Roads that should not all lead to Rome in fact theoretically are postulated to all go exactly there.

Finally, long-term follow-ups raise questions, as fathers on follow-up in reality turn out to be actually mentally healthy when theoretically they should be disturbed, and mothers on follow-up turn out to in fact be absent when according to the theorists they should be engulfing.

THE THEORIES ARE FORMULATED IMPLICITLY

In my experience, some reparative therapists who present their developmental theories as *clinically* determined, such as hypotheses arrived at *observationally*, in actuality have instead come to their conclusions *intellectually* and *implicitly*. They have also subsequently applied them after the fact—then claimed that the facts had preceded the theory. Thus many of the so-called causes of the homosexual illness have not actually been "uncovered" through clinical research as claimed. Instead they have been determined mindfully, even theologically, along the lines of "Because to me it makes sense that . . . I believe that it is true that. . ." It seems to make implicit sense that if a biological male "acts like a woman" in the sexual

act then it follows that his God-given manliness has somewhere along the
line been thwarted and he has reverted to the other, default, womanly
position. And it seems to make implicit sense that traumatic blows to
one's maleness supposedly thwart one's native masculinity, and so heterosexuality. It seems to make implicit sense, at least for those who, in spite
of the evidence, confound gender preference/identity with sexual orientation, that the origin of homosexuality would be in long-buried feelings of
shame related to one's gender, for example, in conflicts about one's masculinity or one's femininity. Yet many homosexuals do not experience dis-
sonance between gender identity and sexual orientation. Many gay men
consider themselves to be "men and gay," or "men because gay," or "not
not men because gay," and they have no shame about being whatever, and
displaying whoever, they are, and prefer to be. Indeed, contrary to some
theories, many gay men believe that the purity of homosexuality (involving "not being contaminated by woman") counts toward a kind of hyper-
masculinity, with being gay making one's masculinity not weaker, but
stronger. There is much shame to be found in gays about being gay, but
this is related less to gender dysphoria than to *internalized* and *internal*
homophobia. (The latter factor, little emphasized, consists of a primary
self-homophobia arising out of erotophobia, e.g., internal conflict over sex
and sexuality, linked with, but not identical to, external homophobia.)
One can be ashamed of being gay, but one is not gay because one is
ashamed—of oneself.

As an upshot, therapists who concoct theories out of common sense
typically create theories that don't survive scientific scrutiny or produce
palpable positive therapeutic results. Thus the belief that "male homosexuality is the product of being too close to mother and so becoming like
her, or having a weak devalued father, one the son doesn't admire, rendering a son's quest to become manly like Dad unlikely to lead anywhere, and
so be impossible of realization"—while it makes intrinsic sense as a theory,
cannot be validated in the real world; doesn't, as we might expect, tend to
manifest itself, in therapy, in the transference; and doesn't in any way
serve as a useful tool to help a gay person go straight

GRAY AREAS BELIE THE EXISTENCE OF CLEAR-CUT DICHOTOMIES

Illustratively, the many observers who attempt to distinguish early (preoedipal) trauma that does not involve gender identity issues from later
(postoedipal) trauma where the trauma occurs during the latency period
(and so does involve gender identity issues, these being ascendant at the

time) are creating false dichotomies in situations where gross overlap is the rule, not the exception.

THE THEORIES RELY ON THE EQUIVALENCY OF A THING AND ITS OPPOSITE

Theories that depend in any way on the concepts of reaction formation/overcompensation, that is, theories that postulate that A = B, with B the obverse of A, can be, and too often are, advanced to fathom almost anything someone wants to explain by invoking the aforementioned, often sophistic, formulation of the identical nature of opposites. Anything goes when A is said to defend against its opposite, B, so that one is aggressive because one fears being passive, or the other way around. Thus do many homosexuals seek a real he-man because they make a narcissistic object choice—a man who reminds them of, or is identical to, who they always were, are, and will be? Or, doing the opposite, do homosexuals who are in reaction formation/overcompensating seek an idealized masculine type, a real he-man, to make up for what they feel they themselves never were and are lacking now?

THE THEORIES CORRECTLY ELABORATE A FALSE INITIAL PREMISE/POSTULATE

Hypotheses even though false can, once in place, be convincingly elaborated in a way that leads to apparently sensible conclusions, but only if one first accepts the flawed basic initial premise. Paranoid delusional thinking is often so constituted as to reach conclusions that would be true, except for the fact that they start with (and are expansions upon) a false initial premise. For example, once patients accept the initial (false) premise that an alien power has implanted a TV transmitter in their body they can weave manifestly incontrovertible theories/beliefs about aliens and transmitters that are, however, all specious because while they are entirely legitimate elaborations of the primary postulate, the postulate they elaborate is an illegitimate false initial primary hypothesis/fact. Thus reparative therapists convinced (the initial false hypothesis) that homosexuality is a Judy Garland identity disorder can *rationally* seek to explain and prevent homosexuality by intervening early on in a child's pathological identifications by removing all Judy Garland songbooks from the home. This doesn't work, because any such interventions are simplistic, are unwarranted, and because identity issues are not causative of homosexuality but are, if they exist at all, its secondary/trivial/causally inapplicable consequences. In the real world, gender identity, or gender identity disorder (GID), or gender

dysphoria, but not homosexuality can be both described and to an extent understood psychodynamically. Therapy can help gays force a jerry-built gender identity shift. But that therapy won't fundamentally alter a (homo)sexual orientation one bit.

False memory theories exemplify how fantasies (often of a paranoid nature) can, serving as primary false hypotheses, be elaborated to incorrectly alter the perception of reality. The conviction that one has been physically abused early in life can sometimes represent a false memory based on paranoid blame shifting put into place to make "I am gay and ashamed of myself" into the self-excusing "I am not responsible for my being gay, I was made that way by those who attacked (persecuted, raped) me." Therapists who claim that a person becomes homosexual as the result of having been homosexually attacked early in life are sometimes buying into a patient's false initial hypothesis (the assumption of rape), then going on, via elaboration, to postulate a relationship between being gay and an in-reality possibly nonexistent early traumatic life event. (Those individuals who have actually been physically mistreated, even raped, including by same-sex individuals, often do not go on to themselves become bisexual or homosexual. Perhaps more likely they go on to become heterosexual predators.)

EXCEPTIONS ARE NOT ACCOUNTED FOR

Causative theories of homosexuality generally overlook the many exceptions that exist to any given hypothesis. For example, theories that account for homosexuality by postulating a flawed process of identification with a father due to the father's absence fail to explain both why homosexuality can occur when a boy whose father is absent identifies with a parental figure outside the home, and conversely, why the best and most intensive early mentoring creating strong healthy masculine identification with a substitute father still cannot prevent/influence the development of a homosexual orientation.

NONHOMOSEXUALS HAVE SIMILAR/THE SAME PROBLEMS

Specific problems cited as causative of homosexuality, including a lack of identification with a strong father, are found in straights. And many individuals with parental problems cited as significant in causing homosexuality go on to develop not homosexuality but AvPD, or paranoia, or grow up without any significant psychopathology whatsoever.

THE THEORIES REFLECT PERSONAL BIAS

Personal bias creates theories out of established mindsets, especially those involving the belief that homosexuality is of a comparatively low value. Biased theorists find what they are looking for in their patients and/or report only on individuals who fit their preconceived theories. As Nicolosi revealingly puts it, "[my reparative] treatment fits the majority of the homosexual clients in my practice [but] some others are inappropriate for reparative therapy because they show no signs of gender-identity deficit and do not match our developmental model."[4]

THE THEORIES ARE SOCIOPOLITICALLY INSPIRED

Some theories about the cause of homosexuality are unscientific due to being forged for sociopolitical considerations. An example is the social theory that favorable media portrayals of gays and lesbians has a hand in causing homosexuality. Too often sociopolitical affiliations promote a negative social agenda, as in "being gay is a choice, so crimes against gays are not deserving of special protection as are true hate crimes."

THE RESEARCH PROTOCOLS ARE FLAWED

Studies done by questionnaire, interviews of only therapists as a way to evaluate patients, use of convenience samples, and/or use of material gleaned directly from psychiatric sessions but not corrected for transference distortions are all subject to unreliability due to problems with collection and reportage. Gay men who report that their father was distant are mistaken if due to their own narcissistic disappointment in Dad they deny how close he in fact felt and was to them. Gay men who report that their mother was too close might simply be imagining that an appropriately close mother was overwhelming them because the sons, temperamentally shy and withdrawing from Mom as they withdraw from everyone else, misread mother's appropriate desire to get reasonably close as "overly engulfing."

THE THEORIES INCORRECTLY LINK
HOMOSEXUALITY WITH MENTAL ILLNESS

Homosexuality Is Supposedly the Product of Another Mental Illness

According to theorists like van den Aardweg homosexuality is not an isolated sexual "preference" but a symptom of a broader underlying disorder such as some specific "neurotic personality."[5]

These underlying neurotic disorders may conceivably be as described in the following sections.

Depression. Gay patients supposedly become not only depressed because they are homosexual but also homosexual because they are depressed. Gays do get depressed, which they do for both external and internal reasons. External reasons can range from others' disdain to bad luck in connecting with someone to love. Internal reasons can involve: sexual guilt, pathogenic beliefs about oneself saying "I am bad" (e.g., I am sexually compulsive); the belief that homosexual sex is itself bad, disgraceful, embarrassing, and humiliating; the sexual performance difficulties that are supposedly inevitable when two individuals have the same genitals; identity problems (having to do with questions about one's masculinity and femininity); feelings of personal failure; past and ongoing difficulties with parents and other family members; ineffective attempts to couple because of self-defeating interpersonal behaviors with one's partner, for example, due to excessive competitiveness leading to a struggle for supremacy; excessive passivity leading to excessive dependency; or feeling unappreciated for who one is, and so inadequately admired and loved, leading to depressive shame and feelings of inadequacy.

The parents of a gay son caring far less about his well-being than about his being well, that is, not gay, at times dealt with their embarrassment about having a gay son by ignoring him, and at other times by hiding him from the family and the rest of their world. The son's response to being abused this way was to feel shamed, then ashamed of himself, then unloved, then to experience self-hatred, then to become reactively, defensively, overly prideful about being gay—openly expressing this pride in a way that led his family to shame him even more in response by inserting themselves into his romantic and physical affairs, ostensibly to protect him from his going too far in his homosexual ways. That in turn led to his feeling even more guilty about his sexuality than previously, and so even more unappreciated and unloved than before. Ultimately his family wore him down until he became excessively timid and full of self-hatred. His therapist said, "Your being ashamed of yourself and getting depressed is what made you gay." However, as we readily discovered he had had gay fantasies long before he developed his problems with/symptoms of depression.

Another patient's parents' shaming of their son led him to feel broken and unworthy. According to his reparative therapist, his parents' view of him as defective and worthless led to his inability to be sufficiently assertive—and so to his being unable to be a real man with red-blooded heterosexual feelings due to the fear that he might be criticized if he shone in that, or in any other, way. His therapist felt he could cure his patient's

homosexuality by helping him modify his self-view—becoming less depressed by becoming more shamelessly self-assertive, leading him to develop a superiority, as distinct from an inferiority, complex, resulting in his becoming heterosexual. He did as a result of therapy become more assertive, only that resulted in his feeling good enough about himself to seek even more homosexual gratification, and to perform homosexually even more, ever more adequately, and more free of guilt, than previously.

Narcissistic Personality Disorder. Homosexuality is supposedly the consequence of excessive narcissism. This narcissism is characterized by an incapacity to love others, and to instead, in place of loving others, seek attention and love for oneself, by loving another who represents oneself.

Fenichel, tying narcissism to homosexuality, notes that in narcissism the person is the object of his own affections. In homosexuality the man "chooses as love objects young men or boys who, for him, are similar to himself, and he loves them and treats them with the tenderness he had desired from his mother. While he acts as if he were his mother, emotionally he is centered in his love object, and thus enjoys being loved by himself [narcissistic object choice]."[6] He gives "certain pleasures, which the persons are inhibited from granting themselves . . . to others and enjoy[s them] in identification with these others."[7] That "homosexual objects resemble the patient's own person more than heterosexual ones explains the intimate relationship between homosexuality and narcissism."[8] (Unless we buy into the proposition that A = B, the opposite of A, this theory fails to explain why many gays love individuals who are as much as possible the exact opposite of themselves.)

Gender Identity Disorder/Gender Dysphoria. According to some observers a gay sexual orientation is the predictable outcome of problems related to gender identity. Supposedly a causal relationship exists between gender identity (manliness/femininity), gender dysphoria (being at odds with the obvious implications of one's biological gender), gender-identity noncomformity (being a manly girl/girly man), and homosexuality. Accordingly, men and women become gay because they feel less like men and women than they reasonably should feel. Gender identity problems such as effeminacy or tomboyishness occurring early in life will necessarily predict homosexuality appearing later in life.

According to van den Aardweg "the majority of prehomosexual boys are . . . effeminate."[9] These boys play with dolls not soldiers and dress up instead of playing sports. Girls play contact sports and ride bareback. All concerned wish they were members of the opposite sex, and so try to become as much like that as possible. Their parents cause/play into their gender-bending by the unhealthy things they say and do to their children,

as when a father encourages his daughter to join him in sports, leading her to identify with him and so become more manly than she otherwise might have become. A cure for homosexuality would then involve changing homosexual orientation by realigning gender identity with gender. This may require going beyond the imparting of insight and the use of cognitive reorganization techniques to do reparenting via providing a male patient with a corrective emotional experience with a he-male therapist in order to help the male patient properly reidentify with his father, or providing a female patient with a corrective emotional experience with a she-female therapist in order to help the female patient properly reidentify with her mother. Or it may involve a simple remedy: "don't cross your legs like a girl and you won't have sex like a girl." Clearly, gay patients need to talk about the troubled relationships with their parents and, going on from there, require their therapist to affirm them in the gender to which they belong.

While gender identity work can influence certain personality traits/disorders such as excessive passivity and/or excessive aggressiveness and even help heal discrete symptoms such as delusions that contain elements of gender identity disturbance—for example, the delusion that feminizing rays have penetrated the body to destroy a man's genitalia—gender-identity work predictably has no effect on sexual orientation, in the main because sexual orientation issues are separate from GID issues. Even van den Aardweg, who is mostly right only when he disagrees with himself, noting that "cross-gender behavior and interests [do not] necessarily lead to homosexuality"[9] says that "even in cases of strong effeminate behavior in preadolescent boys, called the 'sissy syndrome,' no more than two thirds develop homosexual fantasies in adolescence."[10]

Homosexuality Is Supposedly an Illness in Its Own Right

While some reparative therapists view homosexuality as a product of another mental illness like grief/depression or GID, others see it as a mental illness in and of itself, a psychological disorder in its own right.

Reparative therapists routinely claim that qualities of mental illness exist in homosexuality. They cite as evidence that gay men and women are different personally, socially, and professionally in significant ways from their straight counterparts, being, like other mentally ill patients, equivalently incompetent, unstable, unreliable, unlovable, unappealing, and unhealthy both in their personal and in their professional lives.

Too, as is the case with other mentally ill people, gays' past history reveals specific pathognomonic developmental insults. Gays lack emotional

growth and manifest a psychic immaturity that they never outgrow. In many cases there is psychic infantilism involving serious developmental arrest, especially that which involves a fixation on early unresolved conflicts, to which gays furthermore regress especially at times of stress. Gays are also like other mentally ill individuals in that they are excessively fearful people, and in the case of the homosexual, especially afraid of the opposite sex.

As mentioned throughout, reparative therapists identify so many patterns of disturbed early relationships that the likelihood that any one of them is routinely of central significance to becoming gay is low to nonexistent. One set of research findings seems to contradict the other as researchers blame the development of homosexuality on too much parental control, insufficient parental control, hostile parenting, excessively loving parenting (parental engulfment and seduction), and either parental molestation or total parental abandonment. One can only conclude that specific patterns of family relationships do not predictably cause same-sex orientation, and that the role played by the castrative or passive father and the distant or engulfing mother, while no doubt important to the individual as a whole, is meaningful only for determining the evolution of personality/personality problems/neurosis/psychosis, but not for accounting for the development of homosexuality.

Many reparative therapists specifically make the point that homosexuality is phobic because it develops along similar/the same lines as a phobia (in men it involves a phobia of women and so of heterosexuality). Some, conversely, see homosexuality as a soteria, that is, as the opposite of a phobia. In the latter view men become gay not because they *fear* being real men, and so, unable to comfortably be heterosexual turn away from heterosexuality, but because they *wish* not to be real men, perhaps because of positive earlier interpersonal experiences with a man/men in their lives, and so they turn to homosexuality to express and realize their wish to continue to love and be loved, just as before. Thus according to Fenichel, Freud, disagreeing that homosexuality is posttraumatic in nature instead suggests that it results from the *opposite* of traumatization via attachment injury. According to Fenichel, Freud believed that "the extent of male homosexuality in ancient Greece may have been due to the fact that children were brought up by male slaves."[11] Thus homosexuality might be thought of not as an illness but as a wellness, a conditioned (love) response, one in fact conditioned not by whom one as a child most feared, but by whom one as a child most adored.

For many reparative therapists homosexuality is supposedly a dissociative disorder, that is, it is caused by a mechanism akin to an inability to be

in contact with one's true (heterosexual) self, associated with a need to split off one's true (masculine) identity. One's true (masculine) self and identity is supposedly a heterosexual one, which means that gays are supposedly merely frustrated or repressed straights.

Some reparative therapists believe that homosexuality is primarily a *defense mechanism* akin to reaction formation directed to the heterosexual attraction itself and therefore to one's identity as a real man. Therefore gay is an illness "marked by reaction formation" much as paranoia is an illness "marked by projection." The reaction formation here is against heterosexuality, which gay men believe to be not only undesirable but fearsome as well.

CLASSES OF THEORY

I now present some theories of homosexuality classified according to their main focus of concern.

Childhood Trauma Theories

Many observers postulate that men become gay as a result of *early traumatization*, which can be of an emotional or of a physical nature. Supposedly memories of childhood trauma are suppressed, only to remain active and unconscious, leading to having homosexual relationships whose intent is to help the sufferer mollify/reverse the effects of the early traumatization. The trauma may have been discrete and single, or there may have been generalized and multiple traumata, a one-time trauma or an ongoing traumatization, having been inflicted by fathers and/or mothers through neglect, criticism, seduction, or abandonment. Thus traumatic maternal rejection leads to a man's shyness with women and so his holding back from them as well as his developing a protective (narcissistic) relationship with himself which develops into a "narcissistic sexuality" (= homosexuality), where men love men not so much because men are actually physically appealing but because having been traumatized by women they can but retreat into loving men only (= themselves).

A child's temperament may encourage his or her traumatization because the child is remote and rejecting, thus provoking the attachment injury, that is, the child provokes his or her own attachment victimization.

Some observers believe homosexuality to be a fearful, phobic behavior that occurs specifically as an outgrowth of castration anxiety. Here, according to Fenichel, anxiety about female genitals is associated with the trauma-related fear that one might "become such a being" and the fear that female genitals might be "a castrating instrument capable of biting or

tearing off the penis." So men become homosexual because they "react by refusing to have anything to do with such frightening sights."[12] As such, homosexuality comes not, as does heterosexuality, from the id (as an instinct coming out of the repository of normative instincts) but (more like a neurotic symptom) is created out of conflict between id and fear, that is, it represents a conflict between desire and terror and is a compromise formation where one can still have sex, but not full sex, which is heterosexual, only partial sex, which is homosexual.

Current trauma leads to regression to an earlier, perhaps safer, relationship upon which the man had a "pregenital fixation . . . especially an anal fixation."[13]

Father Theories

Many observers believe that homosexuality is a developmental disorder resulting from specific problems between father and son. Homosexuality can be the result of a failure to bond and identify with the father. The homosexual uses other men (new father figures) to repair a broken relationship with his own father, perhaps one who was absent, distant, remote, and unavailable for shared activities, and was additionally an unempathic man who demanded his son not do what the son was good at but what would make the son more masculine: more like a real man, that is, more like his father. Therefore, the remedy for male homosexuality is for the son to approach his father and ask Dad if they can start doing man things together.

A son can also fail to become a man like his father not only primarily because he doesn't identify and bond with his father, but also secondarily because he develops a too close relationship with his mother, becoming womanly like her, then, as a result, becoming impotent not only personally but sexually as well. A father who denigrates his son, who doesn't recognize him as a valuable manly salient individual, will not only promote the failure of identification with himself but will also indirectly promote/intensify that son's relationship with his mother, thus assuring that the son will never properly identify with his father.

Homosexuality is also an undoing of one's anger at the father through loving men, as well as a reaction formation against being shamed by the father manifest as the seeking of approval from one new, hopefully more accepting, man after another.

When the son *counteridentifies* with a feminine, or otherwise, to him, unappealing, father, he becomes gay because he wants to resemble his (straight) father as little as possible.

Mother Theories

Presumably pre-homosexual boys identify with their mothers and remain identified with, and too close to, them, both personally and sexually. Such a boy becomes gay because he can't distance himself from an overly appealing mother; because he loves his mother excessively due to something in his own nature; because his mother discourages him from leaving her safe and pleasurable orbit; because being lonely she keeps him close because she has no one else in her life; because she is natively strong and dominant; because she is especially gratifying, that is, excessively beautiful or particularly kind; because she defeats a boy's attempts at identification with his father by demeaning Dad to the point of making it harder for his son to identify with him; and/or because she is a castrative woman who jealously threatens to hurt/abandon her son if he gets too close to, and becomes too much like, Dad.

Too often a son and his mother get together to devalue the father. They paint him as an unloving, cold, traumatizing, nonbenevolent man, or a passive sissy man too feminine to stop his wife from trashing him for being a weakling. Whatever the reason, a devalued father is now unavailable as a solid identification figure for the son.

Some soon-to-become-gay sons may have not an overly appealing or engulfing but a remote cloth monkey mother. Such a mother weakens (malnourishes) the son to the point that he stops maturing, creating the familiar infantile "passive, sissy, limp-wristed" homosexual (which is, according to some, all male homosexuals).

Both Parents

In some cases both parents conspire to abuse, ignore, or engulf the son in such a way and to such an extent that he becomes too frightened or too paralyzed to mature enough to firm up an avowed clear heterosexual orientation. Dysfunctional families may also create identification confusion possibly leading to homosexuality as an outcome of not knowing who to identify with, leading to sexual uncertainty.

Siblings

In some cases opposite-sex siblings get so much attention in the family that the boy wants to be a girl, or the girl wants to be a boy, to get some of that attention for himself/herself.

Harsh sibling rivalry leading to ganging up on a son can leave the son feeling that he is an inferior person and hence an "inferior man," so he

becomes homosexual as part of his living out his feeling inferior. Intense rivalry between siblings who attempt to heal their rivalry by joining up and ganging up on one sibling as the sacrificial lamb of the family also leads the victimized boy or girl to weaken to the extent of having social problems outside of the family with his or her peers, making it more likely that he or she will be bullied and as a result retreat from heterosexuality (potency) into homosexuality (impotence).

Temperament

Reparative therapists recognize that temperament, which is inherited and therefore of biological origin, plays a part in the development of homosexuality. This is especially true of the "hypersensitive temperament," which can account for why sensitive gays react one way and why others, who are not uniquely sensitive, react another way to the same stimulus, that is, to the same family events/problems. Some temperamental youngsters not only allow parental bullying to affect them excessively (as they allow their bullying peers to affect them in similar ways), they also self-bully, impairing their heterosexuality by forbidding their own normal heterosexual feelings from taking hold, and/or by punishing themselves for being straight by making themselves queer. (Sensitivity theories fail to jibe with those that paint all gay men as insensitive, bitchy bulliers who create as much anguish for others as they do for themselves.)

Biological/Genetic Origin Theories

While being gay presumably has strong *biological/genetic* roots, that one of a pair of identical twins but not the other can be gay suggests, as Besen says, "that there may be more than a simple genetic explanation for homosexuality."[14] As Besen also notes, it is likely that genes do not exclusively determine but only strongly imply outcome and that other, environmental factors are at work.[15] Generally, reparative therapists strongly deny that there is conclusive evidence that homosexuality is at all genetic. NARTH has to take this view for as Besen says "the anti-gay industry knows that if homosexuality is shown to be a genetic trait acceptance of homosexuality will quickly follow: the American people will not penalize people for who they are."[16]

According to Fenichel, while "psychogenic causes [such as] the blocking of heterosexual object choice"[17] are important, constitutional factors (which have genetic, physical, and biochemical implications) are also decisive, particularly (as noted throughout) biologically determined

temperament that creates a biological disposition to the gender deficit seen in primary GID, which can then go on to become homosexuality.

Bizarre Theories

These are too numerous to list all of them. Cohen according to Besen says that the cause of homosexuality involves "avenging, bitter spirits of dead ancestors."[18] Besen also notes that Nunberg believes that circumcision "supposedly leads to less homosexuality among Jews."[19] Some observers see homosexuality as a sexual dwarfism arising out of prohibition against masturbation, a prohibition that can but lead to a reactive desire to constantly touch other men's penises.

Part III

Therapy

11

❖❖❖

Obtaining Informed Consent

In this chapter I set forth my belief that what is so often called "informed" consent is in fact usually seriously uninformed, so that the cry, "we never try to change those who don't want to change" doesn't reflect the hard reality, which is that when it comes to changing their sexual orientation, few, if any, gays truly know what they want, and are not fully aware of what they are getting themselves into.

Nalbandian says that "no health care provider should force any patient to accept treatment he or she does not want. Patients have the ultimate right . . . it is their body and they have a right to make decisions about their own treatment."[1] But Nalbandian doesn't elaborate on the exact nature of "force" or the precise meaning of "want." He doesn't discuss the possibility that "force" may be subtle and that patients can be inwardly conflicted about what they "want." Indeed, as Besen says, "no matter how much outreach the GLBT community does or how many people are educated to the follies and failures of the ex-gay myth, some individuals will attempt to change. The best that we can hope to accomplish is arming people with the information they need to make informed choices."[2]

Clearly there are complexities involved in determining what constitutes the making of informed choices and so the giving of informed consent. Truly informed consent is in fact difficult to impossible to obtain. To obtain it reparative therapists must satisfactorily answer such questions as "what is intellectually and what is emotionally informed?", "what is

voluntary and what is coerced?", and "what is decided by and what is decided for their patients?"

Few patients who give even somewhat "informed" consent truly know
what they are consenting to. Most are in fact in some ways being brain-
washed without their, and even their reparative therapists, actually knowing what is happening. Many patients whose judgment is clearly impaired
are nevertheless offered the opportunity to give informed consent, and in
turn the consent is accepted as informed as if it reflects adequate judgment
on the patients' parts. How many patients, as Throckmorton and Yarhouse
note, are given, and can actually digest "advanced informed consent" that
reflects a full understanding of "what is causing their distress"; have full
awareness of conflicts about their same-sex attraction; possess a complete
knowledge of "professional interventions available including [reported]
success rates and definitions and methodologies used to report and define
success" also including an appreciation of the relative lack of empirical
support for claims of success; and get information about "alternatives to
therapy [and the] possible benefits and risks of pursuing treatment at this
time [and the] possible outcomes with or without treatment (and alternative explanations for possible outcomes?)"[3]

I personally entered a five-year analysis premised on changing me from
gay to straight. But at no time was this premise itself subject to the analysis I entered. I wanted to go straight through analysis less to be straight
than to please my parents and to be accepted into a psychoanalytic institute. My analyst, however, never worked with me on my all-too-
pathological need to go straight, based to a great extent on my need to
defer to others ranging from my parents to officialdom. She also didn't tell
me, for approximately the first four or so years of the analysis, that being
analyzed made no sense at all for one of the important reasons I was doing
it, for she would see to it that even if I should somehow become ex-gay, I
would still never be admitted to her, or to any other, orthodox psychoanalytic institute. And, moreover, she never helped me grieve over that and
reset my course to become something else more acceptable to me than my
present, second best, default career. Many reparative therapists believe
that they only treat patients whose judgment about acceding to the treatment is significantly unimpaired. My analyst, one of these, was clearly
denying that my neediness and desperation made me impressionable to
the extent of dangerously trumping my good judgment—and that her responsibility to me included lending me *her* good judgment, based on the
insider knowledge she must certainly have had. (Just recently I was told
that she had another agenda in mind, one that was not necessarily homophobic in nature. As a training analyst, that is, one who both analyzed

young analysts medically and as well reported on their progress academically, to some her goal in life seemed to be to protect the other patients of this world from her own patients, all of whom she presumably saw as necessarily dangerously incompetent simply because in the first place they were in treatment with her!)

Many reparative therapists without knowing it obtain consent that is only partially informed. Some assume that it is enough to merely ask and have answered the question, "Are you happy being gay or do you want to change?" They assume that patients' *claims* that they want reparative therapy are reason enough to offer to treat them reparatively. They assume that in their patients, unconscious motivations are not widely in play. As a result, as if motivation is a black or white issue, they downplay the important nuances of, and fail to analyze, exactly what their patients want based on why exactly their patients don't seem satisfied with what they already have. Then, failing to obtain the details of why patients want to change, they forge ahead without considering the differential diagnosis of their patients' dissatisfaction and especially whether or not internalized/internal homophobia is involved in their patients' decision to apply for, and accept, reparative therapy. They especially overlook it when their patients' wish to go straight is an expression of personal psychopathology consisting of neurotic conflicts and their equally neurotic resolution, and so is a product not of true need and legitimate desire but of deep emotional pressures creating equally deep inappropriate urgency. In so many cases, the wish to go straight really represents more a fear of being gay than an actual desire to be heterosexual. Often it is anxiety about who one is that leads to defensive detachment from one's true self. Alternatively, the wish to go straight can be the product of an unfortunate identification with a misguided parent, or with an antagonistic peer, or submission to some other bigoted onslaught. It can be part of an unconsciously determined guilt-ridden Spartan philosophy that recommends routinely forgoing immediate pleasure and postponing immediate satisfaction to the distant future for some greater good presumably yet to come. Under these circumstances "I want to go straight" is more determined *for* patients than it is a byproduct of their free will, the latter having been co-opted to the point that patients have become too judgmentally challenged to effectively recognize their true desires and look out for their overall, and ultimate, best interests.

Though reparative therapists claim that they never force someone to change, though they routinely stress the importance of not coercing patients and only treating patients who really want to go straight and are fully educated about what that change involves, and though they claim

that they give each of their patients enough information to assure that their consent will be fully informed (for supposedly they tell them of all the pros and cons of what they plan to do), what reparative therapists don't realize/say is that patients almost always lack the knowledge and experience they need to absorb/understand what their therapists are telling them. This is in part because to make enlightened decisions about reparative therapy, patients have to make what amount to enlightened *medical* decisions and to make them about themselves, in effect having to be their own doctors and to make their own medical decisions when they are personally involved, and at a peak time of emotionality that renders them far too overwrought to be "well informed" in the sense of dispassionately knowing themselves. Even gay patients, though they are actually gay themselves, nevertheless can have a misguided overall view of gay life, for example, one obtained from media that present an overly distortive picture of what life is like as a gay person, a picture sometimes painted too negatively, resulting in a dark portrait of what being gay is all about without noting its upsides.

Often specific personal life crises, however temporary, and only remotely related to *being* gay, can affect gays' judgment about *remaining* homosexual. To illustrate, many gay men try to go straight when they have just been dumped by a lover or been infected by an STD. As a consequence of allowing their grief to drive their intellect, men sometimes think that they will have better romantic luck only if they turn from men to women.

The information some reparative therapists share with their patients is too often based on wish fulfillment of the therapist's own. Some tout their treatment as being efficient, helpful, and cost effective because they hope that's how it will turn out. Some even wistfully claim that their treatment works for patients who don't want it but accede to undergo it against their will, such as patients who have been forced to get treatment by parents who threaten to stop paying their bills—to cut off their tuition or to throw them out of the house—if they don't agree to convert, or patients who have been referred by the courts demanding that they undergo rehab involuntarily because they are hurting others or destroying themselves. These reparative therapists fancifully conclude that all gays need a firm guiding hand because all gays are at bottom immature and, like all adolescents, in such turmoil that they don't know what is good for them—and only a reparative therapist can help them figure that out. Fancifully, too, they deny that SOCE can actually *disrupt* the natural (spontaneous) progression from gay to straight that can occur, so that watchful waiting rather than active intervention may be the best idea. Instead they

advocate "catching being gay in time" and forcing therapy on even young children, reasoning, again wishfully, that they must do so before it's too late to alter the child's inevitable progress toward becoming homosexual. In this they overlook how reparative intervention can have the reverse effect, for most young people don't much like to be told what to do and tend to become stubborn and resistant in response to feeling/being ordered to do anything at all.

Untrained reparative therapists wishfully deny that their academic credentials and practical experience are inadequate to the task of knowing and so imparting the core information gays need in order to make enlightened decisions about entering reparative therapy. Ultrareligious reparative therapists wishfully equate being religious with being knowledgeable, competent, and honest, even when their honesty is corrupted by their zeal. Many wistfully deny that peer pressure determines and validates their position in any meaningful way even though they are in actuality a tool of their establishment—having allowed peers in their in-group to seriously influence their judgment about the wisdom of undergoing, and the effectiveness of receiving, reparative therapy. Having consensually validated each other's theory and practice, all concerned in groupthink reach the point that they become so convinced of the merits of their personal positions that they run roughshod over the positions of their patients—over their patients' personal ideals, goals, and objectives.

And mostly they wistfully deny that the prognosis of ex-gay therapy is guarded, that negative side-effects and adverse effects are common, and that this is not a pessimistic viewpoint constituting a self-fulfilling prophecy but a reality. For reparative therapy is often a long, expensive, time-consuming, and painful process almost always ending in disappointment and failure, which reparative therapists routinely excuse by saying, "of course it's hard, takes a long time, is generally only partially successful, and not for everybody, yet it is still always definitely worth a try."

In conclusion, informed consent can never be truly informed because it can never be meaningfully independent of unconscious factors/determinants and immune to the unconscious manipulation that is almost always involved when reparative foxes construct the guardhouse from which to guard the hens. One's sexuality is a highly emotional matter. How one responds to one's own sexuality is never neutral. Passion almost always disrupts dispassion. This is particularly true at the beginning of any and all forms of treatment, and especially so at the start of reparative therapy, so that a paradox pertains: that truly informed consent can only exist after therapy has proceeded long enough to tackle the issue of informed consent in an insightful meaningful way; yet therapy shouldn't proceed at all

until the issue of informed consent has been tackled satisfactorily, honestly, and insightfully.

So, potential patients considering reparative therapy must depend on informed *others* to guide them. They must pick honest, well-intentioned, well-trained practitioners who are qualified as psychotherapists first, that is, they have a license based on actual, official training and experience in knowing how to do therapy, and as conversion therapists only next. Unfortunately, such people are few and far between, if they even exist at all.

The following section provides a practical guide meant to outline the actual steps needed to obtain consent that is at least as informed as it can possibly be under the circumstances.

INFORMING THE PATIENT OF THE DIFFICULTIES INVOLVED IN CHANGING SEXUAL ORIENTATION

Reparative therapists must freely admit that change is difficult to accomplish psychotherapeutically beyond consciously abandoning a gay or lesbian identity and choosing to either live *as if* one is straight or celibately. As Kort says, "I have helped many men and women who have a 'homosexual' orientation but do not wish to adopt a gay or lesbian identity or lifestyle to live a heterosexual or asexual lifestyle. For them, coming out as lesbian or gay would mean more trauma. However I do not believe that these individuals can—or do—change their innate sexual and romantic orientation. They simply change how they live and whom they love, just like a heterosexual could do if he or she so chose."[4]

INFORMING THE PATIENT OF THE THERAPIST'S PERSONAL "SCIENTIFIC" ORIENTATION

Therapists should tell their patients if they believe that homosexuality is a mental disorder and as such accessible to psychoanalysis/broadly subject to change through techniques developed for treating other mental disorders. They should also tell their patients that few mainstream therapists view homosexuality as a mental disease—thus, homosexuality as something that both needs to, and can be, cured by therapy is all talk.

INFORMING THE PATIENT THAT REPARATIVE EXPERTISE IS NOT ABSOLUTE

While reparative therapists in the main respect their patients' right to decide for themselves whether or not they wish to undergo therapy to

change, they likely, if only unconsciously, manage to choose their patients' path for them by what they communicate to their patients, if only nonverbally, as "being the one and only right road to traverse, take it from me, the expert." We are all to some extent primed to believe identified experts, that is, we are all exposed to an early stimulus having to do with expertise that leads us to respond in a given way to a later, related, one in the similar or same realm. And those who go for therapy, though they are not at first believers, often soon get "hooked on the transference," which means that they are both prone to believe what the identified experts tell them and fear that if they don't go along with the experts, their therapists, their therapists will reject/abandon them. Therefore, therapists owe it to their patients to downplay their own "unfailing expertise" and to not play at all to their patients' universal tendency to be submissive and overly trusting in the face of "those who, knowing best, must be obeyed."

CONSIDERING AND CORRECTING FOR COUNTERTRANSFERENCE DISTORTIONS

It is especially important for reparative therapists to know themselves, which may require their dispassionately exploring the reasons why they advocate (and do) conversion therapy in the first place even though all the evidence points to its not working well, or at all. They should ask themselves if they have a delusional conviction that it works; if their religious faith trumps their science; if they have control issues in general and in this special circumstance in particular; if they need to learn more because they only think themselves as well informed/trained; and if they need to identify and cure their own Stockholm syndrome having bought into the teachings of a reparative cult whose members are more rabid than rational, meaning that as individual therapists they have become unable to extricate themselves from outside group collective collaborative influence. (Or, as is sometimes the case, do they treat for a much simpler reason—because they need to have something to do, and, not incidentally, could use the money?)

Many fail to consider and correct for personal grandiosity. Reparative therapists who are grandiose do not formally even go through the motions of seeking consent that is both voluntary and informed. Rather they simply demand that their patients give their consent, no questions asked. They often do this for personal reasons, of which there are several. Self-preoccupied, they have difficulty being as neutral as many people believe

that the most competent of therapists actually are. Emotionally involved, they lack the ability to stand aside from their selfish interests, making them so thoroughly unable to foster their patients' individual spirit, goals, desires, and tendency toward self-realization that they fail to consider other approaches/goals for their patients based on what their patients' specific needs and desires actually happen to be. All told, they effectively force their patients into becoming passive recipients of a plan for cure mainly originating in the therapist's own predetermined notions of what their patients ought to be doing for themselves. As such, they don't so much therapeutize their patients as they coerce them into pursuing goals that are not the patients' own, but instead belong almost entirely to their patients' therapists.

For grandiose reparative therapists like this the inherent difficulty in accomplishing reparative goals simply represents a challenge to proceed with greater and greater impetus. With this in mind, therapists too readily side with gays' manifest wish to change by convincing themselves that their patients' wish to change is conflict free, that is, that their patients are not at all ambivalent. Though for most gays if not ambivalence then a gray area between "I want to change" and "I want to stay the same" pertains, these therapists, going into denial, automatically side completely with the homonegative aspect of their gay patients' conflicted mentality because they themselves also feel homonegatively: that being gay is bad, wrong, sinful, and criminal, and that that's why gays need to be in reparative therapy. Such a grandiose homophobic view is espoused by Nicolosi who quoting a grandiose patient with whom he clearly agrees "scoff[s] at the American Psychological Association's idea that homosexuality is equivalent to heterosexuality" and sees homosexuality instead as variously "demeaning to a man's dignity [as well as] unhealthy [and] maladaptive"[5] (and is presumably *totally* that way without any mitigating, adaptive factors whatsoever).

Grandiose therapists actually come to believe that what they do has few to no adverse consequences. They support their own notion that reparative treatment has only benefits by convincing themselves that bad things will happen only if nothing at all is done, compared to the good things that will no doubt come to pass if they and their patients at least give reparation a try.

Overly confident and overly zealous, they assume the mantle of benign therapeutic heroism that is in fact an admixture of excessive expectations and unreachable goals based entirely on their being overly fond of a particular theory and form of therapy—a position that can but originate in their being overly fond of themselves. This can shade over into true

pomposity, so that as therapists they forge ahead because they feel like saints or gurus based on the fantasy of being a providing (although entirely enveloping) mother whose love heals, or of being an omnipotent father who is that guy who can fix everything. Then, like many equally hypomanic individuals, should their patients question them, they blow up because they view being questioned at all as the exact same thing as being given a serious, and seriously rude, argument.

Some grandiose reparative therapists actually try to convey to their patients that they are in the presence of a superhuman figure, one with supernatural powers who can use their therapeutic powers (though these mainly amount to suggestion, advice giving, and manipulation) to great effect and who have powers even greater than those possessed by other less able, less potent, "fellow" traditional therapists.

Grandiose therapists are in fact often secretly hostile to gays. Their wanting to change them is actually based on the belief that they "should change because they are inferior, since homosexuality is not on a par with, or the equivalent of, heterosexuality." Ultimately this hostile attitude deflates their patients in proportion to how inflated their therapists think heterosexuality is. As a result of what they hear, many patients experience a fall in self-esteem and become depressed due to in effect having been asked to compare themselves (generally unfavorably) to their therapists—therapists who view their own sexual orientation as being grander, more powerful, and more divine than that of their patients—and see homosexuals, being broken, as not matching up to the straights of this world.

Grandiose therapists being autocratic rarely seek to learn what they truly need to know about their patients. Talking instead of listening, formulating instead of treating, they fail to handle the patient as a participant ally or joint explorer on the virtuous path to wisdom. Not acting in concert with their patients and instead handing down one-way edicts, they fail to hear their patients out, instead cowing them to the point of suppressing their naturalness, authenticity, and individuality. Failing to take into account their patients' self-generated positions, they fail to respect their patient's originality and so, instead of eliciting growth from their patients, stifle their innate tendency toward maturation, thus seriously inhibiting their patients' personal development, doing so in the very guise of helping them grow. Being self-centered this way they tend toward one-size-fits-all responses to their patients, which responses they formulate out of some shaky theory of their own, one they hold especially dear. They might fail to distinguish sexual orientation from gender identity and fail to account for the variability attributable to their patients' age. In the

latter case, they often become seriously remiss in distinguishing between young homosexuals just discovering their sexuality and their more mature homosexual counterparts for whom having a happy future necessarily entails not change but settling down, and for the rest of their lives, still gay, but now in a functioning, perhaps monogamous, gay relationship.

The grandiose therapists' need for world renown can seriously affect their judgment. For example in *Anything but Straight* the author Besen raises the possibility that Spitzer's study results were guided by his need "to see his face on the little screen."[6]

It can be *religious* grandiosity that leads some reparative therapists to feel that they have a "calling," one perhaps directed by a higher authority who may actually, as they see it, have appointed them to cure gays, and in turn can be counted on to fully participate in the healing process. For some grandiose religious therapists it is their faith that gives them the excuse they need for bypassing formal training, for, as they see it, religious people are not *trained* (to heal) but *chosen* (to save). Some allow religion rather than science to enhance their narcissism to the point that they become like God Himself, all-knowing, absolutely aware of what exactly constitutes maladaptive symptomatic behavior, and what in contrast constitutes good asymtomatic functioning. And so they know how precisely to go about diminishing maladaptive (gay) and enhancing adaptive (straight) behavior.

I recommend that grandiose therapists, before attempting to take over their patients' lives, should secure personal consultation/supervision/analysis for themselves. They should also consider referring their patients on to a therapist whose values, positions, and professional identity, being not excessively gay-deaffirming/excessively conservative/Christian, leave room for their patients to have, and express, a preferred direction of their own.

DETERMINING IF GAYS WANT TO BE IN, OR ARE BEING PRESSURED INTO GOING FOR, REPARATIVE THERAPY

The discussion of the motivation to change should include a determination of "Is it something you want to do or something your wife, or family, or your boss is forcing upon you?" Too often the individuals who enter therapy are not the ones who need it most. In many cases it is those demanding others change who are the ones most in need of change themselves.

DETERMINING IF GAY MEN WANT REPARATIVE THERAPY BECAUSE THEY HAVE EMOTIONAL PROBLEMS

Gays who apply to change through therapy often do so because they have emotional problems—making their wish to change a symptom of an underlying emotional disorder, one that is partly or entirely responsible for the severe difficulty they have in accepting themselves as they are. The initial interview with all gays applying for change must therefore explore the possibility that their motivation to undergo is derived from their personal psychopathology.

Powerful emotions, often existing outside of individual awareness, typically determine what gays think they want, meaning that gays like everyone else operate without fully knowing why they do what they do, as they make their decisions based not on free will but on what is determined for them emotionally—that is, they in effect make their decisions for neurotic reasons. These in turn can be traced back to developmental, intrapersonal, and interpersonal psychodynamics. While these are not factors creating homosexuality, they do affect decisions about remaining homosexual. In my experience, an avowal "I want to go straight" that looks autonomous, unlike homosexuality which is genetic and is thus incapable of being influenced, is in fact acquired and is thus capable of being altered psychologically through assessing, uncovering, and dealing with the developmental, interpersonal, and intrapersonal factors that go into making that avowal up. These factors especially include old and new parent-child interactions marked by problematic issues around dependency, control, and competition that lead to the development of such specific emotional disorders as depression with its self-abandonments, and obsessive-compulsive personality disorder with its incessant, harping, self-questioning. Such patients become like automatons who do not chose to enter therapy as much as they, unconsciously at least, choose to yield themselves up to its unhealthy spell.

Specific emotional problems that affect patient's sense of reality relative to giving informed consent are described in the following sections.

Dependency

Dependent patients make the decision to go straight to please, or, perhaps more accurately, to refuse to displease those they admire/love and/or are, or at least feel, dependent upon. So they submit to reparative therapy even though they basically wish to remain who and what they are.

Passivity

Patients who consent to undergo reparative therapy often do so out of personal passivity. Those patients who after beginning in reparative therapy choose to continue in it are often those who have developed a modified Stockholm syndrome in their relationship to their reparative therapist. These patients are not so much informed as they are ensnared. They *seem* very agreeable as they ostensibly willingly work toward what even they know is an unreachable goal. But they are merely denying that they are being manipulated, their false hope tweaked by their emotionally fraught transference love for their therapists.

Passive individuals like this have often been bullied, rejected, and abused earlier in their lives to the point that later in life they tend to too readily permit reparative therapists to bully, reject, and abuse them all over again and in like manner. They allow reparative therapists to tell them what to do and why they should do it, then follow their therapists' marching orders without demanding that they as patients retain a say in their own fate—giving in just so that they won't be re-exposed to traumatic experiences similar to those out of their past. Many having developed serious erotophobic conflicts about homosexuality manifest as sexual guilt presenting as fear, in this case the fear that if they don't agree to convert they will get into trouble at school, at work, and with the bullies of life out to ruin their existence.

Passive-Aggressiveness

The obverse of passives are passive-aggressive (indirectly assertive/active/activist) patients who are generally too reluctant/too individualistic to go for treatment in the first place, and who, if they do go, are unable to commit to it and so generally quit early in the process. These individuals accept therapy at first, but they only show superficial compliance in order to be in a favorable position to express deep reluctance. They apply to be cured but almost "purposely" stay the same as their way to express their resentment toward those attempting to make them over. While these patients attend sessions they nevertheless do not cooperate with their therapists. They act as if they are doing what they can to change, yet underneath they actively resist all attempts to make them something they feel that they are not. Such individuals often manifest a sadistic bent leading them to stay in therapy and stay gay in part to prove to the reparative establishment that that establishment has nothing to offer them. And they say exactly that in testimonials in order to

publicly embarrass their therapists and/or the reparative therapy industry.

Histrionic

These individuals want and seek to change because they feel *excessively*, tragically hopeless about being gay.

AvPD/Schizotypal

These individuals want and seek change as part of their more general tendency to remove themselves not only from gay life but also from much about everything else personal and sexual around them.

Schizophrenic

For these individuals delusions and hallucinations lead them to think entirely unrealistically about reparative therapy. Schizophrenic gays often enter reparative therapy because, delusional and/or hallucinating, they hear God's voice commanding them to do so, e.g., to do something about their being sinners. They even believe God talks to them and gives them specific advice on how, as sinners, they can become more like saints.

Psychopathic

These patients *say* they want to change just to manipulate others and their environment for practical gain, such as to advance on the job, to meet some conditions parents or society set down for them which they must meet if they are to avoid losing emotional and financial support, or to stay, or get, out of jail.

Paranoid

Paranoid gays believe that everyone, being homophobic, is out to get them for being homosexual, so that for them being gay is dangerous, no matter how closeted they stay and/or how much they live their gay lives quietly. Some feeling satanically persecuted try to change to appease, in order to defang the Devil in order to avoid obliteration and annihilation. They try to go straight because they feel intimidated by the bullies in this world—some of whom are real, but many of whom are, at least for these patients, mostly, if not entirely, imaginary.

Masochists

Masochistic gays are self-abusive individuals who after deriding themselves for being gay willingly enter reparative therapy not in spite of its being, but because it actually is, hurtfully abusive and brutally destructive. Such gays deliberately undertake a reparative program to avoid enjoying life and to instead create some unnecessary suffering for themselves—misery they welcome as, for them, the best part of living. Eagerly seeking a form of therapy that promises to be long, arduous, expensive, and unsuccessful, they discover that reparative therapy nicely fits that bill. Longing for punishment, they welcome the reparative offer to participate in at best nothing, and at worst in something seriously personally destructive. (When the masochistic need to suffer and self-defeat leads them to become suicidal, oppressive reparative therapy can in a paradoxical way be a somewhat healthy substitute for making an actual suicidal attempt.)

Hypomanic

These individuals, feeling overly hopeful about getting over being gay, are all too willing to embark (or too much in denial not to embark) on an endless journey to nowhere. The denial, however, being fragile, sooner or later breaks down and the patient becomes depressed (as described below).

Patients who tried ex-gay therapy and left in angry disappointment may have done so because they initially allowed themselves to be misled by their therapists then woke up after unconsciously misleading themselves by agreeing, or actually arranging, to be bamboozled by the experts, doing so out of their own pathological excess of hope based on euphoric feelings about the effectiveness of available help. These patients deal with feeling "all is lost if I stay gay" by refusing to listen to reason—precisely so that they may hopefully, but irrationally, conclude that "all will be well in the end when my therapist makes me straight."

Depressive

Depressed gay patients characteristically feel "there is something wrong with me, I need to undergo repair." What better fits that bill than a form of therapy that says, "you are broken, and you need to be fixed," and a therapist who says, "I am here to do exactly that to, and for, you"? Depressed individuals, down on themselves, unconsciously want to be misled because they are so self-punitive. They even deliberately arrange to fail by having their hopes shattered. So, even after a period of therapy (predictably) not working, they stay in it even after they come to realize (as if they didn't know it all along) that they are "incurable." They stay in treatment because they believe that the more therapy they undergo, the more self-punishing redemption they can soak up.

They also toy with changing not only to satisfy that part of themselves that hates or despises who they are, but also to compromise a part of themselves to satisfy the behests/demands of others, especially those who are as negatively inclined toward them as they as patients are toward themselves.

Gays who have recently been dumped try in their grief to go straight so that they can have a better, more loss-free life in what they imagine to be the much more stable straight relational world. They believe that if they can succeed at being straight they will feel less like failures because overall they will actually fail far less.

Many gays are depressed due to *internalized* homophobia. They view themselves negatively because others view them that way. Also they are depressed due to *internal* homophobia. Here they have become antagonistic to their own instincts, an antagonism which leads to self-disgust characterized by an enmity to their own sexual and bodily functions. Homosexual desire floods them, but a punitive conscience condemns them. And their observant ego, losing the struggle to sort things out in a healthy manner and make peace within themselves, instead resolves the tension between sexual wish and reactive fear not by compromising, for example by tithing, but by giving in completely to that side of themselves that is excessively, cruelly, unhealthily, rigidly self-depriving.

For depressives, as for patients with OCD, the desire to go straight can be a ritualistic cleansing obtained through reparative therapy as a self-condemnatory, self-punitive self-deaffirmation that represents an atonement, which for them "makes it all right."

Borderline/Multiple Personality Disorder

Borderline gays indulge in an as-if identification involving wishing to become "like straight." This desire is, however, unstable, readily shifting back to their wishing to remain "like gay." Should they suffer from a gender identity disorder/gender dysphoria, they come to believe that it is *as if* they were born in the wrong body or, the "as-if" having dropped out, that they are *actually* a girl in a boy's body, or the other way around.

Gays who develop multiple personality disorder go straight in one of their two (or more) identities.

Obsessive-Compulsive

These gays, being generally ambivalent, are equally ambivalent about their sexual orientation identity. As a consequence they shift between trying to go straight then regretting it then once again trying to go straight, then regretting that. As well, gender dysphoria alternates with gender ac- ceptance as they think too much about how manly or not they are in their own eyes and as others see them.

For many OCD gays being gay = dirty and being straight = clean, with sinfulness versus saintliness another concern of theirs. Now conscience driven, and now guiltlessly hedonistic, then once again conscience- driven, then guiltlessly hedonistic once more, they consent to change, only to then fight those who are trying to help them. Alternately rebel- lious and submissive, they feel coerced into changing their sexual orienta- tion, then rebel, then regret their rebelliousness and settle down once again to doing the reparative work that they request of themselves and others want them to do.

Paraphiliac

Gays who are paraphiliac can be less unhappy about being gay than about being "perverse." So they seek therapy not to change their sexual orienta- tion but to change their "perverted" sexual behavior. Distressed about be- ing paraphiliac they would accept being gay, and stop seeking to become ex-gay, if only they could be gay in a more traditional way, that is, not exhibitionistic, fetishistic, pedophiliac, or sadomasochistic, but instead involved in a happy satisfying, perhaps monogamous marriage.

Post-Traumatic Stress Disorder

Gays suffering from posttraumatic stress disorder attempt to undergo change in order to further/completely distance themselves from the persistent ef- fects of earlier gay-imbued trauma, often ones revived by similar-seeming or duplicate traumata in the here and now. Typically such gays, wounded by earlier attachment traumas, decide to go straight after a relationship breaks up—particularly when they have been dumped—or after they have devel- oped an STD. Such individuals need not ex-gay therapy to go straight, but gay-neutral crisis intervention to help them go on: to first cope and then to move to a higher plane.

Of course, many gays try to change for reality-based reasons, for exam- ple, out of concern for the occupational limitations that even today exist

for gays as a result of antigay bias. But the motivation is to change their sexual orientation to improve their lives, when it should be to improve their lives so that they can comfortably retain their sexual orientation.

In conclusion, not all emotional problems in homosexuals are a product of their homosexuality. Gay people can and do have essentially the same emotional problems as anyone else and for essentially the same, gay-neutral reasons; and these in turn often influence how they are gay, who they are gay with, and whether and why they do or do not wish to change, thus infusing informed consent with emotionally involving conflicts within the self and with others, such as conflicts over being submissive to, versus being rebellious toward, authority. The wish to change from gay to straight almost always has some roots in problems of early childhood, especially those associated with early parent-child pathological interactions around issues of dependency, control, and competition, lived out later in life not in homosexuality per se but in a homosexual's wish to become heterosexual so that he or she can be loved; become dependent on others; better control others or be more readily subject to their control; or get ahead in the competition of life by embracing heterosexuality, here less for moral than for practical reasons—to get new, better, and more stuff.

Even patients without emotional problems, especially when they feel desperate, can, due to transference, become sufficiently emotional to buy into promises of hope, however misguided the premise, and offers of help, however unavailable the assistance. In a typical transferential response, patients' feelings about their therapy and therapist affect the decision-making process. Therefore, consent, in order to be informed, ideally ought to be obtained independent of as many transference distortions as possible so that the patients do not make major decisions based on seeing the therapist as an accommodating or punitive father or as a loving, welcoming or unloving, remote, and rejecting mother, thus believing everything their therapists tell them about the process of going straight, or, for patients having control problems, submitting when they should be rebelling. All these as invalid reasons for changing should not determine consent to undergo therapy for that purpose.

Reparative therapists should additionally also identify the characteristic pathological beliefs (cognitive distortions) about self ("if I am not all good then I am all bad") and others ("if some people don't like me than nobody does") that are the root both of their patients' wish to change, and constitute too much of the real reason for applying for and accepting help.

The wish to change is also infused by significant constitutional (biological) factors involving a patient's temperament. *Hypersensitive* gays

readily form a devalued self-concept based on what others think about and wish for them. That in turn leads them to seek out and agree with abusive persecutors, thence to incorporate such people to become part of their identity, as they, accepting being coerced, agree to continue in therapy out of fear that if they quit they will incur their therapist's wrath.

Just as the final choice about pursuing treatment has to be up to the individual himself or herself (to some extent even if that individual is an adolescent), and just as one must make such a decision as completely dispassionately as possible, so the choice to *drop out* of ongoing therapy should be left up to the individual. But too often therapists keep their patients coming by inducing guilt in them about leaving, or by scaring them by warning them of the dire consequences of not continuing in therapy and of remaining gay instead.

Actually punishing the patient for being uncooperative is common in inpatient settings and though it can result in consent to be treated can still lead to a very bad treatment outcome. Here therapists respond punitively to the resistant patient by retaliating, becoming especially retaliative when patients slip back on any level to their old homosexual ways. They punish their patients by threatening to force medication on them or to put them into the seclusion room. Generally they don't admit that they do such things punitively. Rather they concoct other reasons for thusly responding, such as, "He is a danger to self and others."

Achieving informed consent may take days, months, or even years. During this time the therapist must offer the patient general support as well as insight to help the patient decide what he or she really wants to do with his or her own life, and in therapy.

Reparative therapists must not obtain informed consent by inflating their credentials in an attempt to convince their patients that they, their therapists, are both talented and well trained in the art of doing reparative therapy. Therapists who claim having been once gay and now cured often tout themselves as the best ones for the job. But other professional qualifications are necessary, especially those that involve relevant training along with supervised experience. One doesn't have to be like one's patients (ex-gay) to help patients become similarly ex-gay. If one had to be like one's patients to help them, straight practitioners couldn't help gays, and the only therapist who could help a patient with OCD would be a therapist suffering from obsessions and compulsions—something likely to be as much of a liability as an asset. Therapists who are ex-gay and doing ex-gay therapy should never forget that they are, or ought to be, therapists first and therapists who are ex-gay second, not the other way around. There might be some advantages to being ex-gay oneself,

including enhanced empathy and so the ability to better understand others through a process of sharing. But if only to be consistent reparative therapists should argue exactly the opposite: that since being gay is a mental illness, the ex-gay therapist is unlikely to have overcome all his mental problems and likely to be still too mentally ill to be a safe, effective healer.

GOALS

Goals of therapy need to be kept in focus throughout. Concerns extend to having a discussion about whether to continue in treatment or to stop early, including a discussion of objectives (should the objective be a change in one's sexual orientation perhaps involving a change in one's fantasies from homosexual to heterosexual, or an expansion of one's overt functionality regardless of whether sexual orientation changes or stays the same?). Therapists should also initiate a discussion of procedures to be employed in order to best achieve established goals (e.g., will insight-oriented, cognitive, behavioral, and/or interpersonal treatment—or an eclectic amalgam of all these—be employed and which will be most effective). As well there should be a discussion of time frame: whether weeks or months or years are needed to accomplish outlined objectives. Also a discussion of whether the focus of therapy should be exclusively on sexual orientation or should expand to cover other aspects of the patient's life needs to take place. The therapist must remind the patient over and over again that in reality the prognosis for radical change is poor, for there are few if any patients who can rid themselves to any great extent of strong gay attractions, which will at the very least linger. The possibility of damaging side effects and untoward occurrences leaving gays worse off must be discussed honestly.

INFORMED CONSENT FOR CHILDREN AND ADOLESCENTS

Can children and adolescents give valid informed consent? Some legislatures suggest calling this form of therapy unethical and making it illegal for all 18 years of age and younger. But age is not a reliable indicator of maturity and so of suitability for reparative treatment. Some adolescents are more mature than some adults, while some who are adults in years are actually children in their degree of maturity.

Generally speaking, children are too undeveloped and adolescents too much in transition to definitively decide once and for all what they want

out of life in the realm of being gay or straight. Adolescent sexuality is notoriously chimerical; is invariably accompanied by shifting motivation, erratic judgment, and cognitive turmoil; and is usually associated with acute/chronic emotional difficulties that predictably make today's informed choice into tomorrow's regret. In the young, even so-called firm positional statements are really often but temporary reactions, more in the nature of knee-jerk responses than of accepted permanent alignments. Conflicts, passing whims, irrational philosophies, transient identifications, and long-term plans alternately submissively adhered to and rebelliously discarded are in the very nature of adolescence. In children and adolescents, an identity made up of loyalty to any single self inside is, and should be, ephemeral, and full and final self-realization considered premature. So often wishes are really fears, as "what I prefer to be" is often "what I fear actually being and becoming." Too often what looks like principle is really passion. Therefore, therapists should always consider the possibility of simply monitoring young people, with watchful waiting the cry until there is more and better clarity and greater cloture. Because homosexual enactments are normal at this time of life and so many patients go on to spontaneously morph into heterosexuality, therapy, if it must proceed, is best undertaken for reasons other than those related to sexual orientation change and should rather focus not on homosexuality per se but on more general issues related to a patient's overall emotional health and well-being.

12

❖

Destructive Therapeutic Mindsets

In this chapter I describe some misguided, distortive therapeutic mindsets commonly found in reparative therapists.

MINDSETS DUE TO (PATHOLOGICAL) DEFENSIVENESS

Excessive Sublimation

Reparative therapists are often simultaneously homophobic and at odds with their homophobia. Struggling with their own homophobia they feel as guilty about their own homosexuality as they feel antagonistic to that of others. Then they deal with their homophobia neither by expressing it openly nor by suppressing it completely but by converting, or sublimating, it into manifest excessive caring/loving. This excessive caring/loving is, however, contaminated by a return of the repressed, that is by surfacing homohatred along the lines of "I want to fix you (loving) because you are broken (hating)." Such reparative therapists have typically sublimated their sadism into pity. So, acting like a savior and persecutor combined, they hurt gays in the very act of overassisting them. A bonus of this defense for the therapist is that it maintains the reparative therapist's self-view as a nice person, not out to hurt or harm anybody but pledged to do good, and especially for all who seek their help. A downside for the patient is getting caught up in therapeutic excessiveness.

Projection

Reparative therapists in conflict about their own gay tendencies often attempt to resolve this conflict within themselves by outsourcing the conflict onto their patients—condemning them as they condemn, and instead of condemning, themselves. They act out their self-hatred by taking it out on their patients, as they demand that their patients go straight as an extension of the unconscious revulsion they feel about themselves being gay, which is in turn the product of their own guilt-driven inside pressures. For such therapists telling their patients "I want you to change" is tantamount to telling them "I dislike you the way you are because you embody what I most dislike about myself." Fired up by their assessment of homosexuality as bad and their assessment of their homosexual patients as evil—an assessment largely motivated by their own need to douse the flames within—they deal with their patients' being gay not like a doctor healing a medical disorder but like an exorcist removing a satanic possession.

Pathological Identification

Identification, much like projection, involves doing to, or for, another what one might wish to do to, or for, oneself. Reparative therapists who say that their only wish is to meet their patients' needs are often in fact mainly concerned with meeting needs of their own, with their "empathy" more like externalized reattributed self-concern than true concern for others, their patients. Putting themselves in their patients' shoes and treating their patients as they would want to be treated themselves, they try to help gays become heterosexual not so much for the patients' sakes, as they claim, but vicariously, for their own sakes, as they deny. Pressure to heal themselves definitively translates into becoming coercive toward their patients, as they demand that their patients accomplish what they as therapists personally feel that they themselves must do. As a result, they become controlling individuals along the lines of "you better comply with my demands that you stop being gay—exactly as I ought comply with the same demands I make on myself."

Displacement

In displacement reparative therapists focus on redemption for their patients to vicariously deal with their own personal redemptive needs—variously revisiting and hopefully solving their own sexual/relationship problems by helping their patients attend to theirs. In particular, reparative

therapists who are themselves avoidants advocate that gays distance themselves from other gays to become as isolated as they themselves would like to become as their personal way to resolve their problems with their own (infantilizing) parents, particularly with parents who kept them, as children, at home because they were afraid of losing them to outside competition or unduly feared their getting hurt.

These projecting, identifying, displacing therapists have in common their inability to respect the boundary between therapist and patient. Having created their patients in their own images, they fail to acknowledge that their patients are separate, distinct individuals with unique, very personal needs and wants. Having turned the therapist-patient relationship into a self-object relationship, with their own wants, needs, goals, objectives, dreams, and nightmares commingled with those of their patients, they have created a pathological mesh that fails to differentiate out and so respect their patients' healthful needs so that their patients as individuals can maturely separate and individuate themselves from others, including, or perhaps especially, from their therapists.

Rationalization

Many reparative therapists rationalize their *being controlling* toward their patients by claiming that they are offering gays seeking therapy a choice among several alternatives—when in fact, overtly or covertly, subtly, often unconsciously, they are guiding their patients' choice—influencing their patients nonverbally by turning "informed" into "infused" consent. Some fail to respect their patient's decision to not undergo SOCE by sophistically blaming "the patient's transference negativity," or by citing the supposedly poor judgment that all gays have (for why else would they choose to be, and stay, gay?), giving these things as their explanation for discounting what gays say they want and desire. Others for similar controlling purposes cite their personal belief that since gays are "sinners" they can't really be expected to make any choice that isn't itself sinful—in the sense of being morally and ethically deficient/wrong. Some cite their belief that since gays are criminals it is invariably ordained that they will predictably seek to extend their criminality into making antisocial choices, and in particular the choice to go their own, antiestablishment way. To justify taking over their patients' lives these therapists find ways to convince themselves that all gays have a markedly higher level of mental health problems than do heterosexuals, with being gay a product of another emotional disorder or constituting an emotional disorder in and of itself, typically depression.

Reparative therapists also rationalize their psychoanalytic predilections and practices by emphasizing the universal curative aspects of uncovering forms of therapy, in this case uncovering the so-called developmental pathology that causes gays to become homosexual. (Since all of us regardless of our sexual orientation have some developmental difficulties, reparative therapists can always point to so-called pathognomonic developmental problems in homosexuals, and build bridges between cause and effect, however tenuous or nonextant are the connections they make.)

Furthermore, reparative therapists rationalize their advocating celibacy by claiming it is not a disease in and of itself but a cure for one—although celibacy is in fact not a solution to a problem but a new difficulty constituting a symptom substitution/fresh symptom development, characterized by an antierotic, antipleasure, anti-self-fulfillment mindset generally based upon reparative therapists' not atypical guilt about enjoyment, and as such clearly related to such guilt-infused mental disorders as OCD with its punitive conscience ascendant, and such anxiety-infused mental disorders as AvPD with its fearful removal from others, in particularly others who wish to get too close.

Reparative therapists further typically rationalize their premise that homosexuality is problematic—that is, unnatural, immoral, and sinful—by deeming this belief unchallengeable simply due to its being a matter of faith.

Manipulativeness

Reparative therapists who suffer from mild psychopathy routinely defend questionable behavior by citing acceptable reasons to be manipulative, such as, "being passive and accepting gets you nowhere compared to seeing to it that you have things go your way." These therapists place personal gain in importance above truth and the collective good, and consider winning the only thing worthwhile, to the point that they tie themselves up in knots attempting to validate theories that prove useful to them and justify their subsequent actions. In typical fashion, to satisfy their vested interests in doing reparative therapy they deliberately discard contravening facts that seem to disprove their points, thus allowing themselves to think whatever they wish and do whatever they like. They twist their logic in the direction of some desirable objective of theirs, in this case proving the validity and usefulness of reparative therapy by cherry-picking their theoretical formulations and clinical observations, doing so in such subtle and skillful ways that even they come to believe in the truth of distortions created by their selective inclusion and/or selective omission. They deliberately interpret data in a way that satisfies their professional, including their financial, objectives.

They thus readily find proof that straight life is the default and gay life the deviation so that they *can* come around to viewing gay life as being of secondary quality, and then come forth, guilt and self-recrimination free, with a bag of remedies that they aver will bring about the sexual-orientation changes of which they approve and in which they are invested. Not usually they say that they only wish to improve gays' quality of life by enhancing gays' personal and professional status, when in fact the only personal and professional status they set out to enhance is that of their own.

MINDSETS DUE TO CULTURAL DISTORTIONS

Cultural distortions account for the common failings many reparative therapists bring to their therapy. Some have difficulty treating gays because gays are from a subculture that is in some respects different from the one that their therapists come from. These therapists treat gays not as if they are from a unique, just-as-valid, parallel, subculture but as if they are a bad example of someone from the only good culture, the one to which the therapist belongs. Those therapists who are white Christian straight men often have a minimum of experience with (and so little respect for) anyone else to the point that they fail to appreciate the social disapproval other groups suffer through on a regular basis. This leads these therapists to attribute all gays' problems to individual development/personal lacks, while disregarding the important effects of the social oppression particularly common for minorities.

MINDSETS DUE TO SAMPLING ERRORS

As the result of the sampling errors behind the distortive mindsets of many reparative therapists, these therapists equate "my gay patients" with "all gays," and so come to believe that all gays have meaningful emotional problems like the ones that propel their own patients into needing and applying for treatment. Thus reparative therapists claim that gays and lesbians "have a markedly higher level of mental health problems than do heterosexuals" simply because they only work with gays who come to them feeling anxious, or having become depressed.

MINDSETS PROBLEMATIC FOR ALL FORMS OF VERBAL PSYCHOTHERAPY

Not all problems with reparative therapy are the product of destructive mindsets unique to the genre. Some are the product of problems endemic

to virtually all, or actually all, forms of verbal psychotherapy, reparative and traditional alike. Therapeutic amateurism is not confined to reparative therapy. Patients feel disappointed in their therapists for reasons other than the disappointment gay patients feel about their therapists' not keeping their promises to turn them straight. Most patients respond in the transference to what their therapists say in ways that don't match what the therapists actually meant. Patients project their self-criticisms onto their therapists to become "my therapist's criticism of me." And of course even therapists who always "say the right thing" can antagonize *paranoid* patients, some of whom, itching for a fight, are too readily likely to get angry and depressed enough to quit treatment based upon their own innate suspiciousness of what the therapist intends. Most patients being hypersensitive perceive any exploration of their psychopathology as critical, leading them to feel antagonistic to any therapist who digs for truths. The forthright therapist who eschews euphemisms and "tells it like it is" especially antagonizes *narcissistic* patients who rather than accept medical diagnoses as the purely scientific formulations they in fact are, predictably view them as critical personal attacks, as if their therapists are being insulting by pinning unnecessarily harsh labels on those who would prefer to be not condemned but given a pass for, or even approved in, their troubled thinking and behavior. Excessively *passive* patients almost always resent being total-pushed into being active in any way, even when their being pushed is for their own benefit. The therapist who tells patients to try to be something they are not may readily court resentment—not only over "become heterosexual!" but also about seemingly "putting too much pressure on" an unwilling patient to *become* anything at all, that is, to comply in any way, to anything. Even therapy that is too affirmative can scare avoidant patients who fear closeness, and masochistic patients who find being actively supported to be threatening. Although it's generally a useful technique, attempts on therapists' part to make a point by revealing their personal struggles via speaking of their personal triumphs and difficulties, triumphs and difficulties of theirs that they believe to be similar to those of their patients, hoping thusly to teach their patients something, routinely threaten avoidant patients who feel they are being forced into more intimacy than they can comfortably handle. Depressed patients routinely respond to tales of what the therapist accomplished by feeling themselves personally unaccomplished, by comparison. A reparative therapist who was otherwise a helpful healer took a patient on a tour of his 27-room mansion, his intent being to show his patient what could be achieved with hard work and dedication. Only then the patient complained that that started him thinking, "my therapist is so much better than I am."

13

❖

Reparative Therapeutic Mantras

In the following chapter I describe some of the standard therapeutic beliefs reparative therapists have as to how exactly to make a homosexual into a heterosexual.

CURE INVOLVES IMPROVING RELATIONSHIPS WITH STRAIGHT MEN

Reparative therapists believe that helping gay men feel comfortable in relationships with straight men, that is, less uncomfortable about forming friendships with them—friendships where sexual interest does not prevail, and where mutual acceptance is the main or only goal—will not only help improve gays' mentality and life through enhancing their comfort in straight relationships, but will also walk back to reduce/completely overcome their SSA.

Nicolosi seems to see homosexuality as a form of acting out involving a desperate attempt to receive male affirmation. He views homosexuality as a distraction "from the pain of deep alienation that exists in the absence of authentic emotional attachments."[1] And so he feels that decreasing gay alienation of necessity diminishes homosexual attraction.

CURE COMES FROM EXPANDING SOMATIC AWARENESS THROUGH ENHANCING AWARENESS OF CORE FEELINGS

Many reparative therapists focus on analyzing their gay patients' "core feelings." They expose core feelings in the presence of what Nicolosi calls "therapist resonance,"[2] that is, in the presence of feedback from the therapist-as-empathic-listener. Some focus on increasing awareness of core feelings because they believe that exposing feelings effectively exposes the split off (shamed) aspects of the client's self, diminishing shame in general, and in specific reducing the shame attached to heterosexual feelings, thus enhancing one's heterosexuality by reducing shame-based conflicts about being heterosexual. Some conflicts about sexuality and heterosexuality may be reduced or eliminated this way, but there is little evidence to suggest that this will in any way extend to reducing homosexuality itself.

CURE COMES FROM ENHANCING THE TRUE SELF

Many reparative therapists buy into the postulate that enhancing the true self is curative. Only in gays the heterosexual self is not, as many reparative therapists believe, the true self, for it is the homosexual self that is the true self. Therefore, catalyzing the emergence of a true self is as likely to enhance as it is to diminish homosexuality. Besides, in some individuals, aspects of a true self should be not enhanced but suppressed, even shut down—in particular those "true self avoidances" that distract from/interfere with the individual's ability to sustain relationships with other people.

Reparative therapists believe that since homosexuality is the result of repression of the true (heterosexual) self, lifting this repression will cure homosexuality. They also believe that this repression is the unfortunate outcome of the struggle between being and not being a man, with the latter the winning (unrepressed) position. Therefore enhancing manliness by relieving the repression of "being a man" cures homosexuality.

"True self" theories, often with "self" defined according to gender-based stereotypes, routinely fail to recognize that it is possible to be true to oneself only in those rare instances where there is merely one self inside, meaning that the self inside is not a self-divided, that is, that there are not *selves* (plural) inside that are in conflict with each other due to mixed feelings about who one is and should be—especially conflicts of a "manly versus unmanly" or of a "straight versus gay" nature. Pressing conflicts between selves makes self-realization difficult because the many "mes"

inside each longing to come out struggle intensely with each other for primacy and recognition, making it hard for the patient to form an overall superordinate (umbrella) "self," or "identity." In particular, conflict between manly and unmanly is likely to put inner harmony out of reach. Instead, the following tension-producing conflicts tend to remain active: conflicts between self-expression (manly) versus excessive self-control (not manly); challenging society (manly) versus being socially sensitive and cooperative (unmanly); challenging the establishment (manly) versus cooperating with the establishment (unmanly); rebellion (manly) versus assimilation (unmanly); taking risks (manly) versus remaining safely (and productively) behind in the shadows (not manly); close-fisted aggression (manly) versus hands-down abdication (not manly); and a lot of sex as a good thing (manly) versus too much sex as a bad thing (unmanly). When such conflicts remain, and they likely always will, one cannot, among other things, sufficiently firm up a singular identity leading to that overall strength of character that is a supposed requirement for conquering one's homosexuality.

CURE COMES FROM VIEWING HOMOSEXUALITY AS AN ILLNESS AND TREATING IT AS SUCH

Reparative therapists conflating being gay with being psychologically ill portray homosexuality not as a healthy state but as a symptomatic condition, not as a fundamentally instinctual desire or urge but instead as a compromise formation installed to resolve conflicts such as the one between masculine aggression and shame. In this view, homosexuality is not a primary urge but a form of acting out—propelled by anxiety and guilt—not a healthy search for a connection with others yearning for completion and gratification, but an attempt to reduce/resolve inner personal, very unhealthy, psychic chaos. So, reparative therapists try to work out the forces that supposedly have produced this symptom/malady, forces they routinely identify as the very same ones that produce other psychological symptoms such as phobias. Thus they treat homosexuality as mainstream therapists treat other "emotional symptoms": they analyze it, or modify it behaviorally, or deal with it cognitively, airing and correcting cognitive distortions such as those about women who are rejecting and those involving fear of being mutilated by a vagina with teeth (vagina dentata); fear of being castrated by father; a need to suffer manifest as depriving oneself of guilty heterosexual gratification; shame that leads to diminished assertiveness and hence in men to feminine passivity and sexuality; and so on. They do this though the homosexual instinct is a basic force of

nature. And as such it can no more be reduced to its component parts (psychoanalyzed), or changed cognitively, or influenced behaviorally using psychological techniques than can the heterosexual instinct be so broken apart into its component parts then modified in like manner.

A corollary is that the forces that produce the *homosexual* symptom do not stop there but readily spill over into producing contiguous undesirable *personality traits*—those stereotypically undesirable ones supposedly regularly associated with homosexuality, and thus presumably a part of the personality of all homosexuals. Thus shame about masculine assertiveness, which reparative therapists postulate to be a main cause of homosexuality, presumably bleeds over into effeminacy creating the mincing, passive-dependent queen afraid of his own shadow, which is what all homosexuals are like. Or the self-object relational component that is so central to making the homosexual sexual object choice a narcissistic one predictably spills over to become the personal self-centeredness, selfishness, and narcissism supposedly characteristic of all homosexuals, most of whom can be expected to preen in front of a mirror, plucking their eyebrows and arranging and rearranging their hair.

Moreover, since homosexuality is a symptom like any other psychic symptom, or a mental illness in its own right, as well as supposedly being predictably associated with other, related or unrelated mental problems, reparative therapists (and the lay public) ought not to accord it the same social status as heterosexuality, just as they ought not to accord schizophrenia the same social status as normalcy.

This view of homosexuality as in itself a sort of mental illness or as a product of another mental illness overlooks how homosexuality is a vital force in and of itself, not a symptom, but a primary sexual urge/goal, so that clearly the approach to homosexuality would be not to analyze it as a symptom, e.g., in men as a phobia of women, or in women as a phobia of men, but to respect it as authentic and primary to the self, in preparation for then leaving it be.

CURE INVOLVES INTEGRATING UNINTEGRATED OLD TRAUMATA

Since homosexuality is the product of having experienced trauma early in life, in particular trauma due to gender shame, identifying and dealing with earlier trauma—and coping with and integrating these trauma—predictably represents a major step toward curing homosexuality.

Some reparative therapists employ trauma theory to recommend recovered memory treatment. This treatment usually starts with a search for

earlier traumata that occurred at the hands of gays' parents. But such an approach (unfortunately) leads patients to blame, and so to hurt their parents emotionally, especially parents who, though misguided, are nevertheless positively motivated and, being as sensitive as most parents are, easily wounded by a therapeutic approach that uncritically accepts a child's misguided allegations about how bad the parents were to their child(ren).

Actually, techniques meant to recover and mitigate the later effects of early trauma can, if judiciously employed, be potentially useful for patients with certain homosexually related psychological issues, such as compulsive homosexuality. But such techniques are wasted on ordinary healthy persons who happen to be gay. While early traumata might influence a person's psychology, and therefore trauma recovery-and-reduction therapy might help treat a symptom such as a phobia, recovered trauma techniques are misapplied when used to treat homosexuality, for being gay is not the product of an earlier traumatic encounter. Besides, even if there were a relationship between early trauma and the development of homosexuality as a post-traumatic response, trauma recovery after the fact is unlikely to help much anyway, for, as I learned from my work with veterans, the intractability of post-traumatic stress disorder exists because reversing the effects of early trauma entirely, that is, unseeing a trauma that has already been experienced, is a difficult to impossible task.

CURE INVOLVES CONSCIOUSLY REJECTING BEING GAY

Many reparative therapists believe that one can cure homosexuality by consciously rejecting being gay, so that as van den Aardweg suggests, one can at least start getting better by just "saying 'no' [to homosexuality]," that is, saying the equivalent of "I shall not give in."[3] Thus gays should reject all homosexual excitation and all homoerotic enjoyment, and do so totally even if that requires being housebound—which it might, since attractive members of the same sex exist everywhere, and one can only deal effectively with being stimulated by them by never going out in the first place, or, when out, wearing blinders and never turning one's head in the direction of someone appealing. One solution based on a degree of conscious control involves praying the gay away/praying oneself straight as an operative curative. (Praying the gay away also supposedly works in part because submission to a more moral, more prohibitive higher authority occurs with direct intervention from the Almighty helping one refuse to succumb.)

Van den Aardweg also suggests "physical mortifications" that can help "in the battle with sexual obsessions." He suggests that one "Pray to the

Blessed virgin, whose intercession with God is particularly effective" and pray "at the very moment of temptation."[4]

Supposedly it also helps to think of oneself not as gay but as straight. For while thinking of oneself as gay may be satisfyingly pleasurable, it tends to have a self-fulfilling component to it because it catalyzes the vicious cycle between thinking homosexually, acting homosexually, and becoming homosexual (or becoming *even more* homosexual). One supposedly ultimately becomes like what one in the beginning only fantasizes being. And supposedly the reverse is also true, for one becomes no longer what one no longer thinks of oneself as being.

CURE INVOLVES REMEDYING GENDER-IDENTITY DEFICIT

Many reparative therapists believe that all gay men are effeminate, and that all lesbians are masculine, so that all that is necessary to cure being gay is to reverse the gender-deficit patterns which not only accompany but also cause being gay. This theory tells us in effect that, as Twain suggests, "clothes make the man,"[5] and that presumably one can change the man by having him change his clothes, that is, in the case of homosexuality, by having him or her alter superficial (gay) aspects of himself or herself. Thus van den Aardweg suggests one can cure gays by "not feeding the homosexual complex" by not "nurturing moustaches [and] beards."[6] As he goes on to say, to get better gay men must stop "sewing, or arranging flowers"[7] for that = "wallowing in [one's] half-womanly 'nature.'"[8] They must stop avoiding "getting their hands dirty doing manual work" and instead play "a competitive game like soccer or baseball"[9] and otherwise "take . . . on normal gender-linked 'roles' to . . . become "normally assertive."[10] The lesbian woman should learn to love "a nice gown"[11] and becoming more submissive "obey . . . the authority of *men* [and] live with [a] man and give [herself] to him . . . care for him . . . and surrender to his masculinity."[12] "Using a tape recorder"[13] men should learn to change effeminate and women to change hypermasculine speech patterns; for example, men should never put two "s's" as the end of words like "yes" (and not draw out the two "s's" that do end words like miss). And gays should do this all in the service of "taking on normal gender-linked 'roles'."[14]

This works neither as suggested nor in reverse. For example, a straight patient of mine who spent a lot of time dressing up in feminine clothes—heels and gown, makeup with rubber pants/diapers underneath and the like—never showed the slightest inclination whatsoever to turn gay.

14

❖

Serendipitous Positive Effects/ Unintended Positive Consequences

In this chapter, which supplements chapter 9, I discuss specific ways reparative therapists positively affect their patients due to treatment's unplanned aspects and unintended positive consequences, that is for reasons unaligned with their assumed/proclaimed objectives and manifest goals. Many gay patients feel, and are, helped simply because of the human contact involved in reparative therapy. Here the therapist's "merely caring" has on some level led to real, positive results, which go beyond "the placebo effect." Even just agreeing to see someone for help can make that person feel better before he or she actually starts therapy. As some informal studies have shown, just putting a patient on the waiting list can be seriously therapeutic. Furthermore, many reparative therapists without actually recognizing it use methods that bear a strong resemblance to mainstream therapeutic interventions. However misapplied to curing homosexuality these may be, they can relieve other problems their patients have: ancillary problems of the homosexual that keep him or her from feeling happy and being a well-adjusted gay individual.

Reparative therapy serendipitously offers the elements described in the following sections.

Encouragement

All reputable therapists, whether analytically, cognitive-behaviorally, or biologically oriented, can encourage their patients to improve simply by conveying the (spoken or unspoken) wish to have their patients do well. Patients hear "I want you to get better," perhaps for the first time in their lives, and so as much as possible try to, and often do, oblige.

Welcoming

All therapists extend a welcoming hand simply by agreeing to see a patient. This welcoming hand is not much different from the one extended by the club that accepts or the employer who hires. Even patients who don't want to be in therapy (but have, for example, been forced to go by their parents) can experience this "welcoming effect" and will likely react to it in a strongly positive, thankful way.

Hope

The mere act of accepting patients into treatment gives them hope that better days lie ahead. Patients ask themselves, "Would my therapist see me if he or she didn't think something could be done for me?"

Information/Understanding

"All" therapists wittingly or unwittingly tell at least some of their patients at least a few things that they might want or need to know and can use in order to survive, prosper, and hopefully get better. Thus reparative therapists who speak of curing homosexuality itself may in fact be, if only unwittingly, informing patients about the nature of, and problems associated with, as well as instituting the remedy for, their patients' *compulsive* homosexuality/promiscuity—thus helping their patients become better homosexuals perhaps after becoming monogamous in a way that the patients themselves desire, want, and actually have to date tried on their own, although so far without success.

Even the most misguided reparative therapists have some good ideas that their patients can use. Thus van den Aardweg says that gay men and women should "become less ego-centered"[1] and acquire "those human and moral virtues that have a 'deegocentrizing' effect and embrace the capacity to love."[2] He adds "'psychotherapy,' if it is sound, can offer valuable points of insight about the origin and structure of troublesome

emotional and sexual habits, but not discoveries that will cause a change overnight. For instance, no psychotherapy can provide a sudden liberation, as is pretended by certain 'schools', by unblocking repressed memories or emotions."[3] "What is required is much common sense and quiet daily perseverance."[4]

Reparative therapists who explore conflicts about heterosexuality as a prime factor in the development of homosexuality often simultaneously help their patients resolve other troublesome simultaneously extant core conflicts. Closeted or "non-gay homosexuals," for example, while attracted to men sexually but in conflict about or disgusted with being gay because the ideal self (ego-ideal) is at odds with the homosexual self and its yearnings, can benefit from reducing internal anti*human* messages— not because this releases heterosexual desire and capacity, but because that helps them become less conflicted overall about meeting their legitimate needs and honestly and openly validating their sincerest feelings.

Permission

Reparative therapists while inducing guilt in some ways relieve guilt in some other ways. For example, reparative therapists who attempt to cure homosexuality by reducing gays' feeling ashamed about their lack of masculinity often simultaneously help reduce gays' shame about other things as, for example, as a "side-effect" therapists impart forgiveness and absolution for what the patient believes to be unacceptable assertiveness. Paradoxically, by giving patients permission to change they also give patients permission to stay the same, for relieving one self-directed criticism tends to free patients up to feel less self-critical in other ways, about other things.

Reassurance

Therapists' neutrality and calm can by themselves be sufficient to reduce overall patient anxiety especially if the therapist who on the one hand says "it's bad to be gay" also says, "but overall, you haven't ruined yourself completely, there are plenty of good things about you and your life that we can note and start working with."

Organization/Structure

The structure that comes from having regular therapy sessions can have an organizing effect on patients' thoughts and behaviors by providing a

calming effect for gays whose lives are presently less structured than might be desirable—especially so for adolescents whose lives routinely tend to be somewhat chaotic.

A Positive/Loving Therapeutic Relationship

Even reparative therapists who view gays as broken because homosexual can convey that overall they still like their patients, accept them more than they cast them out, respect them rather than demean them overall, and praise them more than they criticize them. Patients on the one hand are told they are broken and on the other hand hear the positive thera-peutic message "you are worthy of being fixed." Offers of love come in many forms, and many gays see the command to change not as a criticism of who they are but, correctly or incorrectly, as a loving inducement to do better, which to patients means that at least someone cares about them and is concerned with their fate.

Real, often positive, aspects exist in most patient-therapist relation-ships to inspire patients and help them feel better. This can be true even in spite of the therapist's strictest attempts to maintain therapeutic neu-trality and avoid crossing boundaries. For some therapists the mere status of "being a patient" doesn't, without violating boundaries, completely dis-qualify the individual from also being perceived as an attractive and wor-thy person overall, one with real, admirable characteristics.

While it is true that the reparative therapist begins from the premise that the patient is broken, it is also true that many reparative therapists regardless of that view respond to their patients, broken or no, with basic respect. And many gays, like many straights, have been so abused to date that a reparative therapist is the only one who has ever held a door open for and closed it behind them, given them an opportunity to participate in deciding how hot or cold the room should be, or offered them a tissue to dry tears of joy or sadness. For some individuals, gay or straight, a thera-pist, reparative or mainstream, is the first person in their lives who lis-tened to them and wanted to help them out, and they are truly appreciative of that. Even if therapists respond critically in the same way others have responded, and still respond, to men and women who are gay, they often do so more constructively than destructively, that is, more in a kind than in a cruel way. Many therapists who tell gays that they are broken, how-ever wrong their hypothesis may be, are still telling them just that for the right reasons: not to criticize but to cure them. That alone makes the re-parative therapist different from others, and in particular separate from parents who tell their gay children the very same thing, but do so less to

help them get better than to simply demean them—without offering them any real suggestions for improvement.

Empathy

Almost all therapists try to put themselves in their patients' place to improve the prospect that they can heal them by seeing them through the patients' own eyes. (On the negative side, empathy can sometimes miss the point: its target is, and wishes to continue to be, an entirely separate person.)

Opportunity to Abreact

Reparative therapy like any other form of therapy encourages patients to talk about what burdens them, helping them get their troubles off their chests.

The Possibility of Celibacy

Reparative therapists can help patients live with an unacceptable/unbearable but unalterable sexual orientation by guiding them toward their becoming celibate, which, however pathological that might be, at least allows *selected* patients (those who can't do any "better") to stay gay without living as gay, fully maintaining their religious beliefs and following their religious calling, while living a life that while limited is also now at least less infused with guilt.

The Opportunity to Redirect (Displace) Blame

Blaming one's parents and society for making one gay, and blaming all one's troubles on being gay, however pathological that may be in and of itself, helps some gays feel less blameworthy overall, and therefore generally less sheepish and depressed.

The Opportunity to Grieve

Reparative therapists' exploration of the role grief plays in "causing homosexuality" can help patients grieve over problematic aspects of their lives they previously have not been able to face and put behind them. That way they can help their patients deal with actual losses such as the loss of a father's love—and get past these losses by viewing them simply as

unfortunate events that happened long ago, won't likely be repeated, and so ought to stop negatively affecting them so much in the here and how. When indicated, reparative therapists can also help patients identify which losses are more perceived than actual, such as father "losses" that never actually occurred, having simply been misinterpretations of what in reality happened in a given father-son relationship.

The Possibility of Giving the Family What They Seem to Want

For families that strongly disapprove of a childs being gay, the sons being in therapy tends to at least temporarily calm the homonegative waters and hold off family predation.

Group Therapy

Group reparative therapy provides the individual with some of the benefits of nonreparative group therapy, including support, understanding, mastery, enhanced resilience, absolution, abreaction to release feelings and impart that sense of relief that comes from unburdening, and consensual validation that gives relief through sharing and positive feedback. Developing any group identity can be temporarily sustaining—if only irrationally so, this being a valid goal when there is a psychological emergency. Biblical studies that are part of some groups' efforts can help individuals reconcile negative biblical teachings with a sustainable positive self-view. These studies can help patients use their relationship with the Almighty not to perpetuate but to resolve earthly inner and interpersonal struggles.

The Illusion That "I Am At Least Trying"

Patients who feel that they are at least trying to go straight can subdue some of their conflicts about being gay by convincing themselves that they are, after all, working on their problems, and so are doing everything that they possibly can to solve them.

Reassurance That No Reasonable Alternatives Exist

Though in the majority of cases SOCE doesn't work to change sexual orientation, some anecdotal reports suggest that it seems to work on rare occasions, though without anyone scientifically knowing why. Therefore, although the odds are long and even impossible, and the process full of

uncertainties, perhaps SOCE should still be tried for those who are in great distress because they don't accept their sexual orientation for personal (perhaps ultrareligious) reasons. The question is fairly asked: how else can anyone help gays who are deeply religious and whose religious beliefs prohibit homosexuality? An argument can be made for both sides: "don't try it, it won't work," or "try it because there is nothing else, and what do you have to lose?" (I clearly favor the former argument and throughout specify the things that *can* be lost.)

Healing through Terminating

Patients' positive responses/testimonials about reparative therapy can be real, that is, they do not always only reflect the impaired judgment characteristic of positive transference. They may not merely reflect a general response to the therapists' offer of help rather than a specific response to anything the therapist actually said or did. They might rather be a sanguine response to the approach to, or to termination itself because termination is so welcome that it stirs up good feelings seemingly about one's therapy, but actually about its finally being over. Such feelings, becoming generalized, very often cast a rosy backward-looking glow on the entire treatment process itself.

Affecting What Traditional Techniques Have to Offer

As discussed in chapter 9, the better (less worse) ex-gay therapists actively advocate and do a form of traditional psychotherapy. More identified with non-ex-gay psychoanalysts/cognitive therapists than with the general run of ex-gay therapists, they, modifying their therapy to become more mainstream, do what amounts to a form of, or at least contains sanguine elements characteristic for, affirmative therapy. Often, hoping to retain their ties to a group dedicated to ex-gay therapists some therapists proclaim they do reparative work but actually set out to help their gay patients stay gay but be better at it. These therapists have found a way to stay identified with the ex-gay movement by doing a little reparative therapy but using rationalizations such as "while you are mulling over converting you need supportive therapy, and while you are working on your sexual orientation we have to develop insight into your lifestyle problems" in order to feel free to be mainstream therapists "when no one is looking." The "while" then becomes if not forever then at least "at some indeterminable tomorrow," so that therapy meant to be reparative in effect begins to look more like traditional supportive and insight-oriented treatment.

We should always remember that conversely many traditional thera-pists do some therapy that can be classified as reparative. Does criticizing and banning reparative therapy apply in these cases, to those therapists who say to themselves, "there is a chink in this gay man's armor so let me just give conversion therapy for him a try"?

—◆—

15

Therapeutic Errors

In this chapter I outline some of the therapeutic errors that reparative therapists often make with their patients.

CROSSING BOUNDARIES

Individual and/or group reparative therapy sometimes degenerates into crossing boundaries where therapists, as Throckmorton and Yarhouse suggest, "engage in dual relationships with clients or provide physical nurturance to clients . . . or become physically close to . . . clients in a group therapy situation [or] refer clients to retreats, support groups or interventions requiring boundary violations as a condition of participation."[1]

OFFERING TRIVIAL TREATMENTS

Some reparative therapists recommend trivial treatments to, and even impose these trivial treatments upon, their patients. These treatments go beyond simply not working to actually doing harm. One way some do harm is by disappointing patients, depressing them by promising them more than can be delivered. Some threaten avoidant patients by being excessively intrusive. Some sidetrack patients from real work that needs to be done. An example of trivial therapy is concrete wishful-thinking-based intervention such as "get a picture of your father and tear it up to prove, once and for all, your independence from him, releasing

you forever from your need to seek men sexually because now you have mastered and triumphed over your father fixation." Another is, "Think of your homosexuality as being in a balloon, then think of the balloon as floating up and out of your head and into the stratosphere." Therapists who recommend such "therapeutic" actions think that they are doing some good, but their patients, sometimes merely mystified, sometimes completely outraged, and mostly finding themselves embarrassed, feel no better or actually feel worse because they come to recognize that they have been snookered.

OFFERING ONE-SIZE-FITS-ALL THERAPY

Reparative therapists, failing to distinguish between different gay patients, deal with homosexuality, not with the homosexual, that is, they fail to tailor their approach to the individual homosexual patient under their care. One-size-fits-all theory forms the basis of one-size-treats-all therapeutic interventions offered to all patients as if all gay and lesbian patients globally fit some specific, in fact procrustean, mold. It is as if different gay patients do not have different personalities/personality disorders/neuroses that in turn influence their need for, and likely responses to, specific therapeutic procedures. All of a therapist's patients might be asked to pray the gay away even though some patients are more, and some less, prayerful because some are more, and some less, religiously oriented (and/or more or less suggestible/hypnotizable) than others. Patients with histrionic tendencies likely buy into the therapist's views about God and His influence more readily than patients who are paranoid, who as a result of their paranoia tend to be far more suspicious/way less trusting of what they hear and are told, including about matters of religion. The *psychoanalytic* approach must be (but often isn't) chosen only for certain patients, and then additionally adapted to them. Thus schizophrenic gays should not be put on a couch and "analyzed" but should rather be offered covering, supportive techniques. Invoked psychodynamics must not be applied across the board, cookbook fashion, to all patients, overlooking individual developmental and dynamic psychology, so that we should hear (if at all) not about "the psychodynamics of homosexuality" (as if these are standard issue) but about "the psychodynamics of this particular homosexual." Some gays are guilty about their homosexuality while others are less alarmed by the homosexuality itself and more concerned with problems they are having in a given homosexual relationship, or relative to the occupational fallout from being gay. (If in the first place we are to believe in the reparative therapists' view of homosexuality as the product of

pathological developmental factors, we should also ask why the identified psychodynamics of homosexuality are so different from reparative therapist to reparative therapist. For example, why do some speak of gays being homosexual because they are in pain about an early sexual traumatic seduction resulting in a developmental fixation at, or a regression to, an unpleasant earlier time, while others speak of being homosexual because of problems related to having had an engulfing mother or having had a weak father.) Gays who want to change because for them being gay is an occupational hazard are different from gays who want to change because as schizophrenics they hear voices telling them that they are bad people who should mutilate their genitals to become not a gay man but a straight woman. Age differences have an impact too, so that gay individuals who are older, more settled, and more set in their ways require a different therapeutic approach from gay adolescents who being young are still in their formative, impressionable years. The approach to bisexuals (or others who have already had heterosexual experiences) ought to differ from the approach to those gays who have never been active sexually with members of the opposite sex. Though in the long run it likely makes no real difference, it is at least intellectually more honest to take the degree of a given patient's psychological sophistication into account, for being psychologically savvy versus psychologically naive will predictably determine the exact approach that should, at least theoretically, be taken to "bring change about." For example, in my experience psychoanalytic approaches are more desired by, and friendly to, patients who are insight oriented than are cognitive-behavioral approaches, which are more desirable for, and friendlier to, patients with a preference not for self-discovery but for being told how to think and what to do.

INDULGING IN INAPPROPRIATE UNCOVERING

Uncovering, whether it involves sensitizing clients to their body responses, or revealing deep inner transcendent intellectual truths, can cause brittle patients to decompensate psychologically and develop an overt psychosis, and more intact patients to become more ashamed of themselves and thereby increasingly depressed. However much intimate revelations (as often recommended) are elicited specifically in order to have them occur in the presence of a caring attuned therapist, brittle gays who heal by repressing feelings that they can't otherwise handle are better off not getting their feelings out in the first place and under any circumstances, therapist in the wings or no, but if they keep them in under most or all conditions. Also, patients urged in therapy to get their feelings out

might learn to do that very thing at inappropriate times and in inappropriate places in real life, for example, with their families, or at work, and these outed feelings might very well be hurtful to others and/or carry along with the rising tide the sexual feelings that gays are at the same time being urged to suppress. The recommended recovery of painful traumatic memories can be detrimental to patients' mental health for many reasons. Especially when the memories supposedly aroused are not of late life but are of early childhood, even of intrauterine, experiences, patients can develop a false memory syndrome consisting of the making of untrue accusations about their parents, or others—accusations that are personally hurtful to all concerned, and as well can additionally have far-ranging seriously negative legal implications.

There is no certain relationship between therapeutically abreacting feelings and becoming straight. Certainly as Nicolosi states, "doing powerful emotive/affective work, which is then followed by narrative reconstruction,"[2] including reconstructing an early emotional experience with a significant other, such as a past emotional betrayal, in the presence of an attuned therapist can help selected patients do better personally and relationally. For example, selected patients might become more aware of here-and-now interactions with others that become troublesome due to the perpetuating of early narcissistic injury, and for some patents dropping dysfunctional defenses directed against core feelings can improve their relationships with others. But in no way or in any respect can such an approach seriously help patients change their sexual orientation.

FOCUSING ON REPARATION INSTEAD OF ON OVERALL SELF-REALIZATION (AND SELF-IMPROVEMENT)

Reparative therapists who focus exclusively on "it's bad to be homosexual" fail to focus, when indicated, on "it's neither bad nor good to be homosexual; but it is good, not bad, to be homosexual in a good way, one that is self-rewarding, not self-destructive."

BEING NARCISSISTICALLY EMPATHIC

Excessive therapist empathy involves narcissism when it extends to assessing patients by projection along the lines of "I can help you overcome your pain because I feel your pain as I feel mine." This can be counterproductive should therapists, being themselves depressed, be projecting their own depressive issues onto their patients, along the lines of "I know what

you are going through because I know what suffering is like, because I am suffering too, and in the same way" and "I got over my suffering by going straight, so I see your need to, and can help you, get over yours, in exactly the same manner." Here "what I want for myself" too readily becomes transformed into a boundary-crossing "what I want for you, and so what you *should* want for yourself."

Narcissistic reparative therapists reliving their own emotional and personal depressive worldview vicariously with their patients see their patients' life as depressing because they are dissatisfied with their own lives, and want to share. They might transfer their self-punitive morality onto their patients so that their own morality becomes what is morally right for others. They might chastise and flagellate their patients as they chastise and flagellate themselves as they displace their own self-directed sadism onto their patients in the form of sadistic straightjacketing therapeutic formulations as to why being gay is sinful and going straight is sacred, accompanied by punitive therapeutic actions (such as recommendations to self-mortify).

Doing these things makes reparative therapists unique among healers in their taking stands when they should, emulating traditional therapists, instead remain morally and intellectually neutral.

BEING EXCESSIVELY ATTUNED

Therapeutic attunement is a valuable asset unless therapists listening with the third ear forget to listen with the first two, doing that to the point that they over-interpret (distort) everything their patients say by reading not the lines themselves but between them: often selectively, and usually according to some theoretical persuasion of the therapists' own, already in place, creating bias.

DEMANDING COMPLIANCE

Reparative therapists who speak of enhancing autonomy simultaneously undermine it by demands they make for patient compliance. Some, becoming reparative martinets, do battle with uncooperative patients by condemning them for wanting to remain homosexual, thus provoking deep, difficult-to-manage negative emotions in their patients, which emotions the therapists go on to wrongly identify as the patient's problem, e.g., as "intractable rebelliousness manifest in your resistances to therapy." The patient protests, "I like being homosexual, it's me," to which the therapist replies, "This is a resistance to my helping you become straight,"

to which the patient replies, "That's not true, it's a position I take, not a protest I make," to which the therapist replies, "I'll pray for you and your soul (but not exactly for your mental health and personal independence—from me)."

BEING OVERLY PARENTAL

Too many reparative therapists unwittingly allow their patients to provoke them to countertransference manifest by their acting like their patients' own mothers or fathers: silent, unavailable, engulfing, controlling, constricting, and competitive. Patients who need to repeat a relationship with an absent mother provoke their therapists to defensively pull back and distance themselves from their patients, while patients who need to repeat a relationship with a punitive father provoke their therapists to shock their patients into opening up by dredging up and avulsing their (the patients') repressed feelings, then excusing opening up their patients' old wounds as part of what is needed to help restore the patients' full mental health.

USING UNPRODUCTIVE OR COUNTERPRODUCTIVE BEHAVIORAL TECHNIQUES

One example of an unproductive behavioral technique is the direct therapeutic attempt to divert a patient's thought processes from something unhelpful or sinister onto something neutral and so calming (e.g., "relax by thinking of green meadows"). This often has a paradoxical effect, for it simply calls more attention to what is being diverted, leaving gays not less preoccupied with their sexual fantasies, but able to think of little else.

16

❖❖❖

Diverse Therapeutic Approaches to SOCE

Not all reparative therapy is done in the same way, in large measure because reparative therapists are from many different schools of psychiatric thought. In this chapter I discuss how individual reparative therapists are thusly defined not only by their goals (changing gay to straight) but also by the specific therapeutic techniques they use to achieve their objectives.

Most reparative therapists agree that a homosexual orientation is not, with the exception of the contribution of individual temperament, biologically determined. Rather, as they see it, the individual acquires homosexuality psychologically over the course of his or her development, with being gay not innate but, according to Nicolosi, the product of "early gender based trauma."[1] Since homosexuality is acquired it can therefore be unacquired, and so, just as a phobia or a depression can be cured by talk therapy, homosexuality can be relieved by the use of various verbal psychotherapeutic techniques. Thus gays who given insight into their past come to understand it thoroughly will no longer be forced to repeat it and so won't have it affect their present. Alternatively, homosexuality can be cured by interpersonal therapy, cognitive-behavioral therapy, or supportive psychotherapy, especially when these involve therapeutic empathy and positive resonance/feedback, depending on the orientation and preferences of the individual therapist.

The different therapeutic approaches are described in the following sections.

BEHAVIORAL

Punitive behavioral methods were once widely used in an attempt to make gays straight. But today if they are used at all (and they still may be) this is often done strictly in secret, behind closed doors.

Essentially nonpunitive behavioral methods exist. These include *reinforcing* heterosexual and *extinguishing* homosexual behavior; *isolation* to avoid stimulation (staying out of restrooms on the highway); cold-shower techniques including but not limited to becoming celibate—the latter involving renouncing all sexuality not only homosexuality; and prayer in order to enlist the powerful assistance of the Almighty. Often used are nonpunitive exhortations to live well, e.g., to eat the right foods and to exercise to enhance good health and well-being—as if the "unsound mind that is homosexuality" cannot exist in the "sound body that is natural heterosexuality."

No behavioral interventions can make a homosexual heterosexual, that is, alter sexual fantasies, for these stubbornly remain untouched no matter what is done. Still, behavioral methods can help compulsive homosexuals become less so, if only by exercising an overall calming effect through offering the patient the structure of regular sessions. Unfortunately, behavioral techniques can have distinctly harmful effects, especially when they involve negative ("don't do that") input from the therapist. Negative input can by itself cause gays to become conflicted not only about their homosexuality but also about themselves overall, thus interfering (as can also happen in heterosexuals) with their ability to perform sexually as each time gays try to have sex they fail because they hear their therapist's little voice telling them to become sexually dysfunctional, and how and why to go about that, making the patient like, as a number of my patients have said, the proverbial centipede who can no longer walk because, newly confused, it thinks too much about the wisdom of getting from here to there and how each leg can best serve that purpose—ultimately rendering all the legs, and the creature itself, entirely immobile.

Yet many reparative therapists claim that behavioral interventions can not only alter the accoutrements of homosexuality (such as its compulsiveness) but can also psychologically alter the homosexuality itself. For example, they view homosexuality as a creation of the "false man" and then intervene by attempting to turn the false into the real (heterosexual) man through having the patient play traditional gender roles, which too often turn out to be stereotypes ("go to the shooting range," "stop sewing," "stop lisping," "go to the sports bar") in order to promote manly identifications by forcing manly and foregoing womanly behaviors. These

therapists claim that such interventions can make sissies into (real) boys because, since being gay and being macho are presumably incompatible, making behavioral changes to a less than macho self (becoming less sissified) is presumably capable of turning a gay sissy into a macho straight man. The confusion between sexual orientation and gender identity that exists for some reparative therapists has now become the actual basis not only of their theory but also of their practice, revealing, among other things, how reparative therapists have little to no realistic firsthand knowledge of what the full range of gays is like, with gays varying considerably in their thoughts and actions from highly effeminate, however rare this is now, to highly masculine. The reparative view of what all gay men are like has likely come about in part because it is the more passive gay man who enters reparative therapy due to having acceded to others' wishes to go straight, while it is the less passive gay man who has chosen to avoid treatment entirely and stick with who he is. If reparative therapists would only attend a gay gym/go into the locker room, they might change their view of all gay men as creatures who do not act like real men, and so they might stop postulating that urging gays to act like real men is the royal road to making them straight.

ABREACTION

Abreaction involves "getting the thoughts and feelings out," often accompanied by offering the patient a corrective emotional experience accomplished by having an attuned therapist validate and help patients understand and integrate their emerging thoughts and emotions. Therapists who use abreaction often attempt to bring gays back in time to dredge up some original, presumably life-altering trauma. They encourage them to relive it, then help them master it through reevaluating its significance, while grieving for any implied or actual loss, potential or real, entailed. Simultaneously, they offer healing feedback in the form of support ("I understand") and reassurance ("You can handle this," "After all, things aren't so bad").

Many reparative therapists ask their patients to air (and work through) the long-suppressed anger assumed to be at the heart of becoming gay, such as anger at an enveloping mother (or passive father). The idea here is that once a man becomes less angry at women he will become less fearful that a woman will retaliate, such as by castrating him, and so more at ease personally and sexually with all women. Some therapists tell gays to act as if their pillow is their traumatizing mother, and then beat up on the pillow in order to discharge negative feelings toward Mom. Supposedly

once discharged, angry feelings will no longer fester to make someone gay. Therefore those so treated will turn straight.

RELIGIOUS

Religious therapists claiming with Nicolosi that "sexual feelings [do not necessarily] take precedence over deeply held religious beliefs"[2] freely use religious beliefs to suppress or lyse (homo)sexual feelings.

Mainstream therapists tend to hold their patients blameless in the hope that they can thereby help them open up and discuss their problems more freely as they come to have less guilt and shame. Religiously oriented therapists, doing the opposite, instead of trying to help gays become less guilty, offer them reasons to become more so, perhaps in the hope that they will feel so guilty about being gay that they will simply have to turn straight.

In effect, all many religious therapists are doing is renaming methods already in use by nonreparative, nonreligious therapists making aspects of religious therapy little more than mainstream (legitimate) therapeutic techniques, now by another name. Thus religious therapists renaming what is essentially a time-tested mainstream process of "suggestion" call it "prayer" and "inspirational reading." They refer to the placebo effect as a "spiritual conversion," and for them the time-tested technique of "exhortation" takes the form of an "appeal to one's better, more highly moral, nature," such as "inculcating obeisance" ("deference"). They call meditation and relaxation therapy "contemplation." Autosuggestion becomes "telling oneself one must do God's will"; "abreaction" overlaps with "confession"; "enhancing empathy" becomes "serving others"; and instead of "positive thinking" religious therapists might recommend "offering gratitudes," which are in some ways the equivalent of repetitive acts that accentuate the positive to avoid thinking and acting on forbidden negative (gay) thoughts and actions. Patients develop "insight" through "religious meditation"; "regression therapy" such as touch therapy becomes for religious therapists the "laying on of hands"; while "environmental manipulation" such as advising patients to stay out of situations that alight gay fantasies in religious therapy terms might involve entering a religious retreat where everyone having become celibate stays away from sex entirely.

INSIGHT-ORIENTED

According to many reparative therapists, homosexuality follows the paradigm of an emotional illness. Therefore the therapeutic task involves

uncovering developmental lags, fixations, and regressions, as well as early traumata, not only gender-based trauma, but also more general trauma arising out of relational problems with one's parents and one's society—basically the same traumata that can lead to the development of other psychological illnesses such as avoidant personality disorder. For example, for some reparative therapists homosexuality is chiefly the product of anxiety as the wish to be straight, e.g., masculine and assertive, comes up against shame over masculine assertiveness, leading to defensive, feminine passivity with its repression of forbidden, feared heterosexuality. Supposedly homosexuality is a phobia of heterosexuality. And that can be cured by working out how "this phobia came to be."

But the parallels between the creation of homosexuality and the construction of a mental illness such as a phobia, while there, are insubstantial. For homosexuality is almost entirely inborn, unlike many mental illnesses which are largely acquired. Homosexuality is largely biological, while neurosis is largely psychological. Being gay is itself a biological instinct not a manifestation of a conflict about one's true, biological urges.

Too often reparative therapists confound sexual orientation and gender identity. They thus view homosexuality as an outgrowth of gender identity discontinuity, and the cure for homosexuality as involving the insightful realigning of gender identity with gender. According to this view gay = sissy and lesbian = butch, while the remedy involves enhancing a man's masculinity and a woman's femininity by discussing the relevant deviations and coming up with ways to reverse them. But Fenichel notes that various "aspects of 'masculinity' or 'femininity' sometimes coincide in the same [normal] individual [and] vary independently of each other, so that there are very active masculine homosexuals and very passive [feminine] ones [and] impulses with active aims as well as with passive ones occur in both sexes." Additionally "what is called masculine and [what] feminine depends more on cultural and social factors than on biological ones."[3] Even so, reparative therapists mistakenly claim that treatment of male homosexuality involves becoming seen as a real man by other men (real men) as if feminine men and masculine women cannot be heterosexual, and masculine men and feminine women cannot be homosexual; as if one can influence masculinity and femininity directly and as if by doing so one will in turn influence sexual orientation.

EDUCATIVE

Reparative therapists educate gay men as to why they should be straight and as to how exactly they should accomplish this. But their educative

process generally derives from ideas that represent a negative view of gays and gay life created by selectively gleaning aspects of gay life according to the therapists' need to affirm their own negative preconceived theoretical notions about what being gay actually entails.

BIOLOGICAL

Reparative *psychiatrists* sometimes use suppressive medications, including antianxiety medications and antidepressants, in an attempt to quell sexual desire. Some prescribe masculinizing hormones like testosterone for gays (under the assumption that gays being sissies lack balls) or feminizing hormones for lesbians (under the assumption that lesbians being butch are flooded by strong masculinizing hormones)—this though in reality the effect hormones have on fantasy life is minimal to nonexistent. (Prescribing ECT and psychosurgery such as lobotomy for the gay man is of course ineffective, dangerous, illegal, and unthinkable.)

Eye movement therapy seems to rely simplistically upon the hypothesis that eyes are connected to the brain which is the seat of the emotions and so altering eye movement can alter the brain and that can alter the emotions that create, thus curing, homosexuality. To me this is, among other things, an example of the cognitive error that avers that since A leads to B, altering B can predictably walk back to affect/cure A. Thus "gay men cross their legs; uncrossing their legs makes gay men straight."

ECLECTIC

Therapeutic interventions from a variety of theoretical orientations are often employed simultaneously by a therapist who believes that using a wide range of interventions collaboratively is the best way to address a given patient's singular needs.

INPATIENT

Reparative therapists may recommend hospitalization, often doing so in seemingly patient-friendly terms, such as calling the hospital a "retreat" to describe what is in actuality a psychiatric inpatient setting, one where gay is presumed to be the illness and the goal of cure is becoming straight, almost no matter what it takes, and this can include isolating, drugging, and intimidating the patient into what amounts to "saying Uncle."

17

---◆❖◆---

Therapy for Children
and Adolescents

For the adolescent patient therapy should be directed as much to the parents as to the patient and should involve enlisting the parents' cooperation to help their child not to become straight but to become happy and well adjusted.

I believe that parents who refer their children for therapy ought always to have a few sessions with the child's therapist, with the discussion geared in large measure to exploring the parents' motivations for requesting (demanding) that their child be repaired.

Most parents need to be educated about homosexuality. They should be told that a young person's having same-sex experiences does not mean that he is, or she is not, a homosexual. A young person's sexual orientation is characterized not by being fixed but by fluidity and unpredictability—with shifting orientation accompanied by serious experimentation. Thus early sexual orientation is generally more whimsical than settled. Parents should be informed that because child/adolescent homosexual activities may or may not be early manifestations of what is to become a fixed sexual orientation, children and adolescents should generally not be prematurely offered, or forced into, intensive invasive interventional deep therapy for what may very well be a passing fancy. Instead, watchful waiting is often indicated; and this would be for another reason as well: since SOCE is rarely helpful anyway, "early intervention" SOCE is no more helpful than SOCE done later.

Adolescents will either grow out of, or into, their homosexuality or remain bisexual mostly depending on forces out of their, or anyone else's, control. Therefore I believe that attempting to influence the course of events "in a timely fashion" is at bottom beyond futile and is instead abusive, virtually a form of sanctioned bullying that is itself at the minimum counterproductive, if only because it can likely lead to rebelliousness, in turn leading to *enhanced* homosexuality as the adolescent, becoming defiant, makes a choice to do not what is therapeutically recommended but exactly the opposite: to defy the establishment by deliberately *not* doing precisely what others expect. If anything, my parents forcing me into therapy had an effect that was the reverse of what they intended: it fixed my homosexuality, once and for all.

Therefore, it's often indicated for all concerned to go into a holding pattern where everyone presides over the situation awaiting clarity. This hands-off approach should especially include a tolerant wait-and-see attitude toward stereotypically defined gender behavior, one that eschews attempting to influence a child in the direction of thwarting "gender inappropriate" and catalyzing socially generated and approved gender stereotypical behavior.

Conversely, *supporting* a child's/adolescent's homosexual tendencies can by itself also be potentially harmful, creating complications that would not have existed if all concerned had just let things go and settle out of their own accord.

Attempts to treat me for being a gay youth had no positive effect on the development of my adult sexual orientation. Forcing me to conform to what my parents wanted for me was an attempt on their part to get me to live their lives for them by getting me to conform to what they believed were, as they saw it, appropriate gender-related behaviors and thus the sexual orientation they desired for me. This only made me uncomfortable, stubbornly resistant, and even defiantly oppositional, causing me to become gayer than before and also generally less attuned to them than ever, and even somewhat unattuned to them forever. Their bullying also made me feel defective because I felt guilty as charged: regretful both for being broken and for not being able to fix myself (as they wanted me to). Forcing me into a mold which didn't fit made me feel so alien to myself that I developed a serious self-image problem, as I thought, "Mom and Dad know what's normal, and it's not me." To me, my parents jumping me and insisting I see a psychiatrist to change meant several things: that they didn't like me as I was, that they were trying to control me to force me to become someone else, and that they would hurt me if I didn't comply by taking away my tuition money, by rejecting me totally, or by disowning me

entirely. This so scared me that for years afterward I felt I might not be able to survive and would suffer permanent emotional damage unless I did what they wanted.

One psychiatrist they made me speak to told me he could make me change, but I didn't want to (and didn't believe him, given how ingrained my homosexuality was). But he didn't care what I wanted, or believed. Instead he kept insisting I come in for my reparative sessions. When I resisted, this further alienated me from him, and ultimately from my parents.

All parents wrong their children who, however, go on nevertheless to forgive them. But my parents had inflicted such a large wound upon me that it seemed as if it would never heal. Also I could now reasonably rationalize all aspects of my adolescent rebellion as being reasonable because "they provoked me," and so, having been provoked, I could freely rebel without concern for whether or not I was being rational and how they might feel or react as a consequence of any negative behavior on my part.

As a result of all this, later in life I constantly struggled with a personal depression. The consequence of my having been thusly abused as a child was that I never felt okay as is, but always felt that I must keep trying to change if I wanted to get even a modicum of parental love and support. Not surprisingly, I was to enter a period of compulsive homosexual sexuality where feeling unloved at home I sought love from strangers, as I simultaneously rebelled against my parents using my sexuality both as a shield to protect me from, and as a sword to get back at, those who wronged me.

In conclusion, I believe that with few exceptions all patients who feel seriously at odds with their sexuality should be offered treatment—not for their sexual orientation, but rather for the distress they might feel about it. Reparative therapists often have considerable talents, but they should use their talents and abilities to treat young patients who apply for or are made to go into SOCE much as other, nonreparative therapists might treat young patients that cross their paths: not for their homosexuality but for emotional responses/disorders that are the cause of their being distressed about being homosexual.

18

❖

Affirmative Therapy

INTRODUCTION

In this chapter I discuss affirmative therapy and distinguish it from other forms of therapy for gays, especially reparative therapy. Affirmative therapy is a mainstream psychiatric technique that many orthodox therapists recommend for gay patients who feel that they need help—not, however, to go straight, but rather to solve emotional problems either related or unrelated to their being gay.

Therapists use the term "affirmative therapy" in several senses. They use it to contrast this form of therapy with insight-oriented therapy. In insight-oriented therapy a goal is self-understanding leading to self-acceptance, while in affirmative therapy the goal of self-acceptance through insight becomes problematical because gay patients, already burdened historically with being deaffirmed, tend to view attempts to impart insight not as "this is how you are, and what makes you that way" but as "this is why you are bad, shouldn't be that way, and so need to change"— making insight therapy, according to some, a potentially misguided approach for gays that can discomfort them as much as it can help them become more comfortable with themselves.

Both insight-oriented and affirmative therapy stand in contrast to reparative therapy, which starts with the premise that "you are unacceptable in your present form" and then attempts to induce not comfort but discomfort—with oneself and with one's homosexuality—doing so in order

to facilitate not self-acceptance, but radical (sexual orientation) self-change triggered by self-questioning and self-deaffirmation.

Some, and perhaps many, affirmative therapists, go beyond merely conveying positivity toward their patients to also employ, albeit in a limited way, and selectively, mainstream therapeutic modalities with the goal of helping gay patients find a definitive solution to their emotional problems/emotional disorder. They aim to reduce gays' anxiety, soften their guilt, and help them better cope using a combination of techniques (an eclectic approach) in order to improve their patients' mental state by improving their patients' relational, occupational, and social functionality. For example, many affirmative therapists add insight-oriented therapeutic techniques selectively, and cautiously, to their armamentarium, imparting insight while trying to make it clear that they are doing so not to be critical but to be constructive. The insight they attempt to impart is, however, only that which is applicable to a specific problem at hand—not an aspect of some general vague objective, such as "to understand yourself," or beside the point, as it can be when it is being used inappropriately to cure homosexuality. Of course, the goal of affirmative insight would not be to change same-sex sexual desire and orientation but to help reduce or eliminate coexisting neurotic symptoms, such as phobias and obsessions, related, or unrelated, to a patient's being gay. Because eclectic affirmative therapists ultimately belong to different schools of thought they have different primary theoretical persuasions, and for that reason they necessarily supplement their affirmation with different therapeutic methods, ranging among insight-oriented, cognitive-behavioral, and interpersonal approaches. Their goal however remains the same: to most effectively treat emotionally troubled gays with all the therapeutic tools at their disposal, the better to help their patients not to go straight, but to be gay in a more salutary way.

As a start, affirmative therapists attempt to convince their gay patients that, generally speaking, those homosexuals who are gay down to their bones, that is, not bisexual, can almost never change over to become heterosexuals. They make it clear that "if I offer you change I will ultimately likely have to deal with your disappointment that things will always remain the same." They state with certainty, upfront, that because change is not a viable possibility it's best not to waste one's time trying to bring it about, for that will only mean having to get over the disappointment of being promised more (that one can be made straight) than can be delivered. For even gays who badly want to go straight will most likely always remain gay, so that what they need to do is to accept that right now, and get over it immediately, and, hopefully, once and for all.

Affirmative therapists view their gay patients' wish to change as itself potentially suspect. They see it as likely the product of fear, guilt, lack of self-respect, and failed self-compassion, a consequence of outside parental and social negative pressures, due to a lack of knowledge about what being gay can and does involve, and the spin-off of a masochistic need to deny how good things are in favor of planted false hope about how much better they could be. Gay men who claim they *wish* to change so often in actuality *fear* what they already are. So, affirmative therapists instead of accepting at face value their patients' wish to change, seek to understand their patients' inability to stay the same. They may attempt to understand this in terms of conflict between who one is and what one feels one should be. That is, the focus is on the self divided, often a self rent asunder by self-questioning and self-punitiveness associated with self-destructive negative goal-directed behaviors as gays deliberately inflict psychological pain upon themselves to remonstrate for being gay, and court self-defeat to punish themselves for being homosexual. In this regard Reich expressed what many gay men need to do (although in another context) when in his book *Character Analysis* he said: "This is our great obligation: to enable the human animal to accept nature within himself, to stop running away from it and to enjoy what now he dreads so much."[1]

This said, much of what we call affirmative therapy *is* primarily focused on giving patients positive feedback. Affirmative therapists make it clear that they see their gay patients, even, or especially, ones with emotional problems, as overall inherently good. According to affirmative therapists, if their gay patients suffer from feeling bad about themselves, they mainly do so not because they *are* bad but because they are instead unable to accept their goodness, and treat themselves accordingly. Along with their reparative therapists, these gays *themselves* view themselves as broken. They remain convinced that being homosexual means that they are suffering from an illness, and even possibly going beyond ill are instead sinners, or actual criminals.

I advise all patients seeking SOCE to first consult with an affirmative therapist before committing themselves to a course of reparative therapy—and to discuss if their wish to undergo reparative therapy is a rational one, or if it is a product of an irrational need, such as a need for self-suppression or self-abnegation. My hope is that the affirmative therapist will inform gay patients thinking about attempting to go straight that they will likely work hard and suffer a lot in therapy only to have nothing to show for it in the end but a continuance of sexual desire as before in spite of anything they do to make alterations to their sexuality—predictably leaving them feeling depressed and unfilled. Reparative therapy perhaps will cause them

to become alcoholic and drug addicted in order to deal with their lack of change, as well as to reduce the pain of having disavowed who they are. And now they will feel not only morally bankrupt, sinful, criminal, socially unacceptable, and inherently pathological—but also like patsies for having wasted so much time and energy trying to do something so inherently undoable, and attempting to accomplish something so inherently unworthwhile as converting to heterosexuality.

SPECIFIC CONTRASTS BETWEEN REPARATIVE AND AFFIRMATIVE THERAPISTS

Reparative therapists believe that homosexuality is a regression from heterosexuality. They then encourage their patients to become straight by becoming less regressed through becoming more mature by comprehending the fears associated with growing up personally and hence sexually. They advocate this growing up through resolving applicable conflicts within oneself and between oneself and other people, as well as between oneself and one's society, community, and church. They advocate this however much these conflicts, though they may have something to do with personality development, have little to do with the development of homosexuality.

Seeing homosexuality as a symptom of another, more fundamental disorder such as depression, or as a disorder in and of itself (e.g., the "illness homosexuality") they attempt to reverse the "disease of being homosexual" with its "symptom of homosexuality" by using talk therapy. This talk therapy generally involves going back in time to reveal the origins of one's same-sex attraction, hoping and searching for that "aha" moment that will, leading to just the right curative revelation, break up the homosexual structure and allow one's natural heterosexuality to break through.

Reparative therapists in effect buy into what Reich (if he were aligned with today's reparative therapists) might have said about homosexuality: that it is the outcome of a failure to develop into a genital character, with homosexuality a regressive form of heterosexuality, a sliding back into pregenitality, a kind of primitive manifestation that is the result of failed genitality, thus constituting a regressive disorganized heterosexuality. Thus:

> The genital character has fully reached the post-ambivalent genital stage . . . the wish for incest and the wish to eliminate the father [or mother] have been given up, the genital interests have been transferred to a heterosexual object which does not, as in the case of the

neurotic character, actually represent the incest object but has taken its place. *The Oedipus complex no longer exists in actuality*; it is not repressed, but free of cathexis [e.g., is no longer energized]. Pregenital tendencies, such as anality, oral eroticism, voyeurism, etc., are not repressed but are partly anchored in cultural sublimations, and partly gratified directly in the forepleasure acts; at any rate, they are subordinated to genitality. The sexual act is the most important sexual goal and that which provides the greatest pleasure. . . . The fewer pregenital demands [that] are repressed, that is, the more the two systems of pregenitality and genitality communicate with each other, the more complete is the satisfaction and the less the pathologic stasis of libido.

The neurotic character [for example, the homosexual], on the other hand, is incapable of orgastic discharge of his free, unsublimated libido . . . . He is always more or less orgastically impotent, for the following reasons: The incestuous objects have an actual cathexis or the corresponding libido is consumed in reaction formations. If there is any sexual life at all, its infantile nature can be readily seen: the woman represents the mother or sister and the love relationship carries the stamp of all the anxieties, inhibitions and neurotic peculiarities of the infantile incest relationship. Genital primacy is either not established or, as in the hysterical character, genital functioning is disturbed by the incest fixation. There is either abstinence, or sexual activity is largely confined to forepleasure acts.[2]

Reparative therapists along with Reich assess homosexuality as "mere forepleasure." And like other mental illnesses they see it as being built out of a fixation on this forepleasure. In general for them homosexuality, causally speaking, involves a developmental lag—an arrest (fixation upon), and/or a backsliding (regression) to specified fixation points, making the homosexual act itself pregenital, that is, immature and regressive.

Interpersonally speaking, the homosexual symptom/illness is supposedly created out of the same interpersonal forces that cause hysteria or depression—in particular: parenting that lacks depth and sensitivity and/or is aggressive, abusive, and nonaffirmative. Patients now continue their interpersonal struggles on the level on which they have originally been traumatized, so that gay men fear all women because they represent the castrative mother; or they never leave the mother's orbit either because she has overgratified them or was too punitive to allow them to mature into normal beings. Or, failing to identify with a distant flawed father,

they cannot become desirably manly. For reparative therapists, reasons like this mean homosexuality is the proper target of curative interpersonal insight whose goal is symptom removal leading to "normal heterosexuality" by releasing the individual from a mother trauma, or a father fear, facilitating this detachment through understanding, allowing for the re-creation anew of healthier identifications. In the man a new healthier identification would involve the following: after the man reconsiders his earlier view of his father as weak and ineffective, he instead newly comes to see his father as more manly, thence helping the gay man become a real man, a he-man, at last the full-blooded heterosexual, the one which he himself aspires to be.

In contrast, *affirmative* therapists do not claim that while heterosexuality is maturity (genitality), homosexuality is a regression from heterosexuality (genitality) to homosexuality (pregenitality). Partly as a consequence of their disbelieving this, they also disbelieve that they can change gays through psychologically oriented talking approaches that have in common the idea that "gays who don't understand their personal history are doomed to stay gay." Affirmative therapists who are psychoanalytically oriented do not believe or claim that psychoanalytic approaches can help the homosexual become heterosexual. They do, however, apply the psychoanalytic approach selectively for other purposes: to cure neuroses, but never to cure homosexuality.

In denying that homosexuality is psychologically pathological any more than is being left-handed or red-headed, affirmative therapists avoid thinking and saying that "since heterosexuality is a mature form of expression of sexuality, therefore homosexuality must = an immature or regressive form of sexuality"—just as hysteria is an immature fixation upon/regression to a boy's Oedipal conflict, and OCD is an immature fixation upon/regression to anality. In short, while affirmative therapists accept the existence of psychodynamics, they deny the relevance of psychodynamic explanations in this particular realm—that is, they don't see psychodynamics as pertaining to the development of homosexuality, and insight into one's psychodynamics as potentially curative of one's gayness.

So, instead of focusing on imparting insight, affirmative therapists recommend a supportive therapeutic approach. For them, insight into core problems doesn't magically move the patient from being developmentally arrested and so from being gay, to being developmentally mature, and so on to becoming straight. Instead, when treatment is at all indicated, they offer a corrective positive here-and-now emotional experience, one that takes place between therapist and patient—not one that is reliant on healing intellectual or emotional uncovering, but one that is reliant on a

healing patient-therapist resonant interaction—a new affirming relationship with real qualities (short of crossing boundaries) to fill in what is currently missing, and to present a new, more positive remedial, healing alternative to what has come before.

Reparative therapists believe that homosexuality is acquired. They cite how no certain biological/genetic causes of homosexuality have ever been found, and that biological/genetic findings that claim significant validity, e.g., the assertion that the brain is different in gays and straights, contain little truth. They conclude that, not being inborn, homosexuality has no biological basis and therefore, being of psychological origin, is not immutable, but can, like any other psychological symptom, be changed through talk therapy.

Reparative therapists inform gays that, furthermore, being gay is psychologically speaking a choice. They say this although saying so has the effect of enhancing gay patients' self-blame via causing them to view themselves as being "entirely responsible for what has happened to them." Many reparative therapists form consensuses with colleagues who *want* being homosexual to be a "psychological disorder" so that with one voice they can claim to reverse it through talk therapy.

In contrast, *affirmative* therapists generally believe and inform gays that they were born that way, so that the "choice" of being gay was made for them. They tell gays that because homosexuality is inborn and gays are born gay, there is little to no sense in trying to change one's sexual orientation, and certainly little hope that talk therapy will have any "remedial" effect upon something so ingrained.

This said, there is a certain amount of hedging of bets here. Reparative therapists, though they see homosexuality as a purely psychological developmental problem, nevertheless by implication include biological factors in what they perceive to be its causation when they aver that temperament (which is inborn) is a causative factor (e.g., because "temperament sensitizes patients to respond to intrapersonal and interpersonal insults"). Also, reparative therapists focus on analyzing deep feelings because feelings are unique in that they, like the homosexuality to whose understanding feelings open the door and contribute, consist of a *somatic* (likely inborn) as well as a *mindful* (likely acquired) component. For their part, affirmative therapists, though they see homosexuality as inborn, sometimes hint, if only in a roundabout way, that psychological factors do play a role in its causation. Thus they note that in some identical twins only one of the twins is gay, a phenomenon that they account for by developmental (that is, nongenetic) factors, especially those involving interactions within the family orbit.

After searching the literature I find few indications that *reparative* therapists use *cognitive* explanations/interventions to attempt to understand the causes of homosexuality and cure it by normalizing pathological cognitions—an approach that, for poorly understood reasons, doesn't seem to be commonly employed by this group of therapists.

In contrast, *affirmative* therapists often use cognitive approaches to help gay patients think better of themselves. The cognitive affirmative therapeutic approach cannot help gay men turn straight, but it can help them rethink how they view their being gay, as they reframe what it is about being gay that is bothering them, and why they feel compelled to criticize, disown, deaffirm, and devalue themselves (unnecessarily) just because they are homosexual. For example, many gay patients can benefit from rethinking the hypothesis that heterosexual = worthy and homosexual = devalued, and from no longer thinking along the all-or-none lines of "not all good = no good at all" that lead them to develop completely negative out of partially positive self-views.

Affirmative therapists often profitably employ cognitive-behavioral techniques to help gay individuals who are depressed. They do this by helping them identify, understand, and correct depressive cognitions especially the typical depressed gay individual's tendency to feel devalued due to an excess of self-criticism modeled upon, and so an extension of, the homonegative criticisms coming in at them from others, ranging from peers to reparative therapists themselves

Positivity/support can be a feature of *reparative* therapy, but only so long as the therapeutic relationship is not predominantly focused on negative/adversarial issues—as too often happens when a reparative therapist attempts to convince gays who are unwilling to change that their fate depends on their going straight and that the reason for going straight is that, being broken, they need to be fixed. Mostly however reparative therapists do not set out to be supportive. Generally they choose to be aggressive, feeling that what they do has to involve being sufficiently active for them to get their patients to once and for all "cut out being gay."

In contrast, *affirmative* therapists are calculatedly supportive and are especially so to give those gays who feel lost and alone due to an absence of social support something that at least seems to resemble, and so to replace, what as patients they feel they are missing. Gays often benefit from the proffered affirmative positive therapeutic relationship itself as much as they benefit from anything specific that might be said during the therapy hour.

Supportive affirmative therapy is particularly helpful for gays grieving about being gay. It can help them better accept their sexuality through

becoming not only less self-judgmental but also less loss-oriented. While reparative therapists deal with gay grief by working on helping gays become straight so that they have less, or nothing at all, to grieve about, affirmative therapists choose to help gays stay comfortable by grieving for, and once and for all getting over, any lost opportunities or hard times they feel may be the result of their being gay. They can help gays get past the grief of not having a standard straight marriage with children and being victims of the professional limitations, which still exist for gays in today's world. They can help those gays grieving over not being straight by pointing out the real benefits of being gay, including the presence of certain personal freedoms and life-choice possibilities not as readily available to men and women who are straight, as exemplified by the compromises straight men and women must make due to the inevitable (stricter for straights) confinements of heterosexual marriage. Affirmative therapists focus on putting the gay life's admitted disappointing aspects, such as less than full social acceptance, into perspective. They do this so that gays can work around the gay life's admitted limitations, and, after getting over not having things that they will never acquire, start experiencing enjoyment from those things that they can and do possess now, are likely to get in the future, and are destined to retain forever if only they will allow themselves to.

Reparative therapists tend to be homophobic and rationalize their homonegativity as provoked and deserved. In contrast, *affirmative* therapists identify reparative therapists' gay negative responses as symptomatic. This helps gays view homophobia as "It's them, not me," as they view homophobia, not gayness, as that which is emotionally troubled. In turn, that helps gays enhance their positive self-view by being able to counter the negative stereotypes they tend to have about themselves, and others have about them.

Reparative therapists focus almost entirely on gays' sexuality and, overlooking any general problems gays might have, fail to deal with anything more than changing one's sexual orientation. Viewing homosexuality as the central, if not the only, issue in the gay person's life; all-determinative of the gay individual's character; and to blame for all the gay person's problems, blaming homosexuality for everything (bad) that happens to gays; they allow neurotic distress of a more generalized nature to go unidentified and unchallenged and so to persist, as they don't try to help their patients live happier more rewarding and more successful lives by helping them solve those neurotic problems that are unrelated to their homosexuality. Yet many gay patients apply for therapy, even reparative therapy, for the same reasons that all patients do: such as being anxious or

fearful—only to quit when they hear that their suffering is entirely due to their being homosexual, so that to be happier they need to become not ex-neurotic but ex-gay.

In contrast, *affirmative* therapists get beyond sexuality to deal with non-sexual issues, for example, those related as much to personality organization as to sexual orientation. Affirmative therapists put their patients' sexuality into perspective as being only one part of the individual, with gay sexuality an aspect of being gay rather than its full equivalent. For affirmative therapists, interpersonal intercourse is just as important, meaningful, desirable, and pleasurable as the sexual kind.

Affirmative therapists recognize that homosexuals have the same problems heterosexuals have, and in about the same proportions, with many of their problems essentially unrelated to sexuality, gay or straight. Thus gay individuals disrupt loving relationships not because "gays can't maintain relationships with a significant other" with "infidelity in gays in character, and fidelity chimerical" but because some gays, like some straights, suffer from an avoidant personality disorder and experience undue interpersonal anxiety, not because they are gay but because (in a gay context) they are avoidant. Affirmative therapists help their patients handle anxiety, strengthen defenses, deal with low self-esteem, integrate anger with love, and gratify healthy and reduce unhealthy dependency. This is not because these things cause or are exclusively the result of homosexuality. It is because these matters are significant in their own right—in gay, as in straight, settings.

Reparative therapists emphasize that stereotypical gender role dysphoria is at the core of/the cause of homosexual deviancy. Convinced that homosexuality is somehow related, at least in part, to gender dysphoria, they anticipate that homosexuality will disappear and that heterosexuality will appear and supervene if the patient becomes more gender attuned, and so less gender dysphoric.

Many reparative therapists believe that all gays are effeminate and all lesbians are masculine and that forcing oneself to act more appropriately (in a stereotypical gender-defined and approved way) will result in gays becoming straight. This, a form of concrete/magical thinking, in effect states that by altering outcome one can undo cause. ("Being a gay man leads you to become sissified; acting less sissified/more masculine walks back to making you straight.") There is no one gender identity associated with homosexuality; the reparative therapists' convictions of what a man or woman *should* be like are stereotypical; and making corrections in what amount to idiosyncratic gender diversities will have no effect whatsoever on the homosexuality these may, or as likely may not, be associated with.

In contrast, *affirmative* therapists attempt to help gay patients feel more comfortable with their gender identity as it now stands and whatever it happens to be. They do not stereotype men and women, and men versus women, e.g., they do not differentiate manly from unmanly behaviors, or pathologize certain gender-based behaviors as unhealthily manly or unmanly, then demand conformity. They not only don't relate gender identity to sexual orientation, they also don't attempt to alter specific supposedly deviant gender identities/behaviors directly, as if by doing so they can thereby walk back to alter homosexual sexual orientation itself.

Affirmative therapists encourage free gender identity expression in the belief that it is not unique gender identity itself but "guilt about deviant gender identity" and the accompanying social disapproval and isolation that constitute the problem. While affirmative therapists draw no causitive connections between gender identity and homosexuality, they do help their patients live a better, more depression-free life by helping them deal with the gender-based social disapproval so commonly associated with being gay—by not yielding to it emotionally, and so by not responding to it either affectively or behaviorally.

Reparative therapists utilize behavioral modification techniques to induce change. Some of these reparative behavioral methods are primitive to bizarre in nature. Thus claiming that "you are what other people see you as," they accept that calling a gay man "dude" will have a startling palliative, or even curative, effect on homosexuality, somehow, and, all by itself miraculously causing a gay man to change over to becoming straight.

The behavioral modification techniques reparative therapists use are of both a nonaversive and an aversive nature. Nonaversive approaches include biofeedback and hypnosis. Aversive approaches include application of uncomfortable or painful stimuli when a patient is fantasizing about sex, e.g., stimulated by pornography, with patient response measured by various methods including the assessment of penile circumference via the use of the penile plethysmograph. They also use *aversion relief* therapy which relies upon stopping uncomfortable or painful stimuli. Other behavioral techniques reparative therapists use are "covert sensitization, systematic desensitization, shame aversion, orgasmic reconditioning, and satiation therapy [as well as] reframing desire [and] redirecting thoughts." They also use semibehavioral educational interventions such as enhancing "dating skills, assertiveness, and affection training" for homosexuals.[3]

They use these approaches because they view homosexuality as a pathological (and therefore potentially remediable) *behavior* (rather than as a primary instinct), and because their (jerry-built) research protocols have

convinced them that many of the behavioral interventions they use are not superficial or trivial but highly effective, that is, no matter how superficial they may seem to some, they nevertheless produce radical curative effects upon many.

To make these approaches sound more scientific than they happen to be they almost always attempt to construct a believable theoretical framework for what they do. They postulate specific underlying behaviorally reversible psychopathology to explain homosexuality, e.g., "homosexuality is a phobia of heterosexuality," then apply valid behavioral techniques that mainstream therapy uses (to treat phobias) in order to treat "homosexuality, the phobia." So they cure homosexuality by facing it down, that is, advocating having hetero sex, much as traditional therapy faces down a phobia of flying by graduated exposure to flying, and/or uses a comforting (human or animal) companion at one's side during flight to reduce anxiety and enable a given behavior.

While many reparative therapists no longer admit to performing seriously invasive behavioral approaches such as "inducing nausea, vomiting or paralysis [or] providing electric shocks in association with a homosexual stimulus to render that stimulus 'unerotic,'" or even admit to using trivial noninvasive approaches such as "snap[ping] an elastic band around the wrist when the individual became aroused to same-sex erotic images or thoughts,"[4] I suspect that some rogue reparative therapists are still doing some such insupportable things. For example, van den Aardweg's fairly recently offering consists of advancing the to me groan-inducing recommendation that one might use curative "physical mortification" therapeutically.[5]

In contrast, *affirmative* therapists believe that invasive behavioral procedures such as inducing nausea in response to homosexual arousal exemplify cruelty/malpractice by deed (too harsh), while less invasive approaches are not very helpful/ineffectual due to being too trivial. They also believe that in a general way behavioral interventions force patients to relinquish individual freedom. They do agree, however, that behavioral therapy can at least temporarily alter not the quality of one's sexual orientation but the *quantitative* expression of one's sexuality. It can do this through redirecting emphasis—for example, from one's sexuality to one's work. Still, most times even these approaches either work poorly or mostly lead to symptom substitution as patients replace one supposedly pathological behavior, such as homosexuality, with another, actually pathological one, such as alcoholism.

Reparative therapists frequently rely on the use of religious approaches/spirituality. Their image of God is of One who condemns homosexuality

as sinful because it's against what He the Almighty wants, expects, and at times even demands, thus in effect placing God in a central position in the reparative process. Therefore reparative therapists help gays become more religious, as they urge gay patients not to attempt to mesh their sexuality with their religious beliefs, but to simply accept the Bible's view of gays as an abomination (contrasting with the more hopeful, more therapeutic view that gays, though different, are as hale, hearty, and healthy as straights), and so on that basis alone must relinquish their "abominable" gay behavior.

Some reparative therapists attempt to force their patients to accept the therapist's own religiously infused ideals, and to do so in a fashion that is too rigid to leave room for "the compromise" that in reality must take place if one is to mesh the irreconcilable facts of one's homosexuality with the tenets of one's religiosity. They tend to also focus on religious issues at the expense of working on more general, often more troublesome, issues of real life. In doing so they give what are essentially traditional psychotherapeutic interventions a religious moniker. To illustrate, they call the technique of exhortation "prayerfulness" and positive thinking "actively performing gratitudes."

In contrast, *affirmative* therapists help gay men focus on their lives as well as on their religion. They encourage their patients to make their own ethical and moral judgments about, and so moral and ethical choices in, life, and to do so without blindly following the ethical/moral choices a reparative therapist makes for them. Overall, they suggest that their patients stop brooding about religious/ethical/moral matters so much and start living instead. They urge their patients to avoid focusing on imponderables in order to resolve their uncertainties—to stop trying to come up with answers to mostly unanswerable questions such as "If God made me gay, how can He at the same time hate gays?" (Such a question, though it remains valid in and of itself, is not easily answered, in part because it refers to hidden and generally unacknowledged deeper issues such as the obsessive self-ambivalence that consists of self-acceptance on the one hand—"because being gay is God's will it's not my fault"— and self-abnegation on the other hand—"I am bad therefore I deserve to be punished, and by a God who has the same thoughts about my true need for punishment as do I.")

Reparative therapists almost always ask their gay patients to hold immoderate views of themselves. Since they are not all good then they are presumably all bad. They thus affirm gays' tendency to deaffirm themselves *completely* because of a *few* things they don't like about themselves—perhaps maintaining a *full* sense of guilt over what are (often

merely perceived) *partial* deviations from perfection, thus enhancing rather than softening their patients' depressive self-view by encouraging their patients to fail to integrate negative self-appraisals into an overall positive self-assessment involving their being at most bent, but by no means completely broken. In this way reparative therapists embrace LGBTs' own sometimes pessimistic view of themselves created by generalizing from those aspects of themselves gays do not admire after suppressing their favorable aspects following the illogic involved in consciously formulating a negative view of self by making the partial picture into the whole picture; and the baneful side of their lives into the essence of who they are and how they live, leading to a thoroughly depressed un-loving self-view, and so self-identity, which is the problem that so often leads gays to eagerly apply for change.

In contrast, *affirmative* therapists ask their gay patients to create a composite view of themselves as neither all good nor all bad but as all human, synthesizing their positive and negative, appealing and unappealing qualities into a whole, more wholesome overall self-accepting self-view. Affirmative therapists do not subscribe to or encourage the development of the "some of the self = all of the self" self- and world-view. They discourage gays from such all-or-none thinking because they feel that such thinking can but keep them from getting full satisfaction out of life due to failing to neutralize life's dissatisfactions with the satisfaction of knowing that as gays they are not all bad because they are not all good, which means that they don't have to pay a certain price for everything they want and get.

Reparative therapists tend to be one-size-fits-all-oriented workers who attempt to help all gays without first evaluating their personal differences and individual needs, instead demanding gays be true not to their differing selves but to some standardized view that their therapists hold of them, and then change along predetermined lines. In contrast, *affirmative* therapists take a more nuanced view where they see gays as all different individuals, only incompletely united in their homosexuality while disunited in their diversity.

Reparative therapists take gays' currently avowed stances as both unalloyed and paramount. They fail to recognize that gays, like everyone else, have mixed feelings about who they are and ought to be, with many "me's" inside of one individual all of which have to be taken into account when determining who that individual is—me's that jostle with one another for recognition and expression, all hoping, often futilely, to be incorporated into an overall superordinate "self" or "identity" forged out of reconciling extant ambivalences. *Affirmative* therapists recognize that most gay

patients have mixed feelings about who and what they are. They have both a masculine and a feminine side as well as an aggressive and a passive side, and a compliant and a rebellious side, and that these different sides can ideally coexist in inner harmony where self-expression and individuality don't coordinate with self-control and social sensitivity; where activism is modified by assimilation, and the need to challenge society is modified by cooperating with the establishment; where taking risks is modified by remaining safely (and productively) behind in at least some shadows; where close-fisted aggression is modified by open-handed abdication; where firm individuality is modified by compliance; and where excessive rebelliousness is modified by sensible submissiveness. For affirmative therapists every gay person is a complex individual hardly constituted as a *completely* straight person inside of a *completely* gay one with that straight person longing to come out, the real me inside that would emerge, if only given a little push, and half a chance.

Reparative therapists often expect adolescent patients to secure their identity (no changing) once and for all even during their formative years, as if in adolescence things don't move so quickly that loyalty to any one self is, as it should be, ephemeral.

In contrast *affirmative* therapists feel that attempts at final self-realization are premature in adolescents and so can lead to more rather than to less inauthenticity. Affirmative therapists understand that gay adolescents often want to be this because they fear being that, so that any firm, fixed identity that exists in youth is generally less a reflection of true desire than the product of anxiety and guilt, not what "I prefer to be" but "what I fear actually becoming," with rebellion and submission *conflicts* in ascendency and especially prone to keeping adolescents from any sort of conflict-free, thorough self-realization.

When *reparative* therapists refer a patient for group therapy (the group can be an informal support group without a trained leader, or a formal group with a leader trained in doing group therapy) their goal is to help them alter their homosexuality. So they generally refer them to a group that will straightjacket the patients into doing some sort of reparative bidding. Reparative therapists know that groups can at times wield more reparative power through strength in numbers than a single therapist could ever hope to muster.

In contract, *affirmative* therapists' objective in referring patients to group therapy is to help them deal with a wide range of secondary emotional problems such as homophobic abuses on the part of others that enhance the gay individual's own failure to self-affirm. They hope that their patients will get any help they need through group consensual

validation, the sharing of experiences, learning from and within the group, and developing real relationships within the group to reduce loneliness and relieve isolation. *Affirmative* group therapy that is *supportive* is especially potent for helping gay patients develop an identity out of shared pride that can serve to protect them from the all-around negativity that so many gays run across in their lives, virtually on a daily basis.

Reparative therapists are often in secret collaboration with their patients' parents—in the common cause of making gay patients straight. In contrast, with *affirmative* therapists if there is any alliance at all with the parents it involves not a contract to make the patient straight but a compact to make the patient healthy. Thus the affirmative therapist might deal directly with the parents, but only to inform them in a general way about the patient's progress and prognosis, often limiting the parents' right to know to whether the patient is still going for treatment and continuing to benefit from it. Parents will often claim, "I am paying the bill therefore I am entitled to progress reports." But the proper therapeutic response to that is that "you are paying for good therapy but, with few exceptions, good therapy involves confidentiality, that is, keeping a patient's secrets, even though that means shutting you, the parents, out so that you cannot unduly influence/control how things go in treatment."

The affirmative therapist might have a few sessions with the family to keep the lines of communication open, reduce misunderstanding, enhance awareness of what all concerned are currently thinking, and tackle counterproductive thinking and behavior. But the affirmative therapist will not use these sessions to push an ex-gay agenda. Affirmative therapists will not inform the parents what they might do to make their child over, only what they might do to make their child better.

A THIRTEEN-STEP GUIDE: SOME WAYS REPARATIVE THERAPISTS CAN BECOME MORE EFFECTIVE

One. Emphasize the patient's concerns, needs, and goals over those of the therapist.

Two. Encourage patients to put unrealistic ambitions aside in favor of pursuing realistic objectives. There is little to no likelihood of patients who are not bisexual changing their sexual orientation from gay to straight. For a deep-seated homosexual who *must* see definitive change, the most feasible (though not necessarily the most desirable) courses involve not changing one's fundamental sexual orientation but becoming sexually less active, celibate, or living *as if* one is a heterosexual, such as assuming the role of being part of a heterosexual couple, such as a couple

made up of that proverbial gay man with an understanding wife who accepts homosexual activity within the course of the marital relationship, allowing the gay man to present a heterosexual face to an antihomosexual world, and to have and raise children. Validate a patient's wish to change in ways other than those involving sexual orientation, that is, by becoming not less of a gay person but more of an effective individual.

Three. Increase the patients' positive sense of self and the esteem in which they hold themselves by helping them seek *conditional* (not unconditional) self-acceptance, that is, to seek self-acceptance even though there remains room for improvement. Help patients lyse rigid homonegative self-attitudes such as the feeling that being gay necessarily means being less of a person, a half-man, or a defective woman.

Four. Recognize that one cannot fully resolve conflicts between homosexuality and religion; for one will always abrade on the other.

Five. Reduce post-traumatic reactions especially where serious ongoing preoccupying resentment exists about mistreatment past and/or present. In the case of how individuals were, or feel that they were, mistreated in the past, the task is to help patients recognize that the present is different from the past. In the case of how patients are, or feel that they are, being mistreated in the present, the task is to help patients sort out actuality from imagination and then do what they can to in reality improve their circumstances.

Six. Help patients manage grief responses about what can probably never be, such as full social acceptance, or a storybook heterosexual marriage. Dispassionately explore options for compromise/renunciation even if these involve making the sacrifices that compromise and renunciation generally require.

Seven. Treat patients' compulsive sexuality. Reparative therapists can treat compulsive homosexuality effectively by helping their patients become not less of, but a better homosexual—through becoming less of a slave to homosexual enactments, so that they remain gay without being frantic about it. Reparative therapists who also respond *affirmatively* help gays enhance their self-respect and reduce shame thus enhancing their relational capacities and so reducing pressure to constantly find what they feel that they are always missing. When compulsive sexuality is a symptom of a mood disorder, treating that mood disorder might help patients settle down after giving up their compulsive cruising/promiscuity, in which some gays indulge in order to feel less isolated and more alive, and so less depressed through the mechanism of seeking sex as a substitute for approval and acceptance from others in order to substitute for the self-approval and self-acceptance they lack within themselves. Patients

suffering from compulsive sexuality because they seek the love of same-sex partners to suppress their hatred of individuals of the same sex can benefit from insight into how same-sex compulsions can be the product of such reaction formation. Many patients can be helped to reduce compulsive sexuality by focusing on the patterns of the rise and fall of sexual desire (which typically comes in waves) by identifying its triggers.

Eight. Try environmental manipulation to help patients improve their surroundings and thus their lives by leaving a harsh environment behind, such as a too-rigid church or an unwelcoming place in which they live. Patients who cannot make a healthy move can be helped to accept and cope with the negativity in their present community.

Nine. Establish a supportive substitute relationship with patients who need someone to tide them over during times of acute stress and loss. Even within the reparative schema reparative therapists can offer their patients a positive relational experience by giving them noncorrective positive feedback, such as by telling them that they are not completely broken but are overall intact—and avoiding destructive criticism which inappropriately focuses on liabilities instead of assets.

Ten. Avoid seducing patients into replacing actual living with falsely sustaining fantasy-pseudo-living in the form of an excessively long-term ongoing therapeutic relationship that infantilizes the patient creating undesirable dependency. (Some reparative therapists using as an excuse that cure can take a long time engage their patients in an ongoing process that is so lengthy that it impoverishes a vibrant youth, ultimately leaving behind a sad old individual—one who comes broken not into but out of therapy.)

Eleven. Teach by giving advice, involving telling patients something they need to know in order to survive, not something they must know in order to change. This would include giving advice based on personal experience that is not too idiosyncratic to be widely applicable, while avoiding giving advice that focuses on liabilities instead of assets.

Twelve. Avoid breaking down healthy defenses. In selected instances help patients become *more* defensive by helping them wall off aspects of themselves that make them unhappy, for example, helping patients *dissociate* away harsh confining aspects of religious beliefs that force them to submit to orthodoxy.

Thirteen. Avoid isolating gay patients. Gays like to be with other gay men and women just as straight men and women like the company of other straight men and women. Therefore reparative therapists should not demand that their patients give up their social lives completely in an attempt to stay away from gay triggers, only to have those of their patients

who take this less than sage advice retreat from life due to feeling that they don't belong anywhere.

In conclusion, reparative therapists who do not only reparative but also affirmative work with the patients who apply for SOCE, who include affirmative approaches in their reparative therapeutic armamentarium alongside the more purely reparative aspects of the treatment they offer, are in the best position to help their patients: not to change from gay to straight, but to improve their adjustment by changing from problematic to creative homosexuality.

19

❖

Treating Compulsive
Homosexuality

In this chapter I describe how some of the psychological theories and therapeutic practices of reparative therapists would be highly useful if applied not to understanding and treating homosexuality, but to understanding and treating *compulsive* homosexuality. Compulsive homosexual enactments represent a problem for some gay men, one characterized by the overwhelming need to seek out sexual relationships, often in astoundingly large numbers, driven not by what the individual necessarily wants, but by what the individual feels forced to go after. What the reparative literature describes as the "psychodynamics of homosexuality" in fact as often as not helps explain not homosexuality but *compulsive* sexuality, both hetero and homosexual. The so-called triggers for *homosexual fantasies* are more significantly triggers for *homosexual enactments*, that is, they light up not quality (gay) but quantity (urgency). Therefore, coming to terms with the past and its traumas or the like won't help understand or cure homosexuality. But it might help the individual diminish the tensions that press for immediate relief in the here and now in the form of urgency to seek homosexual gratification.

Thus Nicolosi notes that a "low-grade emptiness . . . sets [the man] up for male attractions"[1] and homosexual enactment. Perhaps he is implying that these things cause the *maleness* of the attraction, that is, determine its homosexual nature. I believe that low-grade emptiness does act as

a trigger, but only for an immediate transport of sexual, and so in gays homosexual, desire/urgency. No one challenges that, like anyone else, gays can, and often do, use sex as an antidote for loneliness. But while this is problematic, it is not inherently homoerotic.

In compulsive homosexuality the homosexuality expresses needs and soothes fears. Many individuals act homosexually in order to fill themselves up when they feel empty because something appears to be lacking in their lives, or because they feel pain because someone or something has hurt, or is currently wounding, them deeply. As a result they seek pleasurable homosexual enactment not only because they feel sexual, but also as a balm for feeling sad, lonely, and hurt. Their homosexuality is already in place. They now enact it to reduce anxiety and relieve pain by creating the illusion that someone wants and loves them.

Alternatively, homosexual enactment serves the purpose of firming up a shaky identity of a nonsexual nature, such as "being the successful one"; gaining or remaining in control, or winning in an interpersonal competition much like a conquistador might triumph in some rivalry—his or her way to blow out another's candle in order to make his or her own seem to be burning brighter.

Sexual compulsiveness whether heterosexual or homosexual can also express a masochistic need: constantly cruising not so much *out of* shame but as a way to deliberately *bring shame upon oneself*, thus making the search for sex not an act of joy but an act of self-deprecation, a way to punish oneself in order to induce suffering by pursuing anonymous encounters which, as gays (and straights) already know, generally end up badly by leading to relating to unloving people.

Sometimes compulsive homosexuality can be an enactment of hope: that one will this time not be humiliated but will instead be acknowledged or "seen," appreciated as an attractive person, with significant status in life. For depressed individuals who worry that no one likes them because they are gay, special pressures can exist to find others who don't reject them but instead accept them, if only sexually, as a replacement for accepting them personally. Some depressed individuals suffering from low self-esteem feel pressured to seek conquests in order to attain bragging rights for the purposes of enhancing their self-view as "someone others want."

Or, being homosexually compulsive represents being healthily assertive. Or, it can be a way to court danger by individuals who find it exciting to put themselves in (controlled) harm's way, to tempt fate to see what they, in a sexual game of Russian roulette, can get away with without getting figuratively, if not literally, "killed."

For gay men who feel at odds with an aspect of their feminine selves, compulsive gay sex can, one might think somewhat paradoxically, help them feel more like a man—with constant homosexual enactments restitutive in the sense that they lead to their feeling more potent, more macho, as their coupling with another man makes them feel not only as if they belong, but also as if they have become more manly by osmosis—and validated because another man sees them as desirable because he sees them as a real man.

Or, compulsive homosexuality can be installed to repeat or repair an early trauma, often, according to Nicolosi, one involving "shame" or "attachment loss,"[2] as gays attempt to master the effects of attachment-shame trauma by repeating the traumatic situation over and over again, this time in the sexual mode. As with all forms of PTSD, such gays are reviving trauma repeatedly to relive some original hurtful encounter, this time to have things turn out differently, and, for a change, this time to instead come out right. Reparative therapists emphasize how often the original trauma/shame has involved boys not being seen as real, loveable, authentic individuals who can be appreciated for who they are, or hope to eventually become. According to Nicolosi, the reparative goal then would be to turn a boy's traumatic not having been "seen by a man as a man"[3] into his "being seen as a real red-blooded he-man, by a man" (and thence becoming that exactly).

Often reparative therapists correctly suggest that the original trauma can involve a problem with the boy's relationship with his father. Sometimes fathers of gay men have been abusive and neglectful. Or they have been hypercritical. In either case, they have failed to act as a model around which the son could build his own identity and so as an adult come to feel okay as is, and without having to prove himself over and over again by becoming someone else. The homosexually compulsive son thus tries to develop a better self-view and an enhanced status in life via identification with an ongoing series of new sexual partners who for him represent new and improved father figures.

Perhaps the original trauma/shame didn't involve being criticized/abused overtly but instead consisted of being criticized/abused covertly—by a father who said good things which he simultaneously took back with opposing negative nonverbal messages, which were the ones that then became internalized as "poisonous introjects"—to be gotten rid of later in life by incorporating one after another "good penis" to attain manly power and status for oneself in order to undo internalized self-questioning/criticism and reduce feelings of weakness/powerlessness. In many such cases, a gay man feels forced to (sexually) overpower one man after another to

empower himself and so to make himself feel more potent and therefore less "disenfranchised," thereby healing the original wound.

Perhaps the original trauma/shame involved a mother who infantilized her son due to her fear that he might abandon her. Thus a gay man now feels he must repetitively have relationships that are inherently "as un-committed as all gay relationships are," doing so over and over again in order to finally break free from mother, always trying to get away from her in order to at last succeed at developing a degree of independence for himself. When homosexual enactment represents a later-day attempt to cope with ongoing parental displeasure over achieving independence, the homosexual enactment is reparative in the sense of representing inde-pendent activity installed to subdue unwanted dependent passivity. In such cases homosexual acts signify being "grown up," for one is collecting grown-up trophies and developing what is perceived to be a massively enviable collection.

Compulsive homosexuality can also be a way to master early trauma involving being intimidated/bullied. Here active (masculine) conquest is intended to make a gay man feel more masculine—someone who is battle ready, battle worthy, and successful in waging war.

Compulsive homosexual enactments also attempt to resolve conflicts between assertion and shame about being assertive. Shame about being *unassertive* leads to feeling as if one has little or no value. The goal of ho-mosexual conquest then becomes to become someone valuable—because noticed and appreciated, if not for oneself then for one's equipment and performance—the seeking of a compensatory sense of pride and power along with a feeling of meaningful accomplishment attained through ac-tively seeking and having sex with one new, utterly fantastic person after another.

Sometimes compulsive homosexual enactment represents an expres-sion of anger at men, taken back in a compensatory fashion by loving one after another man in order to subdue the reverse feeling.

Often compulsive homosexual enactment is an attempt to reassure oneself that one is safe from feared heterosexuality, as one hangs out only with men to avoid being shamed and rejected by women.

Homosexuality enactment can also be a way to be unimpeachably manly by doing what men do (having good and plenty of sex).

Finally, compulsive homosexuality can unconsciously be a way to de-crease loneliness in order to cope with and master the feeling that "I must be defective, for if I weren't then I would have somebody, but since I am lonely it must mean that, having nobody, I am not much of a person. But having all these men makes me feel more competent and important."

Of course, conscious factors, such as pleasure seeking, are also opera-tive. In such cases "compulsive" homosexuality is what it is, not a cover for something else, such as the need to repair a problematic lack of mas-culinity or to diminish an increased feeling of detachment from the world and the people in it. Compulsive homosexuality may simply represent a need for physical gratification that for some is so pleasurable that it gets repeated over and over again just to have the most possible fun out of life. As such it involves repetitively having old and new pleasurable relation-ships with someone/many someones not in a (futile) attempt to come up with a satisfaction of some deep-seated need, or a resolution of some deep-seated conflict, but as a gratifying action that not so much re-creates old but creates new pleasures as part of a more or less pure form of pleasure addiction.

Compulsive homosexuality can also be a way to get to know other men the better to learn what makes them great so as to be able to discern and identify their desirable characteristics and make them one's own, both to improve oneself, for oneself, and to become the equal of another, an indi-vidual at first only having been passively idealized, but now, merged with, capable of being actively imitated.

TREATMENT

Treatment of compulsive (homo)sexuality starts with understanding these causes in order to reduce sexual acting out by expanding self-awareness through enhancing appreciation of what drives the individual to act this way.

A core therapeutic technique involves identifying the complex issues/events/problems originating in and established in childhood that persist to act as present-day triggers for pressured erotic enactment, such as envy of the male (father) that pushes the adult male to want to "have" another man to enhance his own macho status, or shame about inner truths that leads one man to desperately seek another man's affirmation.

Reparative therapists can help by not advocating sexual orientation change efforts but by instead setting their sights on helping their patients grieve for/adjust to what cannot be altered and cope with/adjust to the consequences of not being able to make things different. Not infrequently, homosexuality becomes less compulsive when a gay man is finally able to get over some loss, such as the loss of a father's love, or some rejection, such as having felt unloved by his mother when he was very young.

Temporarily holding off the need to act out sexually is a way to stay with the resultant core problematic feelings long enough to deal with

them—instead of dissipating them on the spot. This helps gays go from mindless repetition to vital action—which may involve no action at all. Holding off temporarily gives gays breathing room to work toward an improved, more positive, less despairing self-view and worldview.

Many therapists treat compulsive homosexuality by encouraging the patient to have a corrective emotional experience with the therapist. Here the therapist attempts to act in a way that corrects past emotional betrayals and reduces narcissistic injury through a new, more positive, more respectful patient-therapist interaction consisting of a corrective, this time more sensitive, response on the therapist's part. This allows the patient to feel more self-satisfied through having been affirmed by the therapist as an ally/significant other. Conversely, a therapist's being unsupportive in this or any other realm can have a negative effect on patients by provoking them to seek the support they aren't getting from their therapist through enactment consisting of a series of anonymous sexual encounters outside of the therapy hour as patients cruise and pick up one person after another as a way to act out their transference disappointment in their therapist.

Therapists can often help their patients gainfully reduce the frequent need for homosexual enactment by helping them improve their actual circumstances. For example, when I was an intern I went through a period of compulsive sexuality because in fact I had no other life—for I was lonely, had little money, and had to work every other night, sometimes going without sleep for hours on end—a perfect storm that set me up to compulsively act in a way that involved feeding myself a little pleasure, however transient, however desperately motivated, and in any possible way I could.

To be avoided are ineffective/bizarre approaches. Eating right, a cure popular in some reparative circles, might give an individual more healthy energy but it will not transform the individual from "unhealthily gay" to "healthfully straight." Eating right will likely improve one's physical health. That will likely improve one's self image—but rarely enough to reduce the continuing need for self-validation through repetitive sexual conquest. While eating right might reduce a gay person's compulsive need for sex, it certainly won't, as often claimed, affect his or her basic need to serve this sex up in a homosexual form.

In conclusion, reparative therapy techniques can be diverted to helping gays conquer compulsive sexuality. Helpful elements that can reduce sexual pressures can include establishing a positive relationship with a therapist (leading to a transference cure), and enhancement through insight of the sense of one's value, thereby increasing self-esteem so that one

doesn't feel compelled to seek sexual contact simply in order to affirm one's self-worth/improve one's self-image, but one can instead feel like a better, more valuable person without having to compulsively seek a daily dose of affirmative input coming in from the outside. Diminishing the compulsive need for sexual activity by improving the overall quality of individual interpersonal relationships to make them more gratifying and so more sustaining can help. And so can doing grief work so that one accepts earlier losses without constantly having to seek sex to repair deep sadness over having been hurt and even abandoned. Affirmative approaches can also help reduce the pressure to find sexual companions to undo feelings of having been slighted and ignored throughout life—by one's parents, peers, and community, thus creating pressure to seek sexual contact to solve one's problems instead of restructuring one's life to seek not transient superficial joy but ongoing, deep satisfaction and, finally, contentment.

20

❖

Treating Self-Homophobia
(Ego-Dystonic Homosexuality)

In this chapter I emphasize how the wish to be straight is often largely the product of guilt about being gay, which in turn is the product not only of internalized but also of internal (self-originating) homophobia.

Though the literature contains many studies of self-homophobia, these are essentially limited, even by the experts in the field, to observations of the nature and effects of external homohatred. But a great deal of homohatred is internally based, existing independently of the external homophobia that generally gets all the attention. This internal homohatred, consisting of some form of the feeling that "I am broken," comes not from "because others (e.g., my reparative therapist) see me that way," but "because that is how I see, and have always seen, myself."

Gays who are internally homophobic seek SOCE because their homosexuality is "ego-dystonic," that is, because they don't like/accept being gay. Some don't like/accept being gay as a symptom of one or more emotional problems. Such individuals say they seek relief from their homosexuality when in reality they seek relief from a more generalized sense of guilt that is internal and is a symptom of one or more emotional difficulties. Thus guilt about being gay is often less homosexually derived than neurotically generated, less a gay issue than a more general psychological problem however much it gets expressed in the language of "gay guilt."

Some guilty gays are *bipolars* suffering from self-hatred associated with a masochistic need to punish themselves for who they are. Their excessive self-hatred, at first often depressive, tends to spontaneously switch over to become excessive hypomanic self-love, often manifest as excessive self-acceptance, where gays deny their guilt to the point of becoming entirely one with being gay. Almost too accepting of their homosexuality, they treat themselves not as completely broken but as especially intact, no exceptions, as they become excessively proud of being gay—unfortunately only to revert, also as if spontaneously, to once again feeling guilty and ashamed of themselves, and so being depressed. For such gays "I hate me for being gay" alternates with "being gay makes me very proud," but this is less a gay issue than it is an emotional cycling characteristic of an affective problem.

OCDs comprise a large proportion of what Nicolosi calls "non-gay"[1] gay men. For such gays, homosexuality is ego-dystonic because they have come to reject almost anything about who and what they are as bad, or shameful, or dirty. The behavior of such individuals is typically characterized by doing and undoing taking the form of homosexual enactment followed by guilty regrets followed by temptation anew, followed by new action, followed by fresh regrets: a manifestation in the homosexual realm of an unduly harsh conscience alternating with a conscience that is unduly permissive, only to once again switch over to becoming unduly harsh, so that such individuals alternately accept and cherish being gay and self-condemn for being homosexual, not so much affectively, as do bipolars, but instead intellectually. Such OCD gays alternating as they do between being "id" and "superego-oriented" enroll, often after one or more "id" orgiastic homosexual enactments that don't go well (e.g., after they develop an STD), in reparative therapy hoping to somehow undo what they just did, only subsequently to regret their newly found "attempt at sainthood" and long to become sinners once again.

OCD gays buying into religious prohibitions have difficulty reconciling their sexuality with their religiosity. Then they choose to let the impossibility of reconciling their sexuality with their religion bother them excessively. They brood about it continuously, even to the point that they allow it to replace other, healthier concerns. This matter then becomes, for them, a substitute for living.

An OCD gay man feeling overwhelmed by his homosexual (and often by all his sexual) feelings, as if any form of sexuality was unrepresentative of who he was, what he wanted to be, and what his purposes and possibilities in life were, came to feel completely like an unvirtuous sinner. As such he constantly feared God and His wrath, and dreaded God's abandoning him. He sought reparative therapy not so much as a form of

treatment for his homosexuality but for personal salvation—to be wor-
thier of God's embrace via having been good by applying and going for
treatment and for staying in it even though the treatment wasn't working
(and he knew it never would).

For many OCD gay men like him therapy is a kind of ritualistic confes-
sional cum narcotic, and the therapist is less the healer than the shaman,
the priest, offering salvation and absolution along with some magical for-
mula for redemption. The therapist thinks "God's will be done." And the
patient, going along, thinks "That's a good thing, for my homosexuality is
ungodly and as such needs the undoing."

Some nongay gays are excessively *passive dependent*. These gays accept
what others (especially their reparative therapists) tell them about themselves, and in particular what to disavow about who they are. They then
act accordingly as they submissively and far too willingly yield to such
outside pressures. As individuals who will seemingly do anything to avoid
being ignored or unloved, they spend months or even years in therapy just
to get a little attention, approval, affection, and reward for "being good"
and "doing a good thing." Sometimes their entire course of reparative care
is really a hidden search not for a cure but for shelter where even though
the surroundings are spartan, they at least feel that they have come out of
the cold.

As-if borderline individuals readily form identifications with others after
too readily abdicating who they are and what they should be to outsiders.
They especially yield themselves up to those who assume the mantle of
authority. (However hateful a given authority may be, some gays willingly
abandon being "me" in order to become "more like you," the therapist,
doing so voluntarily to please the virtual strangers their therapists in fact
are.) Desperate to feel that they belong to a straight society, they join
straight groups even though, or just because, these groups don't welcome
them. They imagine, "If I go straight then the A team will like me, stop
bullying me, and take me in." Only what they generally, and too sadly,
ultimately discover is that many straights do not consider *ex-gay* to be as
good as *always has been straight*, any more than they consider ex-alcoholics
to be as upright as lifetime teetotalers. Still they comply and yield to group
pressures even though that means noncompliance with, and so an abandonment of, themselves. They morph into someone else in a futile attempt to belong. They accept the brutality that is reparative therapy just
to be able to have someone, anyone, to cling to.

Gays with *post-traumatic stress* symptoms/disorder, especially those
whose symptoms appear after they have been bullied, turn to therapists
and therapeutic organizations not only to help them heal but also

to intervene to protect them from further physical and emotional abuse/bullying over their being gay. They thus use a therapist not only to protect them from themselves but also to shield them from others, generally from the sadists in their lives. Many such gays react negatively to being gay because they have just been traumatized by having been dumped—bad in and of itself, but worse when it revives early problems with rejection and abandonment. Gays are often in crisis after being dumped. They often wrongly attribute having been dumped to their "sexual orientation" which they blame entirely for why "no one likes me." They often believe that if they were straight they wouldn't have been dumped. Reacting so catastrophically to having been let go, they seek therapy nominally "to go straight" but actually as a way to form a substitute (if transitional) relationship, this time with someone, a therapist, who will never dump them (as long, that is, as they keep coming and paying). Especially when they seek group therapy they do so less to gain psychological insight than to be in a place where there are people who can presumably provide them with support—in the form of a substitute for a real social life, however inadequate that particular, jerry-built, replacement inherently is and turns out to be.

Paranoid gays balk after coming out in a knee-jerk "take-it-back" response to having come out. Soaked in imagined fears of what they might have done to themselves and what others might do to them, as a consequence they enter therapy not so much for redemption as for protection.

Of course, not all gays who ask for/submit to reparative therapy are self-homophobic. Gays whose religious beliefs prompt them to go straight may be simply doing what their authentic conscience (often with substantial ties to the dictates of the church) seems to ask of them, readily and willingly agreeing to hand over their sexuality in favor of fully retaining their relationship with their church/God. Religious nongays feel that they are not being self-homophobic, but are instead acting rationally, heeding a conflict-free calling to be straight, and making a conflict-free decision entirely determined by their free will, not one predetermined by inside or outside emotional pressures feeding into irrational neurotic hopes and fears.

Some gays with *realistic* problems with being gay seek "curative" life changes that lead to their enhanced comfort with who they are. They are not so much guilty about being gay as they are actually discouraged by it, for they feel that the actual difficulties realistically associated with being gay are just not worth it. Some of these men and women have been unsuccessful in their gay relationships. They can't find a partner to love, or have found one who doesn't love them, or have gotten involved with one who

encourages them not to be who they are but to be someone else by demanding that they make significant, radical changes in themselves. Others have occupational problems and can't solve them by getting another job in the same or in a different profession. For such individuals, though they have other, and possibly more significant, problems, the goal of going straight can nevertheless still be a symbolic one, a substitute for another, less readily avowed, more practical objective. These individuals think that the gay life will be less fulfilling for them than the straight life. They believe that the gay life predictably works poorly for them in terms of their individual personality and circumstances. They believe that the rewards of being gay are not commensurate with the anxieties they feel about being that way. They thus conclude that they need to change from being gay when in fact they only need to improve on their gay lifestyle.

When gay patients have identifiable emotional problems, therapy should be focused on repairing not their homosexuality but their emotional difficulties, and especially on resolving any emotional problems they might have that keep them from going through life fulfilled and happy men and women with rewarding same-sex experiences, involved in joyful long-term relationships accompanied by a successful career. These gays would likely be less unhappy about their sexual orientation if they were happier about their lives. They likely would accept being/want to continue being gay if they knew how to be gay in the right way, without addiction, depression, STD, and that loneliness that is what so often drives them to say that they need and to seek reparative help. In such situations reparative therapists should not tell their patients gay life is unrewarding, dangerous, and generally fatal. They should tell them that their problems with gay life are the sequelae of their emotional problems, and that these can be effectively remedied—not by going straight, but, still being gay, now going about being that exactly in a better, more self-fulfilling, less troubled way.

Some gays even though they have deep emotional problems nevertheless get sufficient benefit from simple environmental change via environmental manipulation. They might benefit from simply joining another, more welcoming group of gays, those more friendly to, and accepting of, them, and/or seeking to meet another, better partner, and/or dealing with their parents in a better, more mutually satisfactory, healthier, more healing way. Instead of avoiding the entire culture even though that means avoiding the only culture to which ultimately they do, and will ever, truly belong, they might consider moving to a place, or getting a new job, where the culture is more welcoming.

In conclusion, when the gay individual's application for treatment to go straight is itself symptomatic of an underlying self-homophobic emotional state or emotional problem, the wish for change needs not immediate validation but preliminary analysis to determine if what looks to be consciously desired is unconsciously motivated. This avoids having an excessively hopeful patient locked into therapy with an excessively ambitious therapist with both individuals reluctant to shift focus from changing a patient's sexual orientation to helping a patient retain his or her sexual orientation as it is currently, only now in a way that encompasses a greater degree of self-acceptance and a far higher level of self-love.

21

Parental Issues

Too often parents who demand that their children be treated for being gay claim that they do so only out of concern for their children—when in fact they do so mostly out of concern for themselves. They are ashamed of having a gay child in the family: in the eyes of their relatives, friends, and community. They are afraid that they will be harassed and shunned for having made their child gay. They are concerned that their child will get into trouble, humiliating them and even causing legal problems for them. Religious parents don't know whether to see their child being gay as an act of God or as a defiance of Him and of religious teachings. Some parents of gays have a personality disorder marked by their developing seriously adversarial relationships with their child, with the gay community, or with both. Some parents suffer from OCD consisting of scrupulosity, antieroticism, and fears involving dirt and contamination, for which "being gay" makes a perfect symbol. Still others reveal a histrionic tendency to overestimate the downsides of what they and their child face in life. Imagining the worst, they turn a benign situation, their child being gay, into a perceived catastrophe, their life being over. Some, suffering from a post-coming-out post-traumatic stress disorder, act almost as if they have been physically wounded by their child's revelations, to the point that they are constantly in a state of panic and despair as a response to their having given birth to a gay child, They have constant flashbacks to their presumed traumatization by their child coming out, perhaps to the extent that they can seemingly think of nothing else. Everything seems to

remind them of their so-called plight, and any of life's general problems they have call to mind, and they attribute to, the terrible circumstances that pertain for them as parents of a gay son or daughter. They blame their child(ren) being gay for all their emotional troubles, as if it is the child's homosexuality alone that is what is making them anxious and depressed, causing them to drink heavily, and/or interfering with their ability to be successful professionally.

So they seek preventive therapeutic measures, often a right-now cure for their child being gay, even though their child's sexuality is not yet ripe enough to have even begun to take on anything resembling permanent form.

While some of these parents mainly love their gay children and want the best for them, others instead actually dislike their gay child(ren) at least in a part of themselves and want even their own child(ren) corralled and punished. They reject their child(ren) as they feel rejected by them and blame not only themselves for making their child gay but also their child for refusing to listen to Mom and Dad and go straight.

Typically such parents request what amounts to "punitive SOCE" for their child when instead they might, and perhaps should, request evaluation and treatment for themselves. Perhaps Mom and/or Dad need long-term psychotherapy. Perhaps what they most need is advice on how to help their child change his or her environment by moving to a new place, or by getting a whole new group of best friends; or by culling gay-favorable family members from the family pool and becoming closest only to those, while staying far away from the others; and by seeking new religious affiliations more in line with the unalterable fact of their being gay as they attend a church that is more welcoming to LGBTQs than the one to which their child presently belongs.

But too often parents set the wrong goal for their children, asking them, or demanding, they find acceptance among those who don't accept them, even recommending that their child alter his or her sexual preference to fit in with the personal (sexual) preferences of others.

If a gay child needs therapy for general reasons, parents should start by referring the child to a supportive affirmative psychotherapist who will likely begin by informing all concerned that any therapy that takes place should begin with a consideration of the core family, the parents as well as the child, and deal not solely with the child's homosexuality but with the child as a whole, unique person in need of a new, more welcoming environment.

Above all, parents should recognize that reparative therapy almost never works. While as Throckmorton and Yarhouse say, an "adolescent

shift in self-reported sexual and religious identities can occur primarily to please family, peers, or other adult role models, [meaning that] changes may not persist or reflect a permanent [change] in sexual identity."[1]

Parents must recognize that, as noted throughout, most times reparative therapy not only doesn't work for children and adolescents, it can actually harm them. The mere act of sending a child off for treatment of any sort, and especially for homosexuality, can by itself convey a critical, often shattering message: that "you are unacceptable to us as you are, and that we want to be rid of you as you exist in your present state, and as soon as possible." This unloving message can be devastating for the child and his or her relationship with the parents, especially when it goes on to resonate with the unloving therapy that their child will likely be receiving from a therapist who, by the very nature of reparative therapy, will be offering the child the obverse of consolation, and the opposite of comfort.

So often, even if children accede to go for reparative therapy, it is less because they want not to be gay and more because they want a home: someone who loves them, and someone who can make up for what is seriously lacking in their lives, and especially in their relationship with their parents. The affirmative therapist is in the position to do such a child the most good, while the reparative therapist is in the position to do the gay child the most harm. Parents who send their children off to be cured of who they are run the risk of diminishing who their child is, as well as doing harm not only to their child but to the child-parent relationship. The thing that will go away is not their child being gay, but their child. Such parents give their child a hard time, when instead they should be giving them a home. Such children if they must be in therapy ought be in treatment with someone who accepts them, not someone whose first action is to recoil at their very presence but whose first thought is to tell them, misquoting from the Bible, that they "are not weighed in the balance, and are not found wanting."

Notes

DEDICATION

1. Wilhelm Reich, *Character-Analysis* (New York: Orgone Institute Press, 1949), 508.

INTRODUCTION

1. W. R. Besen, *Anything but Straight: Unmasking the Scandals and Lies Behind the Ex-Gay Myth* (New York: Harrington Park Press, 2003), 274.
2. E. Roach, "Exodus Int'l Closes after Chambers' Apology: The Baptist Press" (June 20, 2013), http://www.bpnews.net/BPnews.asp?ID=40574.

CHAPTER 1

1. E. Jones, *The Life and Work of Sigmund Freud* (New York: Basic Books, 1957), 3:195.
2. A. Ross, "Love on the March," *The New Yorker*, November 12, 2012, 47.
3. Ibid.
4. Besen, *Anything but Straight*,124.
5. M. Miller, "D. Reuben. 30 Years Later, the Subject Is Still Sex." (February 17, 1999), http://articles.latimes.com/1999/feb/17/news/cl-8701.

6. R. Isay, "On the Analytic Therapy of Gay Men," in *Affirmative Dynamic Psychotherapy with Gay Men*, ed. Carlton Cornett (Northvale, NJ: Jason Aronson, 1993), 23–24.
7. Ibid., 24.
8. Besen, *Anything but Straight*, 126–127.
9. G. Santayana, *The Life of Reason*, vol. 1 (Charles Scribner's Sons, 1905).
10. R. Isay, "On the Analytic Therapy of Gay Men," in *Affirmative Dynamic Psychotherapy with Gay Men*, ed. Carlton Cornett (Northvale, NJ: Jason Aronson, 1993), 25–28.
11. H. Pinsker, (2013), Personal Communication.
12. J. Nicolosi, *Shame and Attachment Loss, the Practical Work of Reparative Therapy* (Downers Grove, IL: IVP Academic, 2009).
13. NARTH, "Answers to Frequently Asked Questions about Narth and Homosexuality" (October 11, 2012), www.NARTH.com.
14. Roach, "Exodus Int'l Closes after Chambers' Apology.
15. Ibid.

CHAPTER 2

1. M. Regnerus, "How Different Are the Adult Children of Parents Who Have Same-Sex Relationships? Findings from the New Family Structures Study (2012)," retrieved 4/15/2014 from http://www.sciencedirect.com/science/article/pii/S0049089X12000610.
2. "Report of the American Psychological Association Task Force on Appropriate Therapeutic Responses to Sexual Orientation, American Psychological Association," retrieved 3/14/2013 from http://www.apa.org/pi/lgbt/resources/therapeutic-response.pdf.
3. Ibid.
4. J. Keats, "Ode to a Nightingale: Arthur Quiller-couch, ed., *The Oxford Book of English Verse 1250–1900* (1919), retrieved 3/4/2014 from http://www.bartleby.com/101/624.html.

CHAPTER 3

1. Besen, *Anything but Straight*, 127.
2. G. J. M. Van den Aardweg, *The Battle for Normality: A Guide For (Self-) Therapy for Homosexuality* (San Francisco: Ignatius Press 1997), 59. Quotes used by permission of Ignatius Press.
3. R. Oas, "Panel Calls Reparative Therapy for Homosexuals a Human Rights Violation" (Feb. 15, 2013), retrieved 2/12/14 from www.lifesitenews.com/home/print_article/news/38277/.

4. W. Throckmorton and M. Yarhouse, "Sexual identity practice recommendations" (n.d.), retrieved 3/2/13 from http://www .drthrockmorton.com/sexualidentitytherapyframework0506.pdf.
5. Regnerus, "How Different Are the Adult Children of Parents Who Have Same-Sex Relationships?"

CHAPTER 4

1. "Report of the American Psychological Association Task Force."
2. W. Throckmorton and M. Yarhouse, "Sexual Identity Practice Recommendations" (n.d.), retrieved 3/2/2013 from http://www .drthrockmorton.com/sexualidentitytherapyframework0506.pdf.
3. Ibid., 8.
4. J. Nicolosi, *Shame and Attachment Loss*, 21.
5. Ibid., 20.
6. Ibid., 21.
7. C. Hyberman, "Breaking Bread: Normal Siegel. A Mellower 'Mr. Negative,' But Still Passionate about Free Expression" (July 1, 2013), http://www.nytimes.com/2013/07/01/nyregion/a-mellower-mr-negative -but-still-passionate-about-free-expression.html ?pagewanted=all&0_r=0.

CHAPTER 5

1. B. Brantley, Theatre Review: Children of the World, Unite! 'Matilda the Musical' at Shubert Theater (April 12, 2013), http://www .nytimes.com/2013/04/12/theater/reviews/matilda-the-musical-at -shubert-theater.html?_r=0.
2. "Report of the American Psychological Association Task Force."
3. É Coué, "Every Day, in Every Way, I'm Getting Better and Better," in *Self-Mastery-through Conscious Autosuggestion* (New York: Malkan Publishing, 1922), 118.
4. Helene Deutsch, "Some Forms of Emotional Disturbance and Their Relationship to Schizophrenia," *Psychoanalytic Quarterly* 11 (1942).
5. American Psychiatric Association, *Diagnostic and Statistical Manual of Mental Disorders*, *Fifth Edition* (Arlington, VA: American Psychiatric Association, 2013), http://www.apa.org/pi/lgbt/resources/therapeutic-response.pdf.
6. PDM Task Force, *Psychodynamic Diagnostic Manual* (PDM) (Silver Spring, MD: Alliance of Psychoanalytic Organizations, 2006), 59.

7. J. Nicolosi, *Reparative Therapy of Male Homosexuality: A New Clinical Approach* (Lanham, MD: Jason Aronson/Rowman & Littlefield, 1991), xvii.

CHAPTER 6

1. S. Freud, *Future of an Illusion* (London: Hogarth Press, 1928).
2. D. Cameron, "Beyond Belief," 86, no. 1 *Harvard Medicine* (Winter 2013) (Boston: Harvard Medicine), 37–39.
3. J. Feldman, "Phil Robertson Speaks Out at Church Group: 'I Am a Lover of Humanity, Not a Hater'" (Dec. 22, 2013), http://www.mediaite.com/online/phil-robertson-speaks-out-at-church-group-i-am-a-lover-of-humanity-not-a -hater/.

CHAPTER 7

1. Regnerus, "How Different Are the Adult Children of Parents Who Have Same-Sex Relationships?"
2. R. Spitzer, "Psychiatry Giant Sorry for Backing Gay 'Cure'."(5/18/2012), http://www.nytimes.com/2012/05/19/health/dr-robert-l-spitzer-noted-psychiatrist-apologizes-for-study-on-gay-cure.html?pagewanted=all&0_r=0, 1–3.
3. Besen, *Anything but Straight*, 241.
4. I. Bieber, "Research Summary: Irving Bieber et al. Homosexuality: A Psychoanalytic Study" (n.d.), retrieved 4/26/2013 from http://www.emaso.com/links/REF-Books/REF.4-B.htm.
5. NARTH, Answers to Frequently Asked Questions about Narth and Homosexuality (October 11, 2012), www.NARTH.com.
6. Van den Aardweg, *The Battle for Normality*, 83.
7. Besen, *Anything but Straight*, 109.
8. Ibid., 109.
9. G. J. M. Van den Aardweg, *The Battle for Normality*, 53.
10. Ibid.
11. Ibid., 21.
12. Nicolosi, *Shame and Attachment Loss*,19.
13. Van den Aardweg, *The Battle for Normality*, 125.
14. Nicolosi, *Reparative Therapy of Male Homosexuality*, 4.
15. Van den Aardweg, *The Battle for Normality*, 64.
16. Ibid., 53.
17. Ibid., 61.
18. Ibid., 61.

19. Ibid., 61.
20. Ibid., Dedication.

CHAPTER 8

1. Coué, "Every Day, in Every Way," 118.
2. Nicolosi, *Shame and Attachment Loss*, 31.
3. M. W. Webster, *Webster's Third New International Dictionary of the English Language* (Springfield, MA: Merriam-Webster, 1961), 2174.

CHAPTER 9

1. Besen, *Anything but Straight*, 267.
2. Nicolosi, *Reparative Therapy of Male Homosexuality*, 4.
3. Ibid., 4.

CHAPTER 10

1. N. Rorem, "1961–1972, The Later Diaries" (n.d.), retrieved 1/24/2014 from http://books.google.com/books?id=pmz5XVnBRYMC&0pg=PT 70&0lpg=PT70&0dq=%22one+doesn't+know+how,+one+does%22 +rorem&0source=bl&0ots=EXFJhu4Ohj&0sig=IFNmknTqBSzbOh UpVWGIKbaMSPA&0hl=en&0sa=X&0ei=1aBWU7PrJ-7gsASm0 YLoAQ&0ved=0CBsQ6AEwAA#v=onepage&0q=%22one %20doesn't%20know%20how%2C%20one%20does%22%20 rorem&0f=false.
2. B. Adams, "For Benjamin Britten, Upon the Centenary of His Birth" (December 23, 2013), http://musicologynow.ams-net.org/2013/11/ for-benjamin-britten-upon-centenary-of.html.
3. Van den Aardweg, *The Battle for Normality*, 47.
4. Nicolosi, *Reparative Therapy of Male Homosexuality*, 22.
5. Van den Aardweg, *The Battle for Normality*, 12.
6. O. Fenichel, *The Psychoanalytic Theory of Neuroses* (New York: Norton, 1945), 332.
7. Ibid., 333.
8. Ibid., 337.
9. Van den Aardweg, *The Battle for Normality*, 47.
10. Ibid., 38.
11. Fenichel, *The Psychoanalytic Theory*, 332.
12. Ibid., 331.
13. Ibid., 337.

14. Besen, *Anything but Straight*, 147.
15. Ibid., 149.
16. Ibid., 146.
17. Fenichel, *The Psychoanalytic Theory*, 330.
18. Ibid.
19. Ibid., 331.

CHAPTER 11

1. S. Herskovitz, "Ethically Speaking," UCONN *Today* (Winter 1991–1992), http://alumni.uchc.edu/docs/Nalbandian_story.pdf2, 2.
2. Besen, *Anything but Straight*, 269.
3. Throckmorton and Yarhouse, "Sexual Identity Practice Recommendations."
4. J. Kort, *Gay Affirmative Therapy for the Straight Clinician: The Essential Guide* (New York: Norton, 2008), 14.
5. Nicolosi, *Shame and Attachment Loss*, 19.
6. Besen, *Anything but Straight*, 241.

CHAPTER 13

1. Nicolosi, *Shame and Attachment Loss*, 80.
2. Ibid., 129.
3. Van den Aardweg, *The Battle for Normality*, 123 or 127 saying no.
4. Ibid., 129–130.
5. M. D. Roberts, "Mark Twain: Do Clothes Make the Man" (Feb. 3, 2014), http://www.thehighcalling.org/reflection/do-clothes-make-man.
6. Van den Aardweg, *The Battle for Normality*, 129.
7. Ibid., 137.
8. Ibid., 138.
9. Ibid.
10. Ibid., 139.
11. Ibid., 137.
12. Ibid., 140.
13. Ibid., 137.
14. Ibid., 139.

CHAPTER 14

1. Van den Aardweg, *The Battle for Normality*, 132.
2. Ibid., 20–21.

3. Ibid., 92–93.
4. Ibid., 93.

CHAPTER 15

1. Throckmorton and Yarhouse, "Sexual Identity Practice Recommendations."
2. Nicolosi, *Shame and Attachment Loss*, 148.

CHAPTER 16

1. Nicolosi, *Shame and Attachment Loss*, 148.
2. Ibid., 295.
3. Fenichel, *The Psychoanalytic Theory*, 329.

CHAPTER 18

1. W. Reich, *Character-Analysis* (New York: Orgone Institute Press, 1949), 508.
2. Ibid., 165–166.
3. "Report of the American Psychological Association Task Force."
4. Ibid.
5. Van den Aardweg, *The Battle for Normality*, 130.

CHAPTER 19

1. Nicolosi, *Shame and Attachment Loss*, 32.
2. Ibid., 128.
3. Ibid., 110.

CHAPTER 20

1. Nicolosi, *Shame and Attachment Loss*, 17.

CHAPTER 21

1. Throckmorton and Yarhouse, "Sexual Identity Practice Recommendations."

Index

acquired behaviors, xiv–xv, xvi
Adams, B., 130
affirmative therapy, 204–222; group,
    219; reparative therapists vs.,
    207–219; supportive, 211–212
alcoholism, 51–52, 55, 58, 207
American Psychological Association,
    8, 17, 30–31, 37, 39, 42, 76, 156
antidepressants, 113, 200
anxiety, 23, 35, 48–49, 120–121, 177,
    200, 205, 213
Anything but Straight, 158
as-if borderline personality disorder,
    48, 122, 232
asexuality, 51
auto–suggestion, 120
aversion therapy, 46, 88, 214–215
avoidance/AvPD, 47–48, 82, 136,
    161, 172, 199; homophobia, 86

Baptist Press, xii, 8
The Battle for Normality, xiv, 86
behavioral modification, 120, 214;
    abreaction, 197–198; biological,

200; eclectic, 200; educative,
199–200; in–patient, 200; punitive,
196–197; religious, 198
Besen, W.R., xiii–xiv, 4–5, 20, 66, 75,
119, 145–146, 149, 158
Bieber, I., 4–5, 20, 66–67
bipolar: homophobia, 86–87; patients,
12–13, 78
bisexuals, 9, 17, 24, 26–27, 68–69,
205, 219
blame, 16, 46, 58, 92, 185; reparative
therapy and, 16, 51–52, 108–109;
-shifting, 136
borderline predisposition/personality
disorder, 12, 25–26, 44, 82, 119,
163
Boston Psychoanalytic Institute, 7
brainwashing, 120
Brantly, Ben, 40
Britten, Benjamin, 130
bullying, 34, 40–41, 73, 145, 232

Cameron, D., 62, 75
castration, 3

About the Author

Martin Kantor, MD, is a Harvard-trained psychiatrist who has been in full private practice in Boston and New York City and active in residency training programs at hospitals including Massachusetts General in Boston, MA, and Beth Israel in New York, NY. He also served as assistant clinical professor of psychiatry at Mount Sinai Medical School and clinical assistant professor of psychiatry at the University of Medicine and Dentistry of New Jersey-New Jersey Medical School. Kantor is a full-time author whose published works encompass more than 20 other books, including Praeger's *Now That He's Out: The Challenges and Joys of Having a Gay Son; Now That You're Out: The Challenges and Joys of Living as a Gay Man; Homophobia: The State of Sexual Bigotry Today;* and *The Essential Guide to Overcoming Avoidant Personality Disorder.*

About the Author

JOHN H. BECKSTROM is Professor Emeritus of Law at Northwestern University's School of Law. He holds degrees from the University of Iowa, Harvard University, and the School of Oriental and African Studies, University of London, where he was a Fulbright scholar. He earlier published *Sociobiology and the Law* and, in 1989, *Evolutionary Jurisprudence.*